MATLAB Machine Learning Recipes

A Problem-Solution Approach

Third Edition

Michael Paluszek

Stephanie Thomas

Apress®

MATLAB Machine Learning Recipes: A Problem-Solution Approach

Michael Paluszek
Plainsboro, NJ
USA

Stephanie Thomas
Plainsboro, NJ
USA

ISBN-13 (pbk): 978-1-4842-9845-9
https://doi.org/10.1007/978-1-4842-9846-6

ISBN-13 (electronic): 978-1-4842-9846-6

Managing Director, Apress Media LLC: Welmoed Spahr
Acquisitions Editor: Celestin Suresh John
Development Editor: Laura Berendson
Coordinating Editor: Mark Powers

Cover designed by eStudioCalamar

Cover image by Mohamed Nohassi@Unsplash.com

Distributed to the book trade worldwide by Apress Media, LLC, 1 New York Plaza, New York, NY 10004, U.S.A. Phone 1-800-SPRINGER, fax (201) 348-4505, e-mail orders-ny@springer-sbm.com, or visit www.springeronline.com. Apress Media, LLC is a California LLC and the sole member (owner) is Springer Science + Business Media Finance Inc (SSBM Finance Inc). SSBM Finance Inc is a **Delaware** corporation.

For information on translations, please e-mail booktranslations@springernature.com; for reprint, paperback, or audio rights, please e-mail bookpermissions@springernature.com.

Apress titles may be purchased in bulk for academic, corporate, or promotional use. eBook versions and licenses are also available for most titles. For more information, reference our Print and eBook Bulk Sales web page at http://www.apress.com/bulk-sales.

Any source code or other supplementary material referenced by the author in this book is available to readers on GitHub (https://github.com/Apress). For more detailed information, please visit https://www.apress.com/gp/services/source-code.

Paper in this product is recyclable

To our families.

Contents

CONTENTS

About the Authors

 **Michael Paluszek** is President of Princeton Satellite Systems, Inc. (PSS) in Plainsboro, New Jersey. Michael founded PSS in 1992 to provide aerospace consulting services. He used MATLAB to develop the control system and simulations for the Indostar-1 geosynchronous communications satellite. This led to the launch of Princeton Satellite Systems' first commercial MATLAB toolbox, the Spacecraft Control Toolbox, in 1995. Since then, he has developed toolboxes and software packages for aircraft, submarines, robotics, and nuclear fusion propulsion, resulting in Princeton Satellite Systems' current extensive product line. He is working with the Princeton Plasma Physics Laboratory on a compact nuclear fusion reactor for energy generation and space propulsion. He is also leading the development of new power electronics for fusion power systems and working on heat engine–based auxiliary power systems for spacecraft. Michael is a lecturer at the Massachusetts Institute of Technology.

Prior to founding PSS, Michael was an engineer at GE Astro Space in East Windsor, NJ. At GE, he designed the Global Geospace Science Polar despun platform control system and led the design of the GPS IIR attitude control system, the Inmarsat-3 attitude control systems, and the Mars Observer delta-V control system, leveraging MATLAB for control design. Michael also worked on the attitude determination system for the DMSP meteorological satellites. He flew communication satellites on over 12 satellite launches, including the GSTAR III recovery, the first transfer of a satellite to an operational orbit using electric thrusters.

At Draper Laboratory, Michael worked on the Space Shuttle, Space Station, and submarine navigation. His Space Station work included designing Control Moment Gyro–based control systems for attitude control.

Michael received his bachelor's degree in Electrical Engineering and master's and engineer's degrees in Aeronautics and Astronautics from the Massachusetts Institute of Technology. He is the author of numerous papers and has over a dozen US patents. Michael is the author of *MATLAB Recipes*, *MATLAB Machine Learning*, *Practical MATLAB Deep Learning: A Projects-Based Approach, Second Edition*, all published by Apress, and *ADCS: Spacecraft Attitude Determination and Control* by Elsevier.

Stephanie Thomas is Vice President of Princeton Satellite Systems, Inc. in Plainsboro, New Jersey. She received her bachelor's and master's degrees in Aeronautics and Astronautics from the Massachusetts Institute of Technology in 1999 and 2001. Stephanie was introduced to the PSS Spacecraft Control Toolbox for MATLAB during a summer internship in 1996 and has been using MATLAB for aerospace analysis ever since. In her over 20 years of MATLAB experience, she has developed many software tools, including the Solar Sail Module for the Spacecraft Control Toolbox, a proximity satellite operations toolbox for the Air Force, collision monitoring Simulink blocks for the Prisma satellite mission, and launch vehicle analysis tools in MATLAB and Java. She has developed novel methods for space situation assessment, such as a numeric approach to assessing the general rendezvous problem between any two satellites implemented in both MATLAB and C++. Stephanie has contributed to PSS' *Spacecraft Attitude and Orbit Control* textbook, featuring examples using the Spacecraft Control Toolbox, and written many software User's Guides. She has conducted SCT training for engineers from diverse locales such as Australia, Canada, Brazil, and Thailand and has performed MATLAB consulting for NASA, the Air Force, and the European Space Agency. Stephanie is the author of *MATLAB Recipes*, *MATLAB Machine Learning*, and *Practical MATLAB Deep Learning* published by Apress. In 2016, she was named a NASA NIAC Fellow for the project "Fusion-Enabled Pluto Orbiter and Lander." Stephanie is an Associate Fellow of the American Institute of Aeronautics and Astronautics (AIAA) and Vice Chair of the AIAA Nuclear and Future Flight Propulsion committee. Her ResearchGate profile can be found at https://www.researchgate.net/profile/Stephanie-Thomas-2.

About the Technical Reviewer

Joseph Mueller took a new position as Principal Astrodynamics Engineer at Millennium Space Systems in 2023, where groundbreaking capabilities in space are being built, one small satellite at a time, and he is honored to be a part of it.

From 2014 to 2023, he was a senior researcher at Smart Information Flow Technologies, better known as SIFT. At SIFT, he worked alongside amazing people, playing in the sandbox of incredibly interesting technical problems. His research projects at SIFT included navigation and control for autonomous vehicles, satellite formation flying, space situational awareness, and robotic swarms.

Joseph is married and is a father of three, living in Champlin, MN.

His Google Scholar profile can be found at https://scholar.google.com/citations?hl=en&user=breRtVUAAAAJ and his ResearchGate profile at www.researchgate.net/profile/Joseph-Mueller-2.

Introduction

Machine Learning is becoming important in every engineering discipline. For example:

1. Autonomous cars: Machine learning is used in almost every aspect of car control systems.

2. Plasma physicists use machine learning to help guide experiments on fusion reactors. TAE Technologies has used it with great success in guiding fusion experiments. The Princeton Plasma Physics Laboratory (PPPL) has used it for the National Spherical Torus Experiment to study a promising candidate for a nuclear fusion power plant.

3. It is used in finance for predicting the stock market.

4. Medical professionals use it for diagnoses.

5. Law enforcement and others use it for facial recognition. Several crimes have been solved using facial recognition!

6. An expert system was used on NASA's Deep Space 1 spacecraft.

7. Adaptive control systems steer oil tankers.

There are many, many other examples.

While many excellent packages are available from commercial sources and open source repositories, it is valuable to understand how these algorithms work. Writing your own algorithms is valuable both because it gives you insight into the commercial and open source packages and also because it gives you the background to write your custom Machine Learning software specialized for your application.

MATLAB had its origins for that very reason. Scientists who needed to do operations on matrices used numerical software written in FORTRAN. At the time, using computer languages required the user to go through the write-compile-link-execute process which was time-consuming and error-prone. MATLAB presented the user with a scripting language that allowed the user to solve many problems with a few lines of a script that executed instantaneously. MATLAB has built-in visualization tools that helped the user better understand the results. Writing MATLAB was a lot more productive and fun than writing FORTRAN.

The goal of *MATLAB Machine Learning Recipes: A Problem-Solution Approach* is to help all users harness the power of MATLAB to solve a wide range of learning problems. The book has something for everyone interested in Machine Learning. It also has material that will allow people with an interest in other technology areas to see how Machine Learning, and MATLAB, can help them solve problems in their areas of expertise.

Using the Included Software

This textbook includes a MATLAB toolbox that implements the examples. The toolbox consists of

1. MATLAB functions

2. MATLAB scripts

3. HTML help

The MATLAB scripts implement all of the examples in this book. The functions encapsulate the algorithms. Many functions have built-in demos. Just type the function name in the command window, and it will execute the demo. The demo is usually encapsulated in a subfunction. You can copy out this code for your demos and paste it into a script. For example, type the function name `PlotSet` into the command window, and the plot in Figure 1 will appear.

```
1   >> PlotSet
```

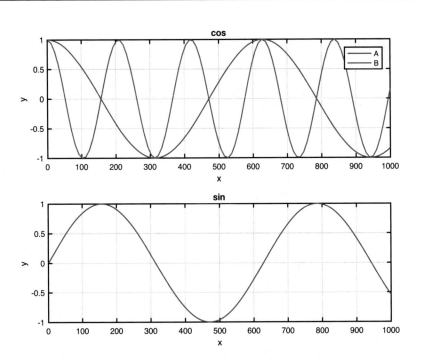

Figure 1: *Example plot from the function PlotSet.m*

If you open the function, you will see the demo:

```
1  %%% PlotSet>Demo
2  function Demo
3
4  x = linspace(1,1000);
5  y = [sin(0.01*x);cos(0.01*x);cos(0.03*x)];
6  disp('PlotSet: One x and two y rows')
7  PlotSet( x, y, 'figure title', 'PlotSet Demo',...
8      'plot set',{[2 3], 1},'legend',{{'A' 'B'},{}},'plot title',{'cos','
         sin'});
```

You can use these demos to start your scripts. Some functions, like right-hand-side functions for numerical integration, don't have demos. If you type a function name at the command line that doesn't have a built-in demo, you will get an error as in the code snippet below.

```
1  >> RHSAutomobileXY
2  Error using RHSAutomobileXY (line 17)
3  A built-in demo is not available.
```

The toolbox is organized according to the chapters in this book. The folder names are Chapter_01, Chapter_02, etc. In addition, there is a General folder with functions that support the rest of the toolbox. In addition, you will need the open source package GLPK (GNU Linear Programming Kit) to run some of the code. Nicolo Giorgetti has written a MATLAB mex interface to GLPK that is available on SourceForge and included with this toolbox. The interface consists of

1. glpk.m

2. glpkcc.mexmaci64, or glpkcc.mexw64, etc.

3. GLPKTest.m

which are available from https://sourceforge.net/projects/glpkmex/. The second item is the mex file of glpkcc.cpp compiled for your machine, such as Mac or Windows. Go to www.gnu.org/software/glpk/ to get the GLPK library and install it on your system. If needed, download the GLPKMEX source code as well and compile it for your machine, or else try another of the available compiled builds.

CHAPTER 1

■ ■ ■

An Overview of Machine Learning

1.1   Introduction

Machine Learning is a field in computer science where data is used to predict, or respond to, future data. It is closely related to the fields of pattern recognition, computational statistics, and artificial intelligence. The data may be historical or updated in real time. Machine learning is important in areas like facial recognition, spam filtering, content generation, and other areas where it is not feasible, or even possible, to write algorithms to perform a task.

For example, early attempts at filtering junk emails had the user write rules to determine what was junk or spam. Your success depended on your ability to correctly identify the attributes of the message that would categorize an email as junk, such as a sender address or words in the subject, and the time you were willing to spend to tweak your rules. This was only moderately successful as junk mail generators had little difficulty anticipating people's handmade rules. Modern systems use machine learning techniques with much greater success. Most of us are now familiar with the concept of simply marking a given message as "junk" or "not junk" and take for granted that the email system can quickly learn which features of these emails identify them as junk and prevent them from appearing in our inbox. This could now be any combination of IP or email addresses and words and phrases in the subject or body of the email, with a variety of matching criteria. Note how the machine learning in this example is data driven, autonomous, and continuously updating itself as you receive emails and flag them. However, even today, these systems are not completely successful since they do not yet understand the "meaning" of the text that they are processing.

Content generation is an evolving area. By training engines over massive data sets, the engines can generate content such as music scores, computer code, and news articles. This has the potential to revolutionize many areas that have been exclusively handled by people.

In a more general sense, what does machine learning mean? Machine learning can mean using machines (computers and software) to gain meaning from data. It can also mean giving machines the ability to learn from their environment. Machines have been used to assist humans for thousands of years. Consider a simple lever, which can be fashioned using a rock and a length of wood, or an inclined plane. Both of these machines perform useful work and assist people, but neither can learn. Both are limited by how they are built. Once built, they cannot adapt to changing needs without human interaction.

M. Paluszek, S. Thomas, *MATLAB Machine Learning Recipes*,
https://doi.org/10.1007/978-1-4842-9846-6_1

Machine learning involves using data to create a model that can be used to solve a problem. The model can be explicit, in which case the machine learning algorithm adjusts the model's parameters, or the data can form the model. The data can be collected once and used to train a machine learning algorithm, which can then be applied. For example, ChatGPT scrapes textual data from the Internet to allow it to generate text based on queries. An adaptive control system measures inputs and command responses to those inputs to update parameters for the control algorithm.

In the context of the software we will be writing in this book, *machine learning* refers to the process by which an algorithm converts the input data into parameters it can use when interpreting future data. Many of the processes used to mechanize this learning derive from optimization techniques and, in turn, are related to the classic field of automatic control. In the remainder of this chapter, we will introduce the nomenclature and taxonomy of machine learning systems.

1.2 Elements of Machine Learning

This section introduces key nomenclature for the field of machine learning.

1.2.1 Data

All learning methods are data driven. Sets of data are used to train the system. These sets may be collected and edited by humans or gathered autonomously by other software tools. Control systems may collect data from sensors as the systems operate and use that data to identify parameters or train the system. Content generation systems scour the Internet for information. The data sets may be very large, and it is the explosion of data storage infrastructure and available databases that is largely driving the growth in machine learning software today. It is still true that a machine learning tool is only as good as the data used to create it, and the selection of training data is practically a field in itself. Selection of data for many systems is highly automated.

■ **NOTE**   When collecting data for training, one must be careful to ensure that the time variation of the system is understood. If the structure of a system changes with time, it may be necessary to discard old data before training the system. In automatic control, this is sometimes called a forgetting factor in an estimator.

1.2.2 Models

Models are often used in learning systems. A model provides a mathematical framework for learning. A model is human-derived and based on human observations and experiences. For example, a model of a car, seen from above, might be that it is rectangular with dimensions that fit within a standard parking spot. Models are usually thought of as human-derived and provide a framework for machine learning. However, some forms of machine learning develop their models without a human-derived structure.

1.2.3 Training

A system which maps an input to an output needs training to do this in a useful way. Just as people need to be trained to perform tasks, machine learning systems need to be trained. Training is accomplished by giving the system an input and the corresponding output and modifying the structure (models or data) in the learning machine so that mapping is learned. In some ways, this is like curve fitting or regression. If we have enough training pairs, then the system should be able to produce correct outputs when new inputs are introduced. For example, if we give a face recognition system thousands of cat images and tell it that those are cats, we hope that when it is given new cat images it will also recognize them as cats. Problems can arise when you don't give it enough training sets, or the training data is not sufficiently diverse, for instance, identifying a long-haired cat or hairless cat when the training data is only of short-haired cats. A diversity of training data is required for a functioning algorithm.

Supervised Learning

Supervised learning means that specific training sets of data are applied to the system. The learning is supervised in that the "training sets" are human-derived. It does not necessarily mean that humans are actively validating the results. The process of classifying the systems' outputs for a given set of inputs is called "labeling." That is, you explicitly say which results are correct or which outputs are expected for each set of inputs.

The process of generating training sets can be time-consuming. Great care must be taken to ensure that the training sets will provide sufficient training so that when real-world data is collected, the system will produce correct results. They must cover the full range of expected inputs and desired outputs. The training is followed by test sets to validate the results. If the results aren't good, then the test sets are cycled into the training sets, and the process is repeated.

A human example would be a ballet dancer trained exclusively in classical ballet technique. If they were then asked to dance a modern dance, the results might not be as good as required because the dancer did not have the appropriate training sets; their training sets were not sufficiently diverse.

Unsupervised Learning

Unsupervised learning does not utilize training sets. It is often used to discover patterns in data for which there is no "right" answer. For example, if you used unsupervised learning to train a face identification system, the system might cluster the data in sets, some of which might be faces. Clustering algorithms are generally examples of unsupervised learning. The advantage of unsupervised learning is that you can learn things about the data that you might not know in advance. It is a way of finding hidden structures in data.

Semi-supervised Learning

With this approach, some of the data are in the form of labeled training sets, and other data are not [12]. Typically, only a small amount of the input data is labeled, while most are not, as the labeling may be an intensive process requiring a skilled human. The small set of labeled data is leveraged to interpret the unlabeled data.

Online Learning

The system is continually updated with new data [12]. This is called "online" because many of the learning systems use data collected while the system is operating. It could also be called recursive learning. It can be beneficial to periodically "batch" process data used up to a given time and then return to the online learning mode. The spam filtering systems collect data from emails and update their spam filter. Generative deep learning systems like ChatGPT use massive online learning.

1.3   The Learning Machine

Figure 1.1 shows the concept of a learning machine. The machine absorbs information from the environment and adapts. The inputs may be separated into those that produce an immediate response and those that lead to learning. In some cases, they are completely separate. For example, in an aircraft, a measurement of altitude is not usually used directly for control. Instead, it is used to help select parameters for the actual control laws. The data required for learning and regular operation may be the same, but in some cases, separate measurements or data will be needed for learning to take place. Measurements do not necessarily mean data collected by a sensor such as radar or a camera. It could be data collected by polls, stock market prices, data in accounting ledgers, or any other means. Machine learning is then the process by which the measurements are transformed into parameters for future operation.

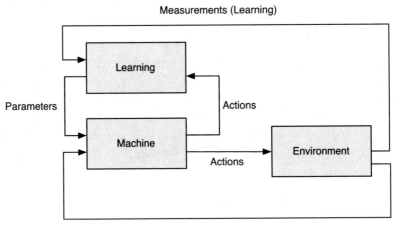

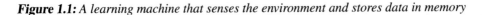

Figure 1.1: *A learning machine that senses the environment and stores data in memory*

Note that the machine produces output in the form of actions. A copy of the actions may be passed to the learning system so that it can separate the effects of the machine's actions from those of the environment. This is akin to a feedforward control system, which can result in improved performance.

A few examples will clarify the diagram. We will discuss a medical example, a security system, and spacecraft maneuvering.

A doctor might want to diagnose diseases more quickly. They would collect data on tests on patients and then collate the results. Patient data might include age, height, weight, historical data like blood pressure readings and medications prescribed, and exhibited symptoms. The machine learning algorithm would detect patterns so that when new tests were performed on a patient, the machine learning algorithm would be able to suggest diagnoses or additional tests to narrow down the possibilities. As the machine learning algorithm was used, it would, hopefully, get better with each success or failure. Of course, the definition of success or failure is fuzzy. In this case, the environment would be the patients themselves. The machine would use the data to generate actions, which would be new diagnoses. This system could be built in two ways. In the supervised learning process, test data and known correct diagnoses are used to train the machine. In an unsupervised learning process, the data would be used to generate patterns that might not have been known before, and these could lead to diagnosing conditions that would normally not be associated with those symptoms.

A security system might be put into place to identify faces. The measurements are camera images of people. The system would be trained with a wide range of face images taken from multiple angles. The system would then be tested with these known persons and its success rate validated. Those that are in the database memory should be readily identified, and those that are not should be flagged as unknown. If the success rate was not acceptable, more training might be needed, or the algorithm itself might need to be tuned. This type of face recognition is now common, used in Mac OS X's "Faces" feature in Photos, face identification on the new iPhone X, and Facebook when "tagging" friends in photos.

For precision maneuvering of a spacecraft, the inertia of the spacecraft needs to be known. If the spacecraft has an inertial measurement unit that can measure angular rates, the inertia matrix can be identified. This is where machine learning is tricky. The torque applied to the spacecraft, whether by thrusters or momentum exchange devices, is only known to a certain degree of accuracy. Thus, the system identification must sort out, if it can, the torque scaling factor from the inertia. The inertia can only be identified if torques are applied. This leads to the issue of stimulation. A learning system cannot learn if the system to be studied does not have known inputs, and those inputs must be sufficiently diverse to stimulate the system so that the learning can be accomplished. Training a face recognition system with one picture will not work.

1.4   Taxonomy of Machine Learning

In this book, we take a larger view of machine learning than is normal. Machine learning as described earlier is the collecting of data, finding patterns, and doing useful things based on those patterns. We expand machine learning to include adaptive and learning control. These fields started independently but now are adapting technology and methods from machine learning.

Figure 1.2 shows how we organize the technology of machine learning into a consistent taxonomy. You will notice that we created a title that encompasses three branches of learning; we call the whole subject area "Autonomous Learning." That means learning without human intervention during the learning process. This book is not solely about "traditional" machine learning. Other, more specialized books focus on any one of the machine learning topics. Optimization is part of the taxonomy because the results of optimization can be discoveries, such as a new type of spacecraft or aircraft trajectory. Optimization is also often a part of learning systems.

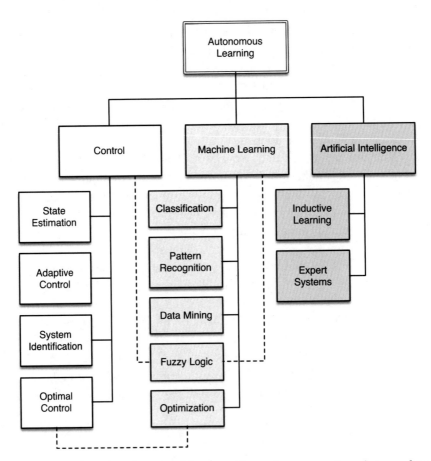

Figure 1.2: *Taxonomy of machine learning. The dotted lines show connections between branches*

There are three categories under Autonomous Learning. The first is *Control*. Feedback control is used to compensate for uncertainty in a system or to make a system behave differently than it would normally behave. If there was no uncertainty, you wouldn't need feedback. For example, if you are a quarterback throwing a football at a running player, assume for a moment that you know everything about the upcoming play. You know exactly where the player should be at a given time, so you can close your eyes, count, and just throw the ball to that spot. Assuming the player has good hands, you would have a 100% reception rate! More realistically, you watch the player, estimate the player's speed, and throw the ball. You are applying feedback to the problem. As stated, this is not a learning system. However, if now you practice the same play repeatedly, look at your success rate, and modify the mechanics and timing of your throw using that information, you would have an adaptive control system, the second box from the top of the control list. Learning in control takes place in adaptive control systems and also in the general area of system identification.

System identification is learning about a system. By system, we mean the data that represents anything and the relationships between elements of that data. For example, a particle moving in a straight line is a system defined by its mass, the force on that mass, its velocity, and its position. The position is related to the velocity times time, and the velocity is related and determined by the acceleration, which is the force divided by the mass.

Optimal control may not involve any learning. For example, what is known as full-state feedback produces an optimal control signal but does not involve learning. In full-state feedback, the combination of model and data tells us everything we need to know about the system. However, in more complex systems, we can't measure all the states and don't know the parameters perfectly, so some form of learning is needed to produce "optimal" or the best possible results. In a learning system, optimal control would need to be redefined as the system learns. For example, an optimal space trajectory assumes thruster characteristics. As a mission progresses, the thruster performance may change, requiring recomputation of the "optimal" trajectory.

System identification is the process of identifying the characteristics of a system. A system can, to a first approximation, be defined by a set of dynamical states and parameters. For example, in a linear time-invariant system, the dynamical equation is

$$\dot{x} = Ax + Bu \tag{1.1}$$

where A and B are matrices of parameters, u is an input vector, and x is the state vector. System identification would find A and B. In a real system, A and B are not necessarily time invariant, and most systems are only linear to a first approximation.

The second category is what many people consider true *Machine Learning*. This is making use of data to produce behavior that solves problems. Much of its background comes from statistics and optimization. The learning process may be done once in a batch process or continually in a recursive process. For example, in a stock buying package, a developer might have processed stock data for several years, say before 2008, and used that to decide which stocks to buy. That software might not have worked well during the financial crash. A recursive program would continuously incorporate new data. Pattern recognition and data mining fall into this category. Pattern recognition is looking for patterns in images. For example, the early AI

Blocks World software could identify a block in its field of view. It could find one block in a pile of blocks. Data mining is taking large amounts of data and looking for patterns, for example, taking stock market data and identifying companies that have strong growth potential. Classification techniques and fuzzy logic are also in this category.

The third category of autonomous learning is *artificial intelligence*. Our diagram includes the two techniques of inductive learning and expert systems. Machine learning traces some of its origins to artificial intelligence. Artificial intelligence is an area of study whose goal is to make machines reason. While many would say the goal is "think like people," this is not necessarily the case. There may be ways of reasoning that are not similar to human reasoning but are just as valid. In the classic Turing test, Turing proposes that the computer only needs to imitate a human in its output to be a "thinking machine," regardless of how those outputs are generated. Systems like ChatGPT appear to easily pass the Turing test. This leads to a need to redefine intelligence. If ChatGPT can produce a decent sonata, is it as "intelligent" as a composer? In any case, intelligence generally involves learning, so learning is inherent in many Artificial intelligence technologies such as inductive learning and expert systems.

The recipe chapters of this book are grouped according to this taxonomy. The first chapters cover state estimation using the Kalman Filter and adaptive control. Fuzzy logic is then introduced, which is a control methodology that uses classification. Additional machine learning recipes follow with chapters on data classification with binary trees, neural nets including deep learning, and multiple hypothesis testing. We have a chapter on aircraft control that incorporates neural nets, showing the synergy between the different technologies. Finally, we conclude with a chapter on an artificial intelligence technique, case-based expert systems.

1.5 Control

Feedback control algorithms inherently learn about the environment through measurements used for control. These chapters show how control algorithms can be extended to effectively design themselves using measurements. The measurements may be the same as used for control, but the adaptation, or learning, happens more slowly than the control response time. An important aspect of control design is stability. A stable controller will produce bounded outputs for bounded inputs. It will also produce smooth, predictable behavior of the system that is controlled. An unstable controller will typically experience growing oscillations in the quantities (such as speed or position) that are controlled. In these chapters, we explore both the performance of learning control and the stability of such controllers. We often break control into two parts, control and estimation. The latter may be done independently of feedback control.

1.5.1 Kalman Filters

Chapter 4 shows how Kalman Filters allow you to learn about dynamical systems for which we already have a model. This chapter provides an example of a variable gain Kalman Filter for a spring system. That is a system with a mass connected to its base via a spring and a damper. This is a linear system. We write the system in discrete time. This provides an introduction to Kalman Filtering. We show how Kalman Filters can be derived from Bayesian statistics. This ties it into many machine learning algorithms. Originally, the Kalman Filter, developed by R. E. Kalman, R. S. Bucy, and R. Battin, was not derived in this fashion. Kalman Filters typically learn about the state of a system, and their learning rate is fixed by a priori assumptions about the system noise.

The second recipe adds a nonlinear measurement. A linear measurement is a measurement proportional to the state (in this case, position) it measures. Our nonlinear measurement will be the angle of a tracking device that points at the mass from a distance from the line of movement. One way is to use an Unscented Kalman Filter (UKF) for state estimation. The UKF lets us use a nonlinear measurement model easily.

The last part of the chapter describes the Unscented Kalman Filter configured for parameter estimation. This system learns the model, albeit one that has an existing mathematical model. As such, it is an example of model-based learning. In this example, the filter estimates the oscillation frequency of the spring-mass system. It will demonstrate how the system needs to be stimulated to identify the parameters.

1.5.2 Adaptive Control

Adaptive control is a branch of control systems in which the gains of the control system change based on measurements of the system. A gain is a number that multiplies a measurement from a sensor to produce a control action such as driving a motor or other actuator. In a non-learning control system, the gains are computed before operation and remain fixed. This works very well most of the time since we can usually pick gains so that the control system is tolerant of parameter changes in the system. Our gain "margins" tell us how tolerant we are to uncertainties in the system. If we are tolerant to big changes in parameters, we say that our system is robust.

Adaptive control systems change the gain based on measurements during operation. This can help a control system perform even better. The better we know a system's model, the tighter we can control the system. This is much like driving a new car. At first, you have to be cautious driving a new car because you don't know how sensitive the steering is to turn the wheel or how fast it accelerates when you depress the gas pedal. As you learn about the car, you can maneuver it with more confidence. If you didn't learn about the car, you would need to drive every car in the same fashion.

Chapter 5 starts with a simple example of adding damping to a spring using a control system. Our goal is to get a specific damping time constant. For this, we need to know the spring constant. Our learning system uses a Fast Fourier Transform (FFT) to measure the spring constant. We'll compare it to a system that does know the spring constant. This is an example of tuning a control system. The second example is model reference adaptive control of a first-order system.

This system automatically adapts so that the system behaves like the desired model. This is a very powerful method and applies to many situations. Another example is ship steering control. Ships use adaptive control because it is more efficient than conventional control. This example demonstrates how the control system adapts and how it performs better than its nonadaptive equivalent. This is an example of gain scheduling. We then give a spacecraft example.

The next example is longitudinal control of an aircraft, extensive enough that it is given its own chapter. We can control the pitch angle using the elevators. We have five nonlinear equations for the pitch rotational dynamics, velocity in the x direction, velocity in the z direction, and change in altitude. The system adapts to changes in velocity and altitude. Both change the drag and lift forces and the moments on the aircraft and also change the response to the elevators. We use a neural net as the learning element of our control system. This is a practical problem applicable to all types of aircraft ranging from drones to high-performance commercial aircraft.

1.6 Autonomous Learning Methods

This section introduces you to popular machine learning techniques. Some will be used in the examples in this book. Others are available in MATLAB products and open source products.

1.6.1 Regression

Regression is a way of fitting data to a model. A model can be a curve in multiple dimensions. The regression process fits the data to the curve producing a model that can be used to predict future data. Some methods, such as linear regression or least squares, are parametric in that the number of parameters to be fit is known. An example of linear regression is shown in the following listing and in Figure 1.3. This model was created by starting with the line $y = x$ and adding noise to y. The line was recreated using a least squares fit via MATLAB's `pinv` pseudo-inverse function.

The first part of the script generates the data.

LinearRegression.m

```
 6  x     = linspace(0,1,500)';
 7  n     = length(x);
 8
 9  % Model a polynomial, y = ax2 + mx + b
10  a     = 1.0;     % quadratic - make nonzero for larger errors
11  m     = 1.0;     % slope
12  b     = 1.0;     % intercept
13  sigma = 0.1; % standard deviation of the noise
14  y0    = a*x.^2 + m*x + b;
15  y     = y0 + sigma*randn(n,1);
```

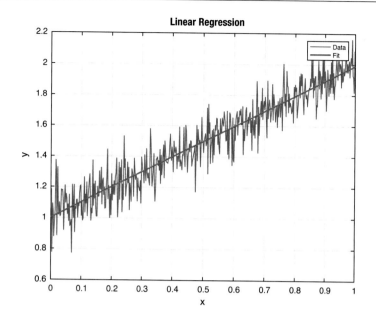

Figure 1.3: *Learning with linear regression when $a = 0$*

The actual regression code is just three lines.

LinearRegression.m

```
18   a      = [x ones(n,1)];
19   c      = pinv(a)*y;
20   yR     = c(1)*x + c(2); % the fitted line
```

The last part plots the results using standard MATLAB plotting functions. We use `grid on` rather than `grid`. The latter toggles the grid mode and is usually OK, but sometimes MATLAB gets confused. `grid on` is more reliable.

LinearRegression.m

```
23   h = figure('Name','Linear Regression');
24   plot(x,y); hold on;
25   plot(x,yR,'linewidth',2);
26   grid on
27   xlabel('x');
28   ylabel('y');
29   title(h.Name);
30   legend('Data','Fit')
31
32   figure('Name','Regression Error')
33   plot(x,yR-y0);
34   grid on
35   xlabel('x');
```

11

```
36  ylabel('\Delta y');
37  title('Error between Model and Regression')
```

This code uses `pinv`. We can solve the problem:

$$Ax = b \tag{1.2}$$

by taking the inverse of A if the length of x and b are the same:

$$x = A^{-1}b \tag{1.3}$$

This works because A is a square matrix but only works if A is not singular. That is, it has a valid inverse. If the length of x and b is the same, we can still find an approximation to x where $x = \text{pinv}(A)b$. For example, in the first case, A is 2 by 2. In the second case, it is 3 by 2, meaning there are three elements of x and two of b.

```
>> inv(rand(2,2))

ans =

     1.4518    -0.2018
    -1.4398     1.2950

>> pinv(rand(2,3))

ans =

     1.5520    -1.3459
    -0.6390     1.0277
     0.2053     0.5899
```

The system computes the parameters, slope, and y intercept, from the data using an algorithm. The more data, the better the fit. As it happens, our model

$$y = mx + b \tag{1.4}$$

is correct. However, if it were wrong, the fit would be poor. This is an issue with model-based learning. The quality of the results is highly dependent on the model. If you are sure of your model, then it should be used. If not, other methods, such as unsupervised learning, may produce better results. For example, if we add the quadratic term x^2, we get the fit in Figure 1.4. Notice how the fit is not as good as we might like.

In these examples, we start with a pattern that we assume fits the data. This is our model. We fit the data into the model. In the first case, we assume that our system is linear. In the second, we assume it is quadratic. If our model is good, the data will fit well. If we chose the wrong model, then the fit will be poor. If that is the case, we will need to try a different model. For example, our system could be

$$y = \cos(x) \tag{1.5}$$

with the span of x over several cycles. Neither a linear nor a quadratic fit would be good in this case. Limitations in this approach have led to other techniques, including neural networks.

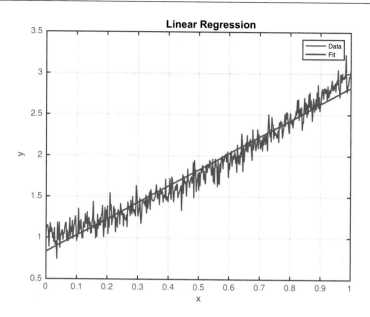

Figure 1.4: *Learning with linear regression for a quadratic model with* $a = 1.0$

1.6.2 Decision Trees

A decision tree is a tree-like graph used to make decisions. It has three kinds of nodes:

1. Decision nodes

2. Chance nodes

3. End nodes

You follow the path from the beginning to the end node. Decision trees are easy to understand and interpret. The decision process is entirely transparent although very large decision trees may be hard to follow visually. The difficulty is finding an optimal decision tree for a set of training data.

Two types of decision trees are classification trees which produce categorical outputs and regression trees which produce numeric outputs. An example of a classification tree is shown in Figure 1.5. This helps an employee decide where to go for lunch. This tree only has decision nodes.

This might be used by management to predict where they could find an employee at lunchtime. The decisions are Hungry, Busy, and Have a Credit Card. From that, the tree could be synthesized. However, if there were other factors in the decision of employees, for example, it is someone's birthday, which would result in the employee going to a restaurant, then the tree would not be accurate.

Chapter 10 uses a decision tree to classify data. Classifying data is one of the most widely used areas of machine learning. In this example, we assume that two data points are sufficient

13

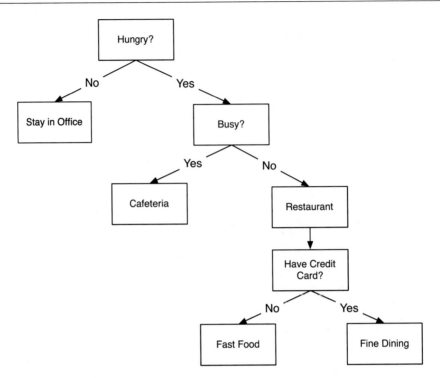

Figure 1.5: *A classification tree*

to classify a sample and determine to which group it belongs. We have a training set of known data points with membership in one of three groups. We then use a decision tree to classify the data. We'll introduce a graphical display to make understanding the process easier.

With any learning algorithm, it is important to know why the algorithm made its decision. Graphics can help you explore large data sets when columns of numbers aren't helpful.

1.6.3 Neural Networks

Introduction

A neural net is a network of neurons designed to emulate the neurons in a human brain. Each "neuron" has a mathematical model for determining its output from its input; for example, if the output is a step function with a value of 0 or 1, the neuron can be said to be "firing" if the input stimulus results in a 1 output. Networks are then formed with multiple layers of interconnected neurons. Neural networks perform pattern recognition. The network must be trained using sample data, but no a priori model is required. However, usually, the structure of the neural network is specified by giving the number of layers, neurons per layer, and activation functions for each neuron. Networks can be trained to estimate the output of nonlinear processes, and the network then becomes the model.

Figure 1.6 displays a simple neural network that flows from left to right, with two input nodes and one output node. There is one "hidden" layer of neurons in the middle. Each node

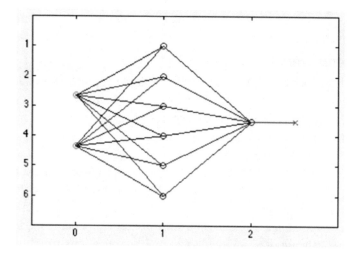

Figure 1.6: *A neural net with one intermediate layer between the inputs on the left and the output on the right. The intermediate layer is also known as a hidden layer*

has a set of numeric weights that are tuned during training. This network has two inputs and one output, possibly indicative of a network that solves a categorization problem. Training such a network is called deep learning.

A "deep" neural network is a neural network with multiple intermediate layers between the input and output.

This book presents neural nets in several chapters. Chapter 7 introduces a neural network as part of an adaptive control system. This ties together learning, via neural networks, and control. Chapter 8 provides an introduction to the fundamentals of neural networks focusing on the neuron and how it can be trained. Chapter 9 provides an introduction to neural networks using multilayer feedforward (MLFF) neural networks to classify digits. In this type of network, each neuron depends only on the inputs it receives from the previous layer. The example uses a neural network to classify digits. We will start with a set of six digits and create a training set by adding noise to the digit images. We then see how well our learning network performs at identifying a single digit and then add more nodes and outputs to identify multiple digits with one network. Classifying digits is one of the oldest uses of machine learning. The US Post Office introduced zip code reading years before Machine Learning started hitting the front pages of all the newspapers! Earlier digit readers required block letters written in well-defined spots on a form. Reading digits off any envelope is an example of learning in an unstructured environment.

Chapter 11 presents deep learning with distinctive layers. Several different types of elements are in the deep learning chain. This is applied to face recognition. Face recognition is available in almost every photo application. Many social media sites, such as Facebook and Google Plus, also use face recognition. Cameras have built-in face recognition, though not identification, to help with focusing when taking portraits. Our goal is to get the algorithm to match faces, not classify them.

The last chapter in the machine learning group employs deep learning to do spacecraft attitude determination.

Generative Deep Learning

Generative machine learning (ML) models are a class of models that allow you to create new data by modeling the data-generating distribution. For example, a generative model trained on images of human faces would learn what features constitute a realistic human face and how to combine them to generate novel human face images. For a fun demonstration of the power of ML-based human face generation, check out [34].

This is in contrast to a discriminative model that learns an association between a set of labels and the training inputs. Staying with our face example, a discriminative model might predict the age of a person given an image of their face. In this case, the input is the image of the face, and the label is the numerical age. Labels can also be used in generative models.

Generative models are used in a wide variety of applications from drug design to language models for better chatbots and autocomplete features. Generative models are also used in data augmentation to train better discriminative models, especially in situations where training data is difficult or expensive to obtain. Finally, generative models are widely used by artists and composers to inspire or augment their work.

ChatGPT is an example. It can produce all sorts of interesting material based on questions that it is asked. A MATLAB interface to ChatGPT is presented in Chapter 2.

Reinforcement Learning

Reinforcement learning is a machine learning approach in which an intelligent agent learns to take actions to maximize a reward. We will apply this to the design of a Titan landing control system. Reinforcement learning is a tool to approximate solutions that could have been obtained by dynamic programming, but whose exact solutions are computationally intractable [4].

1.6.4 Support Vector Machines (SVMs)

Support vector machines (SVMs) are supervised learning models with associated learning algorithms that analyze data used for classification and regression analysis. An SVM training algorithm builds a model that assigns examples into categories. The goal of SVM is to produce a model, based on the training data that predict the target values.

In SVMs, nonlinear mapping of input data in a higher dimensional feature space is done with kernel functions. In this feature space, a separation hyperplane is generated that is the solution to the classification problem. The kernel functions can be polynomials, sigmoidal functions, and radial basis functions. Only a subset of the training data is needed; these are known as the support vectors [9]. The training is done by solving a quadratic program which can be done with many numerical software.

1.7 Artificial Intelligence

1.7.1 What Is Artificial Intelligence?

The original test of artificial intelligence is the Turing test [13]. The idea is that if you have a conversation with a machine and you can't tell it is a machine, then it should be considered intelligent. By this definition, many robocalling systems might be considered intelligent. Another example is chess programs, which can beat all but the best players, but a chess program can't do anything but play chess. Is a chess program intelligent? What we have now are machines that can do things pretty well in a particular context.

As Machine Learning is an offshoot of artificial intelligence, all the Machine Learning examples could also be considered as artificial intelligence.

1.7.2 Intelligent Cars

Our "artificial intelligence" example is a blending of Bayesian estimation and controls. It still reflects a machine doing what we would consider intelligent behavior. This, of course, gets back to the question of defining intelligence.

Autonomous driving is an area of great interest to automobile manufacturers and the general public. Autonomous cars are driving the streets today but are not yet ready for general use by the public. There are many technologies involved in autonomous driving. These include

1. Machine vision: Turning camera data into information useful for the autonomous control system

2. Sensing: Using many technologies including vision, radar, and sound to sense the environment around the car

3. Control: Using algorithms to make the car go where it is supposed to go as determined by the navigation system

4. Machine learning: Using massive data from test cars to create databases of responses to situations

5. GPS navigation: Blending GPS measurements with sensing and vision to figure out where to go

6. Communications/ad hoc networks: Talking with other cars to help determine where they are and what they are doing

All of the areas overlap. Communications and ad hoc networks are used with GPS navigation to determine both absolute location (what street and address correspond to your location) and relative navigation (where you are concerning other cars). In this context, the Turing test would be a success if you couldn't tell if a car was driven by a person or a computer. Now, since many drivers are bad, one could argue that a computer that drove well would fail the Turing test! This gets back to the question of what is intelligence.

This example explores the problem of a car being passed by multiple cars and needing to compute tracks for each one. We are addressing just the control and collision avoidance problem. A single-sensor version of Track-Oriented Multiple Hypothesis Testing is demonstrated for a single car on a two-lane road. The example includes MATLAB graphics that make it easier to understand the thinking of the algorithm. The demo assumes that the optical or radar preprocessing has been done and that each target is measured by a single "blip" in two dimensions. An automobile simulation is included. It involves cars passing the car that is doing the tracking. The passing cars use a passing control system that is in itself a form of machine intelligence.

Our autonomous driving recipes use an Unscented Kalman Filter for the estimation of the state. This is the underlying algorithm that propagates the state (i.e., advances the state in time in a simulation) and adds measurements to the state. A Kalman Filter, or other estimator, is the core of many target tracking systems.

The recipes will also introduce graphics aids to help you understand the tracking decision process. When you implement a learning system, you want to make sure it is working the way you think it should or understand why it is working the way it does.

1.7.3 Expert Systems

An expert system is a system that uses a knowledge base to reason and present the user with results and an explanation of how it arrived at that result. Expert systems are also known as knowledge-based systems. The process of building an expert system is called knowledge engineering. This involves a knowledge engineer, someone who knows how to build the expert system, interviewing experts for the knowledge needed to build the system. Some systems can induce rules from data speeding the data acquisition process.

An advantage of expert systems, over human experts, is that knowledge from multiple experts can be incorporated into the database. Another advantage is that the system can explain the process in detail so that the user knows exactly how the result was generated. Even an expert in a domain can forget to check certain things. An expert system will always methodically check its full database. It is also not affected by fatigue or emotions.

Knowledge acquisition is a major bottleneck in building expert systems. Another issue is that the system cannot extrapolate beyond what is programmed into the database. Care must be taken with using an expert system because it will generate definitive answers for problems where there is uncertainty. The explanation facility is important because someone with domain knowledge can judge the results from the explanation.

In cases where uncertainty needs to be considered, a probabilistic expert system is recommended. A Bayesian network can be used as an expert system. A Bayesian network is also known as a belief network. It is a probabilistic graphical model that represents a set of random variables and their dependencies. In the simplest cases, a Bayesian network can be constructed by an expert. In more complex cases, it needs to be generated from data from Machine Learning. Chapter 15 delves into expert systems.

In Chapter 15, we explore a simple case-based reasoning system. An alternative would be a rule-based system.

1.8   Summary

All of the technologies in this chapter are in current use today. Any one of them can form the basis for a useful product. Many systems, such as autonomous cars, use several. We hope that our broad view of the field of machine learning and our unique taxonomy, which shows the relationships of machine learning and artificial intelligence to the classical fields of control and optimization, are useful to you. In the remainder of the book, we will show you how to build software that implements these technologies. This can form the basis of your own more robust production software or help you to use the many fine commercial products more effectively. Table 1.1 lists the scripts included in the companion code.

Table 1.1: *Chapter Code Listing*

File	Description
LinearRegression	A script that demonstrates linear regression and curve fitting

CHAPTER 2

■ ■ ■

Data for Machine Learning in MATLAB

2.1 Introduction to MATLAB Data Types

2.1.1 Matrices

By default, all variables in MATLAB are double-precision matrices. You do not need to declare a type for these variables. Matrices can be multidimensional and are accessed using one-based indices via parentheses. You can address elements of a matrix using a single index, taken column-wise, or one index per dimension. To create a matrix variable, simply assign a value to it, like this 2×2 matrix a:

```
>> a = [1 2; 3 4];
>> a(1,1)
     1

>> a(3)
     2
```

■ **TIP**   A semicolon terminates an expression so that it does not appear in the command window. If you leave out the semicolon, it will print in the command window. Leaving out semicolons is a convenient way of debugging, without using the MATLAB debugger, but it can be hard to find those missing semicolons later!

You can simply add, subtract, multiply, and divide matrices with no special syntax. The matrices must be the correct size for the linear algebra operation requested. A transpose is indicated using a single quote suffix, A', and the matrix power uses the operator ^.

```
>> b = a'*a;
>> c = a^2;
>> d = b + c;
```

© The Author(s), under exclusive license to APress Media, LLC, part of Springer Nature 2024
M. Paluszek, S. Thomas, *MATLAB Machine Learning Recipes*,
https://doi.org/10.1007/978-1-4842-9846-6_2

Table 2.1: *Key Functions for Matrices*

Function	Purpose
zeros	Initialize a matrix to zeros
ones	Initialize a matrix to ones
eye	Initialize an identity matrix
rand, randn	Initialize a matrix of random numbers
isnumeric	Identify a matrix or scalar numeric value
isscalar	Identify a scalar value (a 1×1 matrix)
size	Return the size of the matrix

By default, every variable is a numerical variable. You can initialize matrices to a given size using the zeros, ones, eye, or rand functions, which produce zeros, ones, identity matrices (ones on the diagonal), and random numbers, respectively. Use isnumeric to identify numeric variables.

MATLAB can support n-dimensional arrays. A two-dimensional array is like a table. A three-dimensional array can be visualized as a cube where each box inside the cube contains a number. A four-dimensional array is harder to visualize, but we needn't stop there! Table 2.1 lists some key functions for interacting with matrices.

2.1.2 Cell Arrays

One variable type unique to MATLAB is cell arrays. This is a list container, and you can store variables of any type in elements of a cell array. Cell arrays can be multidimensional, just like matrices, and are useful in many contexts.

Cell arrays are indicated by curly braces, {}. They can be of any dimension and contain any data, including strings, structures, and objects. You can initialize them using the cell function, recursively display the contents using celldisp, and access subsets using parentheses just like for a matrix. A short example is as follows:

```
>> c = cell(3,1);
>> c{1} = 'string';
>> c{2} = false;
>> c{3} = [1 2; 3 4];
>> b = c(1:2);
>> celldisp(b)
b{1} =
string

b{2} =
     0
```

22

Table 2.2: *Key Functions for Cell Arrays*

Function	Purpose
cell	Initialize a cell array
cellstr	Create a cell array from a character array
iscell	Identify a cell array
iscellstr	Identify a cell array containing only strings

Using curly braces for access gives you the element data as the underlying type. When you access elements of a cell array using parentheses, the contents are returned as another cell array, rather than the cell contents. MATLAB help has a special section called *Comma-Separated Lists* which highlights the use of cell arrays as lists. The code analyzer will also suggest more efficient ways to use cell arrays. For instance:

Replace

```
a = {b{:} c};
```

with

```
a = [b {c}];
```

Cell arrays are especially useful for sets of strings, with many of MATLAB's string search functions optimized for cell arrays, such as strcmp.

Use iscell to identify cell array variables. Use deal to manipulate structure array and cell array contents. Table 2.2 lists some key functions for manipulating cell arrays.

2.1.3 Data Structures

Data structures in MATLAB are highly flexible, leaving it up to the user to enforce consistency in fields and types. You are not required to initialize a data structure before assigning fields to it, but it is a good idea to do so, especially in scripts, to avoid variable conflicts.

Replace

```
d.fieldName = 0;
```

with

```
d = struct;
d.fieldName = 0;
```

We have found it generally a good idea to create a special function to initialize larger structures that are used throughout a set of functions. This is similar to creating a class definition. Generating your data structure from a function, instead of typing out the fields in a script, means you always start with the correct fields. Having an initialization function also allows you to specify the types of variables and provide sample or default data. Remember, since MATLAB does not require you to declare variable types, doing so yourself with default data makes your code that much clearer.

■ **TIP**   Create an initialization function for data structures.

You make a data structure into an array simply by assigning an additional copy. The fields must be identically named (they are case sensitive) and in the same order, which is yet another reason to use a function to initialize your structure. You can nest data structures with no limit on depth.

```
d = MyStruct;
d(2) = MyStruct;

function d = MyStruct

d = struct;
d.a = 1.0;
d.b = 'string';
```

MATLAB now allows for *dynamic field names* using variables, that is, `structName.` `(dynamicExpression)`. This provides improved performance over `getfield`, where the field name is passed as a string. This allows for all sorts of inventive structure programming. Take our data structure array in the previous code snippet, and let's get the values of the field a using a dynamic field name; the values are returned in a cell array.

```
>> field = 'a';
>> values = {d.(field)}

values =
    [1]    [1]
```

Use `isstruct` to identify structure variables and `isfield` to check for the existence of fields. Note that `isempty` will return *false* for a struct initialized with `struct`, even if it has no fields.

```
>> d = struct
d =
  struct with no fields.

>> isempty(d)

ans =
  logical
   0
```

Table 2.3 provides key functions for structs.

Table 2.3: *Key Functions for Structs*

Function	Purpose
struct	Initialize a structure with or without fields
isstruct	Identify a structure
isfield	Determine if a field exists in a structure
fieldnames	Get the fields of a structure in a cell array
rmfield	Remove a field from a structure
deal	Set fields in a structure array to a value

2.1.4 Numerics

While MATLAB defaults to doubles for any data entered at the command line or in a script, you can specify a variety of other numeric types, including `single`, `uint8`, `uint16`, `uint32`, `uint64`, and `logical` (i.e., an array of booleans). The use of the integer types is especially relevant to using large data sets such as images. Use the minimum data type you need, especially when your data sets are large.

2.1.5 Images

MATLAB supports a variety of formats including GIF, JPG, TIFF, PNG, HDF, FITS, and BMP. You can read an image directly using `imread`, which can determine the type automatically from the extension, or `fitsread`. (FITS stands for Flexible Image Transport System, and the interface is provided by the CFITSIO library.) `imread` has special syntaxes for some image types, such as handling alpha channels for PNG, so you should review the options for your specific images. `imformats` manages the file format registry and allows you to specify the handling of new user-defined types if you can provide read and write functions.

You can display an image using either `imshow`, `image`, or `imagesc`, which scales the colormap for the range of data in the image.

For example, we use a set of images of cats in Chapter 11. The following is the image information for one of these sample images:

```
>> imfinfo('IMG_4901.JPG')
ans =
            Filename: 'MATLAB/Cats/IMG_4901.JPG'
         FileModDate: '28-Sep-2016 12:48:15'
            FileSize: 1963302
              Format: 'jpg'
       FormatVersion: ''
               Width: 3264
              Height: 2448
            BitDepth: 24
           ColorType: 'truecolor'
     FormatSignature: ''
     NumberOfSamples: 3
        CodingMethod: 'Huffman'
       CodingProcess: 'Sequential'
             Comment: {}
```

```
              Make: 'Apple'
             Model: 'iPhone 6'
       Orientation: 1
       XResolution: 72
       YResolution: 72
    ResolutionUnit: 'Inch'
          Software: '9.3.5'
          DateTime: '2016:09:17 22:05:08'
   YCbCrPositioning: 'Centered'
     DigitalCamera: [1x1 struct]
           GPSInfo: [1x1 struct]
     ExifThumbnail: [1x1 struct]
```

This is the metadata that tells camera software, and image databases, where and how the image was generated. This is useful when learning from images as it allows you to correct for resolution (`width` and `height`), bit depth, and other factors.

If we view this image using `imshow`, it will publish a warning that the image is too big to fit on the screen and that it is displayed at 33%. If we view it using `image`, there will be a visible set of axes. `image` is useful for displaying other two-dimensional matrix data as individual elements per pixel. Both functions return a handle to an image object, only the axes' properties are different. Figure 2.1 shows the resulting figures. Note the labeled axes on the right figures.

```
>> figure; hI = image(imread('IMG_2398_Zoom.png'))
hI =
  Image with properties:

          CData: [680x680x3 uint8]
    CDataMapping: 'direct'

  Show all properties
```

Table 2.4 provide key functions for interacting with images.

2.1.6 Datastore

Datastores allow you to interact with files containing data that are too large to fit in memory. There are different types of datastores for tabular data, images, spreadsheets, databases, and custom files. Each datastore provides functions to extract smaller amounts of data that do fit in memory for analysis. For example, you can search a collection of images for those with the brightest pixels or maximum saturation values. We will use our directory of cat images as an example:

```
>> location = pwd
location =
/Users/Shared/svn/Manuals/MATLABMachineLearning/MATLAB/Cats
>> ds = datastore(location)
ds =
  ImageDatastore with properties:

        Files: {
```

```
        ' .../Shared/svn/Manuals/MATLABMachineLearning/MATLAB/Cats/
          IMG_0191.png';
        ' .../Shared/svn/Manuals/MATLABMachineLearning/MATLAB/Cats/
          IMG_1603.png';
        ' .../Shared/svn/Manuals/MATLABMachineLearning/MATLAB/Cats/
          IMG_1625.png'
        ... and 19 more
        }
    Labels: {}
   ReadFcn: @readDatastoreImage
```

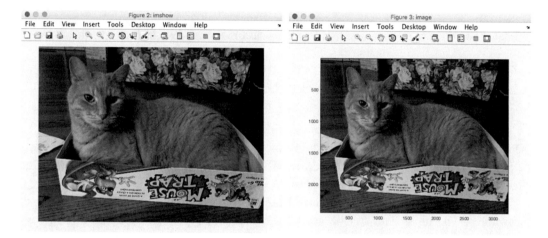

Figure 2.1: *Image display options. A figure created using* imshow *is on the left and a figure using* image *is on the right*

Table 2.4: *Key Functions for Images*

Function	Purpose
imread	Read an image in a variety of formats
imfinfo	Gather information about an image file
imwrite	Write data to an image file
image	Display an image from an array
imagesc	Display image data scaled to the current colormap
imshow	Display an image, optimizing figure, axis, and image object properties and taking an array or a filename as an input
rgb2gray	Write data to an image file
ind2rgb	Convert index data to RGB
rgb2ind	Convert RGB data to indexed image data
fitsread	Read a FITS file
fitswrite	Write data to a FITS file
fitsinfo	Information about a FITS file returned in a data structure
fitsdisp	Display FITS file metadata for all Header + Data Units (HDUs) in the file

Table 2.5: *Key Functions for Datastore*

Function	Purpose
datastore	Create a datastore from a collection of data
read	Read a subset of data from the datastore
readall	Read all of the data in the datastore
hasdata	Check to see if there is more data in the datastore
reset	Initialize a datastore with the contents of a folder
partition	Excerpt a portion of the datastore
numpartitions	Estimate a reasonable number of partitions
ImageDatastore	Datastore of a list of image files
TabularTextDatastore	A collection of one or more tabular text files
SpreadsheetDatastore	Datastore of spreadsheets
FileDatastore	Datastore for files with a custom format, for which you provide a reader function
KeyValueDatastore	Datastore of key-value pairs
DatabaseDatastore	Database connection requires the Database Toolbox

Once the datastore is created, you use the applicable class functions to interact with it. Datastores have standard container-style functions like `read`, `partition`, and `reset`. Each type of datastore has different properties. The `DatabaseDatastore` requires the Database Toolbox and allows you to use SQL queries.

MATLAB provides the MapReduce framework for working with out-of-memory data in datastores. The input data can be any of the datastore types, and the output is a key-value datastore. The map function processes the datastore input in chunks, and the reduce function calculates the output values for each key. `mapreduce` can be sped up by using it with the MATLAB Parallel Computing Toolbox, Distributed Computing Server, or Compiler. Table 2.5 gives key functions for using datastores.

2.1.7 Tall Arrays

Tall arrays were introduced in R2016b. They are allowed to have more rows than will fit in memory. You can use them to work with datastores that might have millions of rows. Tall arrays can use almost any MATLAB type as a column variable, including numeric data, cell arrays, strings, datetimes, and categoricals. The MATLAB documentation provides a list of functions that support tall arrays. Results for operations on the array are only evaluated when they are explicitly requested using the `gather` function. The `histogram` function can be used with tall arrays and will execute immediately.

The MATLAB Statistics and Machine Learning Toolbox, Database Toolbox, Parallel Computing Toolbox, Distributed Computing Server, and Compiler all provide additional extensions for working with tall arrays. Table 2.6 gives key functions for using Tall Arrays.

Table 2.6: *Key Functions for Tall Arrays*

Function	Purpose
tall	Initialize a tall array
gather	Execute the requested operations
summary	Display summary information to the command line
head	Access first rows of a tall array
tail	Access last rows of a tall array
istall	Check the type of the array to determine if it is tall
write	Write the tall array to disk

Table 2.7: *Key Functions for Sparse Matrices*

Function	Purpose
sparse	Create a sparse matrix from a full matrix or from a list of indices and values
issparse	Determine if a matrix is sparse
nnz	Number of nonzero elements in a sparse matrix
spalloc	Allocate nonzero space for a sparse matrix
spy	Visualize a sparsity pattern
spfun	Selectively apply a function to the nonzero elements of a sparse matrix
full	Convert a sparse matrix to full form

2.1.8 Sparse Matrices

Sparse matrices are a special category of the matrix in which most of the elements are zero. They appear commonly in large optimization problems and are used by many such packages. The zeros are "squeezed" out, and MATLAB stores only the nonzero elements along with index data such that the full matrix can be recreated. Many regular MATLAB functions, such as chol or diag, preserve the sparseness of an input matrix. Table 2.7 gives key functions for sparse matrices.

Here is an example:

```
1   %% Sparse matrix
2   % Compares a sparse matrix with a regular matrix
3
4   n = 100;
5   a = sprand(n,n,0.1);
6   b = sprand(n,1,0.1);
7   disp('Sparse matrix: linear equations')
8   tic
9   c = a\b;
10  toc
11
12  disp('Sparse matrix: eigenvalue')
13  tic
14  eigs(a);
15  toc
```

```
16
17   a = full(a);
18   b = full(b);
19   disp('Regular matrix: linear equations')
20   tic
21   c = a\b;
22   toc
23
24   disp('Regular matrix: eigenvalue')
25   tic
26   eig(a);
27   toc
```

```
>> Sparse
Sparse matrix: linear equations
Elapsed time is 0.000433 seconds.
Sparse matrix: eigenvalue
Elapsed time is 0.023754 seconds.
Regular matrix: linear equations
Elapsed time is 0.000158 seconds.
Regular matrix: eigenvalue
Elapsed time is 0.002211 seconds.
```

Sparse is not necessarily faster.

2.1.9 Tables and Categoricals

Tables were introduced in release R2013 of MATLAB and allowed tabular data to be stored with metadata in one workspace variable. It is an effective way to store and interact with data that one might put in, or import from, a spreadsheet. The table columns can be named, assigned units and descriptions, and accessed as one would fields in a data structure, that is, `T.DataName`. See `readtable` on creating a table from a file, or try out the Import Data button from the command window.

Categorical arrays allow for the storage of discrete nonnumeric data, and they are often used within a table to define groups of rows. For example, time data may have the day of the week, or geographic data may be organized by state or county. They can be leveraged to rearrange data in a table using `unstack`.

You can also combine multiple data sets into single tables using `join`, `innerjoin`, and `outerjoin`, which will be familiar to you if you have worked with databases. Table 2.8 lists key functions for using tables.

Table 2.8: *Key Functions for Tables*

Function	Purpose
table	Create a table with data in the workspace
readtable	Create a table from a file
join	Merge tables by matching up variables
innerjoin	Join tables A and B retaining only the rows that match
outerjoin	Join tables including all rows
stack	Stack data from multiple table variables into one variable
unstack	Unstack data from a single variable into multiple variables
summary	Calculate and display summary data for the table
categorical	Arrays of discrete categorical data
iscategorical	Create a categorical array
categories	List of categories in the array
iscategory	Test for a particular category
addcats	Add categories to an array
removecats	Remove categories from an array
mergecats	Merge categories

Here is an example reading an Excel spreadsheet:

```
>> s = readtable('ExcelSpreadsheet','VariableNamingRule','preserve')

s =

  15x3 table

      Number        Value 1        Value 2

    _____     _____     _____


    1.0000e+00     8.0000e+00     6.6667e-01
    2.0000e+00     1.1000e+01     9.1667e-01
    3.0000e+00     1.4000e+01     1.1667e+00
    4.0000e+00     1.7000e+01     1.4167e+00
    5.0000e+00     2.0000e+01     1.6667e+00
    6.0000e+00     2.3000e+01     1.9167e+00
    7.0000e+00     2.6000e+01     2.1667e+00
    8.0000e+00     2.9000e+01     2.4167e+00
    9.0000e+00     3.2000e+01     2.6667e+00
    1.0000e+01     3.5000e+01     2.9167e+00
    1.1000e+01     3.8000e+01     3.1667e+00
    1.2000e+01     4.1000e+01     3.4167e+00
    1.3000e+01     4.4000e+01     3.6667e+00
    1.4000e+01     4.7000e+01     3.9167e+00
    1.5000e+01     5.0000e+01     4.1667e+00
```

2.1.10 Large MAT-Files

You can access parts of a large MAT-file without loading the entire file into memory by using the `matfile` function. This creates an object that is connected to the requested MAT-file without loading it. Data is only loaded when you request a particular variable or part of a variable. You can also dynamically add new data to the MAT-file.

For example, we can load a MAT-file of neural net weights generated in a later chapter:

```
>> m = matfile('PitchNNWeights','Writable',true)
m =
  matlab.io.MatFile

  Properties:
      Properties.Source: '/Users/Shared/svn/Manuals/MATLABMachineLearning
          /MATLAB/PitchNNWeights.mat'
    Properties.Writable: true
                      w: [1x8 double]
```

We can access a portion of the previously unloaded w variable or add a new variable name, all using this object m:

```
>> y = m.w(1:4)
y =
     1     1     1     1
>> m.name = 'Pitch Weights'
m =
  matlab.io.MatFile

  Properties:
      Properties.Source: '/Users/Shared/svn/Manuals/MATLABMachineLearning
          /MATLAB/PitchNNWeights.mat'
    Properties.Writable: true
                   name: [1x13 char]
                      w: [1x8  double]
>> d = load('PitchNNWeights')
d =
       w: [1 1 1 1 1 1 1 1]
    name: 'Pitch Weights'
```

There are some limits to the indexing into unloaded data, such as struct arrays and sparse arrays. Also, `matfile` requires MAT-files using version 7.3, which is not the default for a generic `save` operation as of R2016b. You must either create the MAT-file using `matfile` to take advantage of these features or use the `-v7.3'` flag when saving the file.

2.2   Initializing a Data Structure

It's always a good idea to use a special function to define a data structure you are using as a type in your codebase, similar to writing a class but with less overhead. Users can then overload individual fields in their code, but there is an alternative way to set many fields at once: an initialization function that can handle a parameter pair input list. This allows you to do additional processing in your initialization function. Also, your parameter string names can be more descriptive than you would choose to make your field names.

2.2.1   Problem

We want to initialize a data structure so that the user knows what they are entering.

2.2.2   Solution

The simplest way to implement the parameter pairs is using `varargin` and a switch statement. Alternatively, you could write an `inputParser`, which allows you to specify required and optional inputs as well as named parameters. In that case, you have to write separate or anonymous functions for validation that can be passed to the `inputParser`, rather than just writing out the validation in your code.

2.2.3   How It Works

We will use the data structure developed for the automobile simulation in Chapter 12 as an example. The header lists the input parameters along with the input dimensions and units, if applicable.

AutomobileInitialize.m

```
1   %% AUTOMOBILEINITIALIZE Initialize the automobile data structure.
2   %
3   %% Form
4   %   d = AutomobileInitialize( varargin )
5   %
6   %% Description
7   % Initializes the data structure using parameter pairs.
8   %
9   %% Inputs
10  % varargin:    ('parameter',value,...)
11  %
12  % 'mass'                              (1,1)  (kg)
13  % 'steering angle'                    (1,1)  (rad)
14  % 'position tires'                    (2,4)  (m)
15  % 'frontal drag coefficient'          (1,1)
16  % 'side drag coefficient'             (1,1)
17  % 'tire friction coefficient'         (1,1)
18  % 'tire radius'                       (1,1)  (m)
19  % 'engine torque'                     (1,1)  (Nm)
20  % 'rotational inertia'                (1,1)  (kg-m^2)
```

```
21  % 'state'                               (6,1) [m;m;m/s;m/s;rad;rad/s
        ]
```

The function first creates the data structure using a set of defaults, then handles the parameter pairs entered by a user. After the parameters have been processed, two areas are calculated using the dimensions and the height.

AutomobileInitialize.m

```
30  function d = AutomobileInitialize( varargin )
31
32  % Defaults
33  d.mass          = 1513;
34  d.delta         = 0;
35  d.r             = [  1.17 1.17 -1.68 -1.68;...
36                      -0.77 0.77 -0.77  0.77];
37  d.cDF           = 0.25;
38  d.cDS           = 0.5;
39  d.cF            = 0.01; % Ordinary car tires on concrete
40  d.radiusTire    = 0.4572; % m
41  d.torque        = d.radiusTire*200.0; % N
42  d.inr           = 2443.26;
43  d.x             = [0;0;0;0;0;0];
44  d.fRR           = [0.013 6.5e-6];
45  d.dim           = [1.17+1.68 2*0.77];
46  d.h             = 2/0.77;
47  d.errOld        = 0;
48  d.passState     = 0;
49  d.model         = 'MyCar.obj';
50  d.scale         = 4.7981;
51
52  for k = 1:2:length(varargin)
53    switch lower(varargin{k})
54      case 'mass'
55        d.mass          = varargin{k+1};
56      case 'steering angle'
57        d.delta         = varargin{k+1};
58      case 'position tires'
59        d.r             = varargin{k+1};
60      case 'frontal drag coefficient'
61        d.cDF           = varargin{k+1};
62      case 'side drag coefficient'
63        d.cDS           = varargin{k+1};
64      case 'tire friction coefficient'
65        d.cF            = varargin{k+1};
66      case 'tire radius'
67        d.radiusTire    = varargin{k+1};
68      case 'engine torque'
69        d.torque        = varargin{k+1};
70      case 'rotational inertia'
71        d.inertia       = varargin{k+1};
```

```
72        case 'state'
73           d.x            = varargin{k+1};
74        case 'rolling resistance coefficients'
75           d.fRR          = varargin{k+1};
76        case 'height automobile'
77           d.h            = varargin{k+1};
78        case 'side and frontal automobile dimensions'
79           d.dim          = varargin{k+1};
80        case 'car model'
81           d.model        = varargin{k+1};
82        case 'car scale'
83           d.scale        = varargin{k+1};
84      end
85    end
86
87    % Processing
88    d.areaF = d.dim(2)*d.h;
89    d.areaS = d.dim(1)*d.h;
90    d.g      = LoadOBJFile(d.model,d.scale);
```

To perform the same tasks with `inputParser`, you add either an `addRequired`, `addOptional`, or `addParameter` call for every item in the switch statement. The named parameters require default values. You can optionally specify a validation function; in the following example, we use `isNumeric` to limit the values to numeric data:

```
>> p = inputParser
p.addParameter('mass',0.25);
p.addParameter('cDF',1513);
p.parse('cDF',2000);
d = p.Results

p =

   inputParser with properties:

         FunctionName: ''
        CaseSensitive: 0
        KeepUnmatched: 0
       PartialMatching: 1
          StructExpand: 1
            Parameters: {1x0 cell}
               Results: [1x1 struct]
             Unmatched: [1x1 struct]
         UsingDefaults: {1x0 cell}

d =

   struct with fields:

      cDF: 2000
     mass: 0.2500
```

In this case, the results of the parsed parameters are stored in a Results substructure.

2.3 `mapreduce` on an Image Datastore

2.3.1 Problem

We discussed the `datastore` class in the introduction to the chapter. Now let's use it to perform analysis on the full set of cat images using `mapreduce`, which is scalable to very large numbers of images. This involves two steps: first, a *map* step that operates on the datastore and creates intermediate values, then a *reduce* step which operates on the intermediate values to produce a final output.

2.3.2 Solution

We create the `datastore` by passing in the path to the folder of cat images. We also need to create a map function and a reduce function to pass into `mapreduce`. If you are using additional toolboxes like the Parallel Computing Toolbox, you would specify the reduced environment using `mapreducer`.

2.3.3 How It Works

First, create the `datastore` using the path to the images:

```
>> imds = imageDatastore('MATLAB/Cats');
imds =
  ImageDatastore with properties:

      Files: {
              ' .../MATLABMachineLearning/MATLAB/Cats/IMG_0191.png';
              ' .../MATLABMachineLearning/MATLAB/Cats/IMG_1603.png';
              ' .../MATLABMachineLearning/MATLAB/Cats/IMG_1625.png'
              ... and 19 more
              }
     Labels: {}
    ReadFcn: @readDatastoreImage
```

Second, we write the map function. This must generate and store a set of intermediate values that will be processed by the reduce function. Each intermediate value must be stored as a key in the intermediate key-value datastore using `add`. In this case, the map function will receive one image each time it is called. We call it `catColorMapper` since it processes the red, green, and blue values for each image using a simple average:

```
function catColorMapper(data, info, intermediateStore)

% Calculate the average (R, G, B) values
avgRed = mean(mean(data(:,:,1)));
avgGreen = mean(mean(data(:,:,2)));
avgBlue = mean(mean(data(:,:,3)));

% Store the calculated values with text keys
add(intermediateStore, 'Avg Red', struct('Filename',info.Filename,'Val',
    avgRed));
add(intermediateStore, 'Avg Green', struct('Filename',info.Filename,'Val
    ', avgGreen));
add(intermediateStore, 'Avg Blue', struct('Filename',info.Filename,'Val',
    avgBlue));
```

The reduce function will then receive the list of the image files from the datastore once for each key in the intermediate data. It receives an iterator to the intermediate datastore as well as an output datastore. Again, each output must be a key-value pair. The `hasnext` and `getnext` functions used are part of the `mapreduce` ValueIterator class. In this case, we find the minimum value for each key across the set of images:

```
function catColorReducer(key, intermediateIter, outputStore)

% Iterate over values for each key
minVal = 255;
minImageFilename = '';
while hasnext(intermediateIter)
  value = getnext(intermediateIter);

  % Compare values to find the minimum
  if value.Val < minVal
      minVal = value.Val;
      minImageFilename = value.Filename;
  end
end

% Add final key-value pair
add(outputStore, ['Minimum -  ' key], minImageFilename);
```

Finally, we call `mapreduce` using function handles to our two helper functions. Progress updates are printed to the command line, first for the mapping step and then for the reduce step (once the mapping progress reaches 100%):

```
minRGB = mapreduce(imds, @catColorMapper, @catColorMapper);

********************************
*       MAPREDUCE PROGRESS      *
```

37

```
*******************************
Map    0% Reduce    0%
Map   13% Reduce    0%
Map   27% Reduce    0%
Map   40% Reduce    0%
Map   50% Reduce    0%
Map   63% Reduce    0%
Map   77% Reduce    0%
Map   90% Reduce    0%
Map  100% Reduce    0%
Map  100% Reduce   33%
Map  100% Reduce   67%
Map  100% Reduce  100%
```

The results are stored in a MAT-file, for example, `results_1_28-Sep-2016_16-28-38_347`. The store returned is a key-value store to this MAT-file, which in turn contains the store with the final key-value results:

```
>> output = readall(minRGB)
output =
          Key                                        Value

     _____                      _____

     ''Minimum - Avg Red'      '/MATLAB/Cats/IMG_1625.png'
     ''Minimum - Avg Blue'     '/MATLAB/Cats/IMG_4866.jpg'
     ''Minimum - Avg Green'    '/MATLAB/Cats/IMG_4866.jpg'
```

You'll notice that the image files are different file types. This is because they came from different sources. MATLAB can handle most image types quite well.

2.4 Processing Table Data

2.4.1 Problem

We want to compare temperature frequencies in 1999 and 2015 using data from a table.

2.4.2 Solution

Use `tabularTextDatastore` to load the data and perform a Fast Fourier Transform on the data.

2.4.3 How It Works

First, let us look at what happens when we read the data from the weather files:

```
>> tds         = tabularTextDatastore('./Weather')

tds =

  TabularTextDatastore with properties:

                       Files: {
                              ' .../MATLABMachineLearning2/MATLAB/
                                Chapter_02/Weather/HistKTTN_1990.txt';
                              ' .../MATLABMachineLearning2/MATLAB/
                                Chapter_02/Weather/HistKTTN_1993.txt';
                              ' .../MATLABMachineLearning2/MATLAB/
                                Chapter_02/Weather/HistKTTN_1999.txt'
                              ... and 5 more
                              }
                FileEncoding: 'UTF-8'
    AlternateFileSystemRoots: {}
           ReadVariableNames: true
               VariableNames: {'EST', 'MaxTemperatureF', 'MeanTemperatureF
                              ' ... and 20 more}

  Text Format Properties:
              NumHeaderLines: 0
                   Delimiter: ','
                RowDelimiter: '\r\n'
               TreatAsMissing: ''
                MissingValue: NaN

  Advanced Text Format Properties:
             TextscanFormats: {'%{uuuu-MM-dd}D', '%f', '%f' ... and 20
                              more}
                    TextType: 'char'
          ExponentCharacters: 'eEdD'
                CommentStyle: ''
                  Whitespace: ' \b\t'
     MultipleDelimitersAsOne: false

  Properties that control the table returned by preview, read, readall:
       SelectedVariableNames: {'EST', 'MaxTemperatureF', 'MeanTemperatureF
                              ' ... and 20 more}
             SelectedFormats: {'%{uuuu-MM-dd}D', '%f', '%f' ... and 20
                              more}
                    ReadSize: 20000 rows
```

WeatherFFT selects the data to use. It finds all the data in the mess of data in the files. When running the script, you need to be in the same folder as WeatherFFT.

WeatherFFT.m

```
6   c0 = cd;
7   p = mfilename('fullpath');
8   cd(fileparts(p));
9   secInDay = 86400;
11
12  %% Create the datastore from the directory of files
13  tDS                      = tabularTextDatastore('./Weather/');
14  tDS.SelectedVariableNames = {'EST','MaxTemperatureF'};
15
16  preview(tDS)
17  z = readall(tDS);
18
19  % The first column in the cell array is the date. year extracts the
        year
20  y     = year(z{:,1});
21  k1993 = find(y == 1993);
22  k2015 = find(y == 2015);
23  tSamp = secInDay;
24  t     = (1:365)*tSamp;
25  j     = {[1 2]};
26
27  %% Plot the FFT
28
29  % Get 1993 data
30  d1993     = z{k1993,2}';
31  m1993     = mean(d1993);
32  d1993     = d1993 - m1993;
33
34  e1993     = FFTEnergy( d1993, tSamp );
35
36  % Get 2015 data
37  d2015     = z{k2015,2}';
38  m2015     = mean(d2015);
39  d2015     = d2015 - m2015;
40  [e2015,f] = FFTEnergy( d2015, tSamp );
```

If the data does not exist, `TabularTextDatastore` puts `NaN` in the data points' place. We happen to pick two years without any missing data. We use `preview` to see what we are getting.

```
>> WeatherFFT
Warning: Variable names were modified to make them valid MATLAB
    identifiers.

ans =

  8x2 table
```

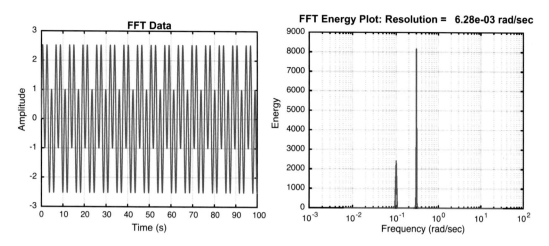

Figure 2.2: *1993 and 2015 data*

EST	MaxTemperatureF
1990-01-01	39
1990-01-02	39
1990-01-03	48
1990-01-04	51
1990-01-05	46
1990-01-06	43
1990-01-07	42
1990-01-08	37

In this script, we get output from FFTEnergy so that we can combine the plots. We chose to put the data on the same axes. Figure 2.2 shows the temperature data and the FFT.

We get a little fancy with plotset. Our legend entries are computed to include the mean temperatures.

WeatherFFT.m

```
42  lG = {{sprintf('1993: Mean = %4.1f deg-F',m1993) sprintf('2015: Mean =
        %4.1f deg-F',m2015)}};

43

44  PlotSet(t,[d1993;d2015],  'x label', 'Days', 'y label','Amplitude (deg-
        F)',...
45    'plot title','Temperature', 'figure title', 'Temperature','legend',lG
        ,'plot set',j);

46

47  PlotSet(f,[e1993';e2015'],'x label', 'Rad/s','y label','Magnitude',...
48    'plot title','FFT Data', 'figure title', 'FFT','plot type','ylog','
        legend',lG,'plot set',j);

49

50  cd(c0);
```

Using MATLAB Strings

Machine learning often requires interaction with humans, which often means processing speech. Also, expert systems and fuzzy logic systems can make use of textual descriptions. MATLAB's string data type makes this easier. Strings are bracketed by double quotes. In this section, we will give examples of operations that work with strings but not with character arrays.

2.5   String Concatenation

2.5.1 Problem

We want to concatenate two strings.

2.5.2 Solution

Create the two strings and use the "+" operator.

2.5.3 How It Works

You can use the + operator to concatenate strings. The result is the second string after the first:

```
>> a = "12345";
>> b = "67";
>> c = a + b

c =

    "1234567"
```

2.6   Arrays of Strings

2.6.1 Problem

We want an array of strings.

2.6.2 Solution

Create the two strings and put them in a matrix.

2.6.3 How It Works

We create the same two strings as before and use the matrix operator. If they were character arrays, we would need to pad the shorter with blanks to be the same size as the longer:

```
>> a = "12345";
>> b = "67";
>> c = [a;b]

c =

  2x1 string array
```

```
   "12345"
   "67"

>> c = [a b]

c =

  1x2 string array

   "12345"    "67"
```

You could have used a cell array for this, but strings are more convenient.

2.7   Substrings

2.7.1 Problem

We want to get strings after a fixed prefix.

2.7.2 Solution

Create a string array and use `extractAfter`.

2.7.3 How It Works

Create a string array of strings to search and use `extractAfter`:

```
>> a = ["1234";"12456";"12890"];
f = extractAfter(a,"12")

f =

  3x1 string array

   "34"
   "456"
   "890"
```

Most of the string functions work with `char` but strings are a little cleaner. Here is the above example with cell arrays:

```
>> a = {'1234';'12456';'12890'};
>> f = extractAfter(a,"12")

f =

  3x1 cell array

   {'34' }
   {'456'}
   {'890'}
```

43

2.8 Reading an Excel Spreadsheet into a Table

2.8.1 Problem

We want to read in a spreadsheet and use it to plot data in a steam table.

2.8.2 Solution

Create a function that uses MATLAB's `readtable]`

2.8.3 How It Works

The following code shows the function to read an Excel spreadsheet. `readtable` reads in the spreadsheet. The corners of the data range are given in `'A1:M384'`. If the spreadsheet's first row is the label, it will use them to denote the names of the columns. Otherwise, it will use column names such as `Var1`. In this case, the table has some duplicate values in the first column, probably due to round-off, so we use `unique` to remove duplicates. The outputs are found using `interp1`. The function `function c = CellToCol(cA)`, at the end of the file, does the conversion from a number in a cell to a number.

TSCurve.m

```
23  % Read the steam table
24  tableT = readtable('Steam_Tables_Temperature.xlsx','range','A1:M384');
25
26  % Convert to double
27  sL = CellToCol(tableT.Var11)';
28  sV = CellToCol(tableT.Var12)';
29
30  % The first path generates the full diagram
31  if( nargin < 1 )
32    t =  tableT.Var1';
33    t  = [t fliplr(t)];
34    s  = [sL fliplr(sV)];
35  else
36    s = zeros(2,length(t));
37    x = tableT.Var1';
38    [~,j] = unique(x);
39
40    for k = 1:length(t)
41      s(1,k) = interp1(x(j),sL(j),t(k));
42      s(2,k) = interp1(x(j),sV(j),t(k));
43    end
44  end
```

The first column will be numbered. The other columns will be cell elements of characters.

TSCurve.m

```
59  function c = CellToCol(cA)
60
```

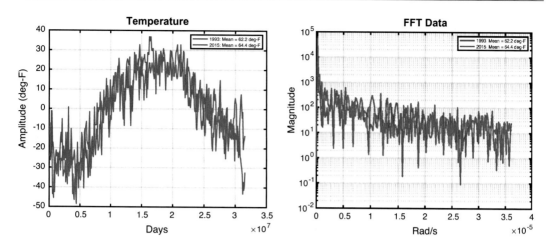

Figure 2.3: *Temperature-entropy diagram generated by* `TSDiagram`

```
61   c = zeros(length(cA),1);
62   for k = 1:length(cA)
63     c(k) = str2double(cA{k});
64   end
```

The function produces a plot if no outputs are requested.

TSCurve.m

```
47   if( nargout < 1 )
48     PlotSet(s,t,'x label','Entropy (KJ/kg K)', 'y label', ...
49       {'Temperature (K)'},'figure title','T-S');
50     [tMax,k] = max(t);
51     p = tableT.Var2';
52     sMax   = s(k);
53     pMax   = str2double(p(end));
54     text(sMax,1.02*tMax,sprintf('Critical Point: %4.1f MPa %4.1f K',pMax,
         tMax));
55     clear s
56   end
```

The plot is shown in Figure 2.3.

2.9   Accessing ChatGPT

2.9.1 Problem

We want to make queries of ChatGPT.

2.9.2 Solution

Create a script using MATLAB's web access tools to access ChatGPT directly through the Internet.

2.9.3 How It Works

The following code shows a script to do a ChatGPT query. You must sign up with ChatGPT to get the api_key. If you don't have an API key, you get the error message:

```
Error using matlab.internal.webservices.HTTPConnector/
    copyContentToByteArray
The server returned the status 401 with message "" in response to the
request to URL https://api.openai.com/v1/chat/completions.
```

It is convenient to set the environment variable to the string "OPENAI_API_KEY".

ChatGPTScript.m

```
1  %% ChatGPT interface
2  % Use the webwrite function to send a question to the ChatGPT API
3  % Save your API key. You need to log on to ChatGPT to create your own
4
5  %% Save your API key. You need to log on to ChatGPT to create your own
6  %% This won't work without a key
7  api_key = "xxxxxxyyyyyyyyzzzzzz"; % Must get your own key
8  setenv("OPENAI_API_KEY",api_key)
9
10 %% Set up the web options for webwrite
11 options = weboptions(...
12     'MediaType', 'application/json','timeout',10,...
13     'HeaderFields', {'Authorization' ['Bearer ' getenv('OPENAI_API_KEY'
            )]});
14
15 %% The ChatGPT query
16 question = 'Write MATLAB code to plot the sine of the sequence from 0
       to 10 pi';
17
18 %% The destination
19 url = 'https://api.openai.com/v1/chat/completions';
20 body = struct(...
21     'model', 'gpt-3.5-turbo',...
22     'messages', {{struct('role', 'user','content', question)}});
23
24 %% This writes to the web
25 response = webwrite(url, body, options);
```

```
26
27  %% Question and response
28  fprintf('Query: %s\n\n',question);
29  disp('Response:')
30  disp(response.choices.message.content)
```

url is the web location. webwrite makes the query at that URL.

The results of the query are shown as follows. Minor changes in the query can make major changes in the response:

```
>> ChatGPTScript
Query: Write MATLAB code to plot the sine of the sequence from 0 to 10 pi

Response:
To plot the sine of the sequence from 0 to 10 pi using MATLAB, follow the
    steps below:

1. Define the range of x values using the linspace command. In this case,
    we will use a range from 0 to 10 pi with 1000 points.

2. Calculate the sine of the x values using the sin command.

3. Plot the sine of x using the plot command.

4. Add x and y labels to the plot using the xlabel and ylabel commands.

5. Add a title to the plot using the title command.

Here is the MATLAB code to plot the sine of the sequence from 0 to 10 pi:

```
% Define x values
x = linspace(0, 10*pi, 1000);

% Calculate sine of x
y = sin(x);

% Plot sine of x
plot(x, y);

% Add labels and title
xlabel('x');
ylabel('sin(x)');
title('Sine of Sequence from 0 to 10 pi');
```

When you run this code, you should see a plot of the sine of the sequence
    from 0 to 10 pi. The x-axis will show the values from 0 to 10 pi,
    and the y-axis will show the corresponding sine values.
```

The resulting ChatGPT code generated plot is shown in Figure 2.4.

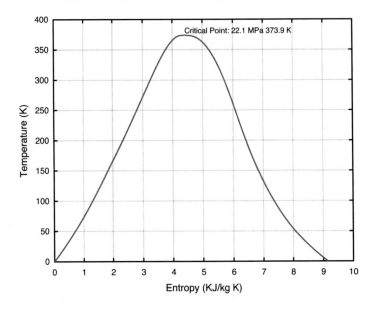

Figure 2.4: *ChatGPT code generates this figure.*

Table 2.9: *Chapter Code Listing*

File	Description
AutomobileInitialize	Data structure initialization example from Chapter 12
catReducer	Image datastore used with mapreduce
ChatGPTScript	Allows queries of ChatGPT
FFTEnergy	Computes the energy from an FFT
Sparse	Example of speed of sparse operations
TSCurve	Generates a temperature-entropy curve from a table
weatherFFT	Does an FFT of weather data

2.10 Summary

There are a variety of data containers in MATLAB to assist you in analyzing your data for machine learning. If you have access to a computer cluster or one of the specialized computing toolboxes, you have even more options. Table 2.9 lists the functions and scripts included in the companion code.

CHAPTER 3

■ ■ ■

MATLAB Graphics

One of the issues with machine learning is understanding the algorithms and why an algorithm made a particular decision. In addition, you want to be able to easily understand the decision. MATLAB has extensive graphics facilities that can be harnessed for that purpose. Plotting is used extensively in machine learning problems. MATLAB plots can be two- or three-dimensional. MATLAB also has many plot types such as line plots, bar charts, and pie charts. Different types of plots are better at conveying particular types of data. MATLAB also has extensive surface and contour plotting capabilities that can be used to display complex data in an easy-to-grasp fashion. Another facility is 3D modeling. You can draw animated objects, such as robots or automobiles. These are particularly valuable when your machine learning involves simulations.

An important part of MATLAB graphics is Graphical User Interface (GUI) building. MATLAB has extensive facilities for making GUIs. These can be a valuable way of making your design tools or machine learning systems easy for users to operate.

This chapter will provide an introduction to a wide variety of graphics tools in MATLAB. They should allow you to harness MATLAB graphics for your applications.

3.1 2D Line Plots

3.1.1 Problem

You want a single function to generate two-dimensional line graphs, avoiding a long list of code for the generation of each graphic.

3.1.2 Solution

Write a single function to take the data and *parameter pairs* to encapsulate the functionality of MATLAB's 2D line plotting functions. An example of a plot created with a single line of code is shown in Figure 3.1.

M. Paluszek, S. Thomas, *MATLAB Machine Learning Recipes*,
https://doi.org/10.1007/978-1-4842-9846-6_3

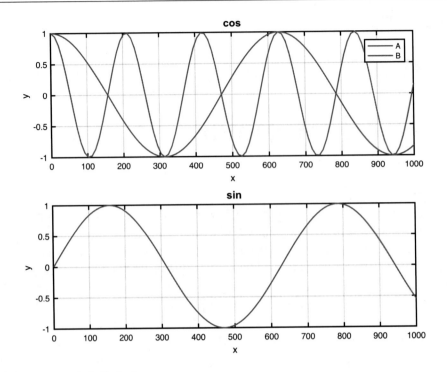

Figure 3.1: *PlotSet's built-in demo*

3.1.3 How It Works

PlotSet generates 2D plots, including multiple plots on a page. This code processes varargin as parameter pairs to set options. A parameter pair is two inputs. The first is the name of the value, and the second is the value. For example, the parameter pair for labeling the x-axis is

    'x label','Time (s)'

varargin makes it easy to expand the plotting options. The function signature is then very simple.

PlotSet.m

```
 1    %% PLOTSET Create two-dimensional plots from a data set.
27    function h = PlotSet( x, y, varargin )
```

The core function code is shown in the following listing. We supply default values for the x- and y-axis labels and the figure name. The parameter pairs are handled in a switch statement. The following code is the branch when there is only one x-axis label for all of the plots. The code arranges plots by the data in plotSet a cell array.

PlotSet.m

```
104    for k = 1:m
105      subplot(m,nCol,k);
106      j = plotSet{k};
107      for i = 1:length(j)
108        plotXY(x,y(j(i),:),plotType);
109        hold on
110      end
111      hold off
112      ylabel(yLabel{k});
113      if( length(plotTitle) == 1 )
114        if k==1
115          title(plotTitle{1})
116        end
117      else
118        title(plotTitle{k})
119      end
120      if( ~isempty(leg{k}) )
121        legend(leg{k},'fontsize',fontSize);
122      end
123      if( k < m )
124        %set(gca,'xtick',[])
125        set(gca,'xticklabel',[])
126      end
127      grid on
128      set(gca,'fontsize',fontSize); % for book images
129    end
130    xlabel(xLabel);
131  else
```

The plotting is done in a subfunction called `plotXY`. There, you see all the familiar MAT-LAB plotting function calls.

PlotSet.m

```
162  function plotXY(x,y,type)
163
164  switch type
165    case 'plot'
166      plot(x,y,'linewidth',1);
167    case {'log' 'loglog' 'log log'}
168      loglog(x,y,'linewidth',1);
169    case {'xlog' 'semilogx' 'x log'}
170      semilogx(x,y,'linewidth',1);
171    case {'ylog' 'semilogy' 'y log'}
172      semilogy(x,y,'linewidth',1);
173    otherwise
174      error('%s is not an available plot type',type);
175  end
```

The example in Figure 3.1 is generated by a dedicated demo function at the end of the `PlotSet` function. This demo shows several of the features of the function. These include

1. Multiple lines per graph

2. Legends

3. Plot titles

4. Default axis labels

Using a dedicated demo subfunction is a clean way to provide a built-in example of a function, and it is especially important in graphics functions to provide an example of a typical plot. The code is shown as follows.

PlotSet.m

```
178  function Demo
179
180  x = linspace(1,1000);
181  y = [sin(0.01*x);cos(0.01*x);cos(0.03*x)];
182  disp('PlotSet: One x and three y rows')
183  PlotSet( x, y, 'figure title', 'PlotSet Demo',...
184      'plot set',{[2 3], 1},'legend',{{'A' 'B'},{}},'plot title',{'cos','
             sin'});
185
186  disp('PlotSet: Two x and two y rows')
187  PlotSet( [x;y(1,:)], y(1:2,:) );
188
189  disp('PlotSet: Two x and two y rows using `plot set`')
190  PlotSet( [x;y(1,:)], y, 'plot set', {[1 2 3],2}, 'legend',...
191      {{'sin','cos A',' cosB'},{'sin vs cos'}} );
```

3.2 General 2D Graphics

3.2.1 Problem

You want to represent a 2D data set in different ways. Line plots are very useful, but sometimes it is easier to visualize data in different forms. MATLAB has many functions for 2D graphical displays.

3.2.2 Solution

Write a script to show MATLAB's different 2D plot types. In our example, we use subplots within one figure to help reduce figure proliferation.

3.2.3 How It Works

Use the `NewFigure` function to create a new figure window with a suitable name. Then run the following script.

MATLABPlotTypes.m

```
6   h = NewFigure('Plot Types');
7   x = linspace(0,10,10);
8   y = rand(1,10);
9
10  subplot(4,1,1);
11  plot(x,y);
12  subplot(4,1,2);
13  bar(x,y);
14  subplot(4,1,3);
15  barh(x,y);
16  ax4 = subplot(4,1,4);
17  pie(y)
18  colormap(ax4,'gray')
```

Four plot types are shown that help display 2D data. One is the 2D line plot, the same as is used in `PlotSet`. The middle two are bar charts. The final is a pie chart. Each gives you a different insight into the data. Figure 3.2 shows the plot types.

There are many MATLAB functions for making these plots more informative. You can

- Add labels

- Add grids

- Change font types and sizes

- Change the thickness of lines

- Add legends

- Change axis limits

The last item requires looking at the axis properties. Here are the properties for the last plot—the list is very long! `gca` is the handle to the current axis. `get(gca)` returns a huge list, which we will not print here. Every single one of these can be changed by using the `set` function:

```
set(gca,'YMinorGrid','on','YGrid','on')
```

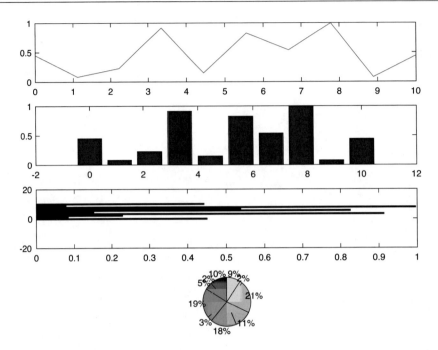

Figure 3.2: *Four different types of MATLAB 2D plots*

This uses parameter pairs just like PlotSet. In this list, children are pointers to the children of the axis. You can access those using get and change their properties using set. Any items that are added to the axis, such as axis labels, titles, lines, or other graphics objects, are all children of that axis.

3.3   Custom Two-Dimensional Diagrams

3.3.1 Problem

Many machine learning algorithms benefit from two-dimensional diagrams, such as tree diagrams, to help the user understand the results and the operation of the software. Such diagrams, automatically generated by the software, are useful in many types of learning systems. This section gives an example of how to write MATLAB code for a tree diagram.

3.3.2 Solution

Our solution is to use the MATLAB patch function to automatically generate the blocks and use line to generate connecting lines. Figure 3.3 shows the resulting hierarchical tree diagram. The circles are in rows, and each row is labeled.

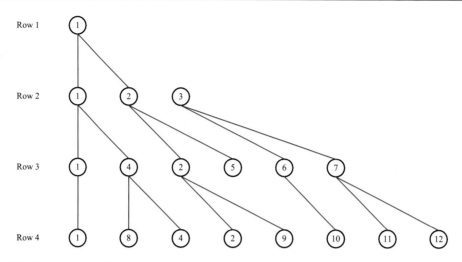

Figure 3.3: *A custom tree diagram*

3.3.3 How It Works

Tree diagrams are very useful for machine learning. This function generates a hierarchical tree diagram with the nodes as circles with text within each node. The graphics functions used in this function are

1. `line`

2. `patch`

3. `text`

The data needed to draw the tree is contained in a data structure, which is documented in the header. Each node has a parent field. This information is sufficient to make the connections. The node data is entered as a cell array.

The function uses a figure handle as a persistent variable so that the same figure can be updated with subsequent calls if desired.

TreeDiagram.m

```
94  if( ~update )
95      figHandle = NewFigure(w.name);
96  else
97      clf(figHandle)
98  end
```

The core drawing code is in `DrawNode`, which draws the boxes, and `ConnectNode` which connects the nodes with lines. Our nodes are circles with 20 segments. The `linspace` code makes sure that both 0 and 2π are not in the list of angles.

TreeDiagram.m

```
136  function [xC,yCT,yCB] = DrawNode( x0, y0, k, w )
137
138  n = 20;
139  a = linspace(0,2*pi*(1-1/n),n);
140
141  x = w.width*cos(a)/2 + x0;
142  y = w.width*sin(a)/2 + y0;
143  patch(x,y,'w');
144  text(x0,y0,sprintf('%d',k),'fontname',w.fontName,'fontsize',w.fontSize,
          'horizontalalignment','center');
145
146  xC  = x0;
147  yCT = y0 + w.width/2;
148  yCB = y0 - w.width/2;
149
150  %% TreeDiagram>ConnectNode
151  function ConnectNode( n, nP, w )
152
153  x = [n.xC nP.xC];
154  y = [n.yCT nP.yCB];
155
156  line(x,y,'linewidth',w.linewidth,'color',w.linecolor);
```

The demo shows how to use the function.

3.4   Three-Dimensional Box

There are two broad classes of three-dimensional graphics. One is to draw an object, like the earth. The other is to draw large data sets. This recipe plus the following one will show you how to do both.

3.4.1 Problem

We want to draw a three-dimensional box.

3.4.2 Solution

Use the patch function to draw the object. An example is shown in Figure 3.4. A single patch is shown in Figure Figure 3.5.

3.4.3 How It Works

Three-dimensional objects are created from vertices and faces. A vertex is a point in space. You create a list of vertices that are the corners of your 3D object. You then create faces that are lists of vertices. A face with two vertices is a line, and one with three vertices is a triangle. A polygon can have as many vertices as you would like. However, at the lowest level, graphics processors deal with triangles, so you are best off making all patches triangles. You will notice the normal vector. This is the outward vector. Your vertices in your patches should be ordered using the right-hand rule, that is, if the normal is in the direction of your thumb, then the faces

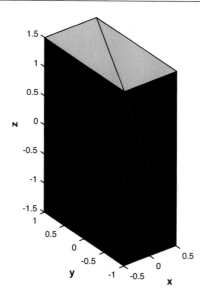

Figure 3.4: *A box drawn with* `patch`

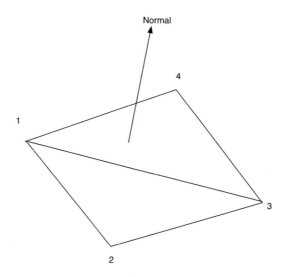

Figure 3.5: *A patch. The normal is toward the camera or the "outside" of the object*

are ordered in the direction of your fingers. In this figure, the order for the two triangles would be

```
[3 2 1]
[1 4 3]
```

MATLAB lighting is not very picky about vertex ordering, but if you export a model, then you will need to follow this convention. Otherwise, you can end up with inside-out objects!

The following code creates a box composed of triangle patches. The face and vertex arrays are created by hand. Vertices are one vertex per row, so vertex arrays are n by 3. Face arrays are n by m where m is the largest number of vertices per face. In Box, we work with triangles only. All graphics processors ultimately draw triangles, so, if you can, it is best to create objects only with triangles.

Box.m

```matlab
18   function [v, f] = Box( x, y, z )
19
20   % Demo
21   if( nargin < 1 )
22      Demo
23      return
24   end
25
26   % Faces
27   f   = [2 3 6;3 7 6;3 4 8;3 8 7;4 5 8;4 1 5;2 6 5;2 5 1;1 3 2;1 4 3;5 6
         7;5 7 8];
28
29   % Vertices
30   v = [-x  x  x -x -x  x  x -x;...
31         -y -y  y  y -y -y  y  y;...
32         -z -z -z -z  z  z  z  z]'/2;
33
34   % Default outputs
35   if( nargout == 0 )
36      DrawVertices( v, f, 'Box' );
37      clear v
38   end
```

The box is drawn using patch in the function DrawVertices. There is just one call to patch. patch accepts parameter pairs to specify face and edge coloring and many other characteristics of the patch. Only one color can be specified for a patch. If you wanted a box with different colors on each side, you would need multiple patches. We turn on rotate3d so that we can reorient the object with the mouse. view3 is a standard MATLAB view with the eye looking down a corner of the grid box.

DrawVertices.m

```
31  NewFigure(name);
32  patch('vertices',v,'faces',f,'facecolor',[0.8 0.1 0.2]);
33  axis image
34  xlabel('x')
35  ylabel('y')
36  zlabel('z')
37  view(3)
38  grid on
39  rotate3d on
```

We use only the most basic lighting. You can add all sorts of lights in your drawing using `light`. Light can be ambient or from a variety of light sources.

3.5 Draw a 3D Object with a Texture

3.5.1 Problem

We want to draw a planet displaying an image of the planet's surface.

3.5.2 Solution

Use a surface object and overlay a texture onto the surface. Figure 3.6 shows an example with a recent image of Pluto.

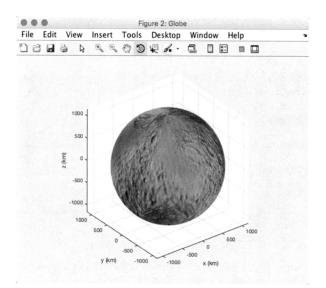

Figure 3.6: *A three-dimensional globe of Pluto*

3.5.3 How It Works

We generate the picture by first creating x, y, z points on the sphere and then overlaying a texture that is read in from an image file. The texture map can be read from a file using `imread`. If this is a colored image, it will be a three-dimensional matrix. The third element will be an index to the color, red, blue, or green. However, if it is a grayscale image, you must create the three-dimensional "color" matrix by replicating the image.

```
p = imread('PlutoGray.png');
p3(:,:,1) = p;
p3(:,:,2) = p;
p3(:,:,3) = p;
```

The starting `p` is a two-dimensional matrix.

You first generate the surface using the coordinates generated from the `sphere` function. This is done with `surface`.

Globe.m

```
1   %% GLOBE Draws a three dimensional map of a planet.
26  function Globe( planet, radius )
35  if( ischar(planet) )
36    planetMap = imread(planet);
37  else
38    planetMap = planet;
39  end
40
41  NewFigure('Globe')
42
43  [x,y,z] = sphere(50);
44  x         = x*radius;
45  y         = y*radius;
46  z         = z*radius;
47  hSurf    = surface(x,y,z);
48  grid on;
```

You then apply the texture.

Globe.m

```
49  for i= 1:3
50    planetMap(:,:,i)=flipud(planetMap(:,:,i));
51  end
52  set(hSurf,'Cdata',planetMap,'Facecolor','texturemap');
53  set(hSurf,'edgecolor', 'none',...
54            'EdgeLighting', 'phong','FaceLighting', 'phong',...
55            'specularStrength',0.1,'diffuseStrength',0.9,...
56            'SpecularExponent',0.5,'ambientStrength',0.2,...
57            'BackFaceLighting','unlit');
```

flipup makes the map look "normal." Phong is a type of lighting. It takes the colors at the vertices and interpolates the colors at the pixels on the polygon based on the interpolated normals. Diffuse and specular refer to different types of reflections of light. They aren't too important when you apply a texture to the surface.

3.6   General 3D Graphics

3.6.1   Problem

We want to use 3D graphics to study a 2D data set. A 2D data set is a matrix or an n-by-m array.

3.6.2   Solution

Use MATLAB surface, mesh, bar, and contour functions. An example of a random data set with different visualizations is shown in Figure 3.7.

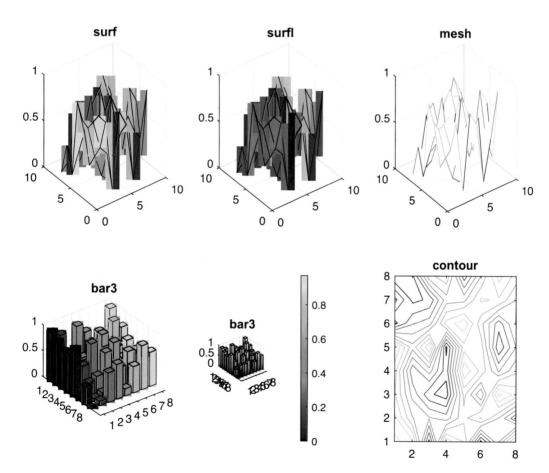

Figure 3.7: *Two-dimensional data shown with six different plot types*

3.6.3 How It Works

We generate a random 2D data set that is 8 by 8 using `rand`. We display it in several ways in a figure with subplots. In this case, we create two rows and three columns of subplots. Figure 3.7 shows six types of 2D plots. `surf`, `mesh`, and `surfl` (3D shaded surface with lighting) are very similar. The surface plots are more interesting when lighting is applied. The two `bar3` plots show different ways of coloring the bars. In the second bar plot, the color varies with length. This requires a bit of code changing the `CData` and `FaceColor`.

TwoDDataDisplay.m

```
10   colormap(h,'gray')
11
12   subplot(2,3,1)
13   surf(m)
14   title('surf')
15
16   subplot(2,3,2)
17   surfl(m,'light')
18   title('surfl')
19
20   subplot(2,3,3)
21   mesh(m)
22   title('mesh')
23
24   subplot(2,3,4)
25   bar3(m)
26   title('bar3')
27
28   subplot(2,3,5)
29   h = bar3(m);
30   title('bar3')
31
32   colorbar
33   for k = 1:length(h)
34           zdata = h(k).ZData;
35           h(k).CData = zdata;
36           h(k).FaceColor = 'interp';
37   end
38
39   subplot(2,3,6)
40   contour(m);
41   title('contour')
```

3.7   Building a GUI

3.7.1 Problem

We want a GUI to provide a graphical interface for a second-order system simulation.

3.7.2 Solution

We will use the MATLAB GUIDE to build a GUI that will allow us to

1. Set the damping constant

2. Set the end time for the simulation

3. Set the type of input (pulse, step, or sinusoid)

4. Display the input and output plot

Note that GUIDE is being deprecated and as of 2023 throws a warning. The new App Designer feature works very similarly, with one tab to lay out the GUI and a second to write the code.

3.7.3 How It Works

We want to build a GUI to interface with the following SecondOrderSystemSim. The first part is the simulation code in a loop.

SecondOrderSystemSim.m

```
26  function [xP, t, tL] = SecondOrderSystemSim( d )
38  omega     = max([d.omega d.omegaU]); % Maximum frequency for the
               simulation
39  dT        = 0.1*2*pi/omega; % Get the time step from the frequency
40  n         = floor(d.tEnd/dT); % Get an integer numbeer of steps
41  xP        = zeros(2,n); % Size the plotting array
42  x         = [0;0]; % Initial condition on the [position;velocity]
43  t         = 0; % Initial time
44
45  for k = 1:n
46    [~,u]   = RHS(t,x,d);
47    xP(:,k) = [x(1);u];
48    x       = RungeKutta( @RHS, t, x, dT, d );
49    t       = t + dT;
50  end
```

Running it gives the plot in Figure 3.8. The plotting code is

SecondOrderSystemSim.m

```
52  [t,tL] = TimeLabel((0:n-1)*dT);
53
54  if( nargout == 0 )
55    PlotSet(t,xP,'x label',tL,'y label', {'x' 'u'}, 'figure title','
         Filter');
56  end
```

`TimeLabel` makes time units that are reasonable for the length of the simulation. It automatically rescales the time vector.

The function has the simulation loop built in.

The MATLAB GUI building system, GUIDE, is invoked by typing `guide` at the command line. This recipe was created using MATLAB R2018a. Note that GUIDE will be deprecated in future releases, replaced by the App Designer, invoked with `appdesigner`. In App Designer, you follow a similar process of laying out the GUI with drag and drop and writing the code in the callback functions.

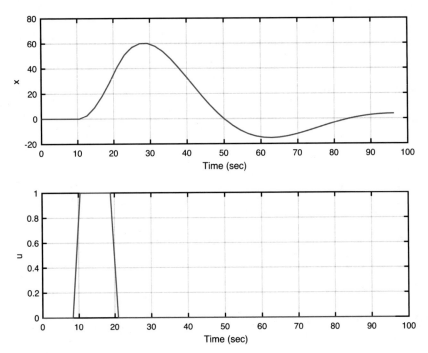

Figure 3.8: *Second-order system simulation*

There are several options for GUI templates or a blank GUI. We will start from a blank GUI. First, let's make a list of the controls we will need from our desired features listed earlier:

- Edit boxes for

  - Simulation duration
  - Damping ratio
  - Undamped natural frequency
  - Sinusoid input frequency
  - Pulse start and stop time

- Radio button for the types of input

- Run button for starting a simulation

- Plot axes

We type "guide" in the command window, and it asks us to either pick an existing GUI or create a new one. We choose a blank GUI. Figure 3.9 shows the template GUI in GUIDE before we make any changes to it. You add elements by dragging and dropping them from the table at the left.

Figure 3.10 shows the GUI inspector. You edit GUI elements here. You can see that the elements have a lot of properties. We aren't going to try and make this GUI slick, but with some effort, you can make it a work of art. The ones we will change are the tag and text properties. The tag gives the software a name to use internally. The text is just what is shown on the device.

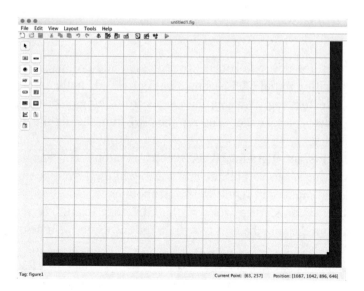

Figure 3.9: *Blank GUI*

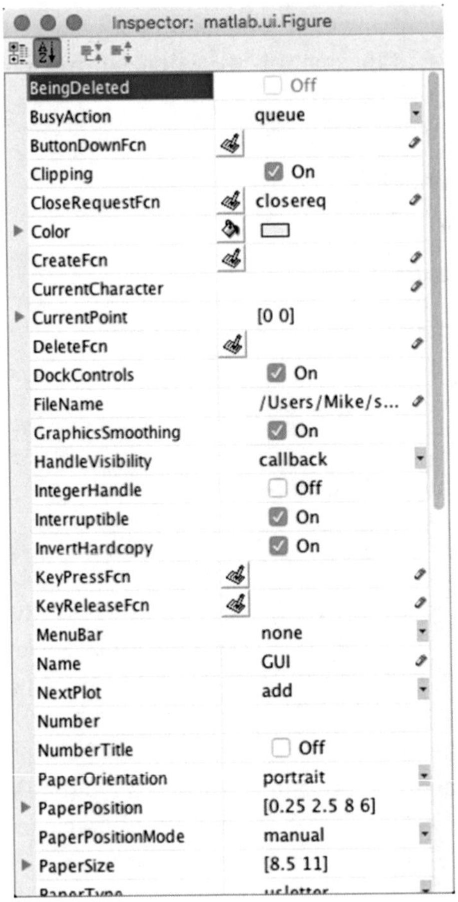

Figure 3.10: *The GUI inspector*

We then add all the desired elements by dragging and dropping. We choose to name our GUI GUI. The resulting initial GUI is shown in Figure 3.11. In the inspector for each element, you will see a field for "tag." Change the names from things like edit1 to names you can easily identify. When you save them and run the GUI from the .fig file, the code in GUI.m will automatically change.

We create a radio button group and add the radio buttons. This handles disabling all but the selected radio button. When you hit the green arrow in the layout box, it saves all changes to the m-file and also simulates it. It will warn you about bugs.

At this point, we can start to work on the GUI code itself. The template GUI stores its data, calculated from the data the user types into the edit boxes, in a field called simdata. The autogenerated code is in SimGUI.

When the GUI loads, we initialize the text fields with the data from the default data structure. Make sure that the initialization corresponds to what is seen in the GUI. You need to be careful about radio buttons and button states.

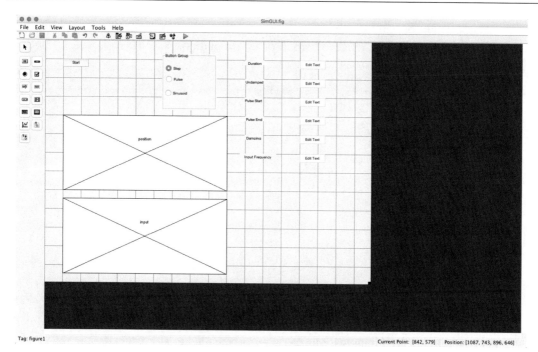

Figure 3.11: *Snapshot of the GUI in the editing window after adding all the elements*

SimGUI.m

```
48  function SimGUI_OpeningFcn(hObject, eventdata, handles, varargin)
49
50  % Choose default command line output for SimGUI
51  handles.output = hObject;
52
53  % Get the default data
54  handles.simData = SecondOrderSystemSim;
55
56  % Set the default states
57  set(handles.editDuration,'string',num2str(handles.simData.tEnd));
58  set(handles.editUndamped,'string',num2str(handles.simData.omega));
59  set(handles.editPulseStart,'string',num2str(handles.simData.tPulseBegin
        ));
60  set(handles.editPulseEnd,'string',num2str(handles.simData.tPulseEnd));
61  set(handles.editDamping,'string',num2str(handles.simData.zeta));
62  set(handles.editInputFrequency,'string',num2str(handles.simData.omegaU)
        );
63
64  % Update handles structure
65  guidata(hObject, handles);
```

When the start button is pushed, we run the simulation and plot the results. This essentially is the same as the demo code in the second-order simulation.

SimGUI.m

```
104  function start_Callback(hObject, eventdata, handles)
105
106  [xP, t, tL] = SecondOrderSystemSim(handles.simData);
107
108  axes(handles.position)
109  plot(t,xP(1,:));
110  ylabel('Position')
111  grid on
112
113  axes(handles.input)
114  plot(t,xP(2,:));
115  xlabel(tL);
116  ylabel('input');
117  grid on
```

The callbacks for the edit boxes require a little code to set the data in the stored data. All data is stored in the GUI handles. `guidata` must be called to store new data in the handles.

SimGUI.m

```
120  function editDuration_Callback(hObject, eventdata, handles)
121
122  handles.simData.tEnd = str2double(get(hObject,'String'));
123  guidata(hObject, handles);
```

One simulation is shown in Figure 3.12. Another simulation in the GUI is shown in Figure 3.13.

3.8 Animating a Bar Chart

Two-dimensional arrays are often produced as part of machine learning algorithms. For situations where they change dynamically, we would like to animate a display.

3.8.1 Problem

We want to animate a 3D bar chart.

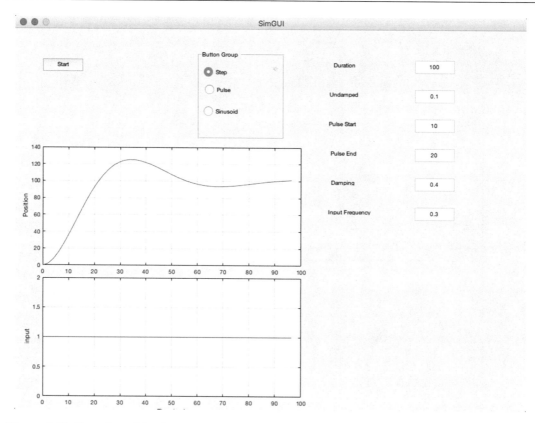

Figure 3.12: *Snapshot of the GUI in simulation*

3.8.2 Solution

We will write code to animate the MATLAB `bar3` function.

3.8.3 How It Works

Our function will set up the figure using `bar3` and then replace the values for the length of the bars. This is trickier than it sounds.

The following is an example of `bar3`. Look at the handle `h`. It is length 3. Each column in `m` corresponds to a `surface` data structure.

```
>> m = [1 2 3;4 5 6];
>> h = bar3(m)
h =

  1x3 Surface array:

    Surface    Surface    Surface
```

Figure 3.14 shows the 3D bar graph.

69

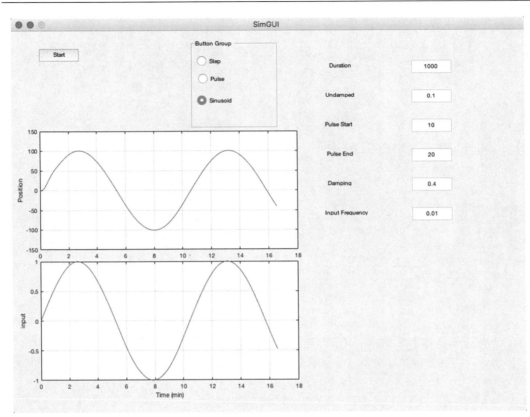

Figure 3.13: *Snapshot of the GUI in simulation*

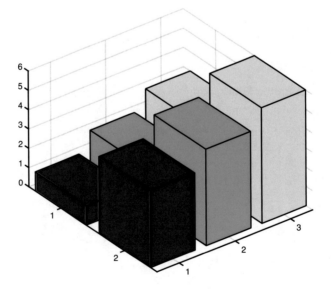

Figure 3.14: *Two by three bar chart*

70

We use the handle h to get the z data. The following is the data for the first column of m:

```
>> z = get(h(1),'zdata')
z =

    NaN      0      0    NaN
      0      1      1      0
      0      1      1      0
    NaN      0      0    NaN
    NaN      0      0    NaN
    NaN    NaN    NaN    NaN
    NaN      0      0    NaN
      0      4      4      0
      0      4      4      0
    NaN      0      0    NaN
    NaN      0      0    NaN
    NaN    NaN    NaN    NaN
```

There are six rows in the z array for each 3D bar, one row for each face, and four values across. Note that each value in m, in this case 1 and 4, appears in the z data four times, twice each in two rows. This defines the height of each bar. We will need to replace all four values for each number in m.

The code is shown as follows. We have two actions, 'initialize', which creates the figure, and 'update' which updates the z values. Fortunately, the z values are always in the same spot, so it is not too hard to replace them. colorbar draws the color bar seen on the right of Figure 3.15. We use persistent to store the handle to bar3.

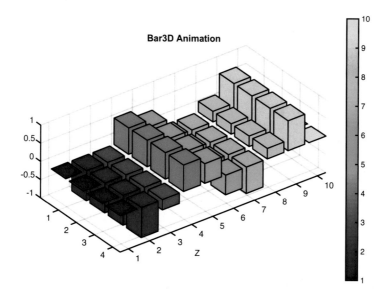

Figure 3.15: *Sinusoidal bar chart and the beginning (left) and end (right) of the animation*

Bar3D.m

```
18   function Bar3D(action,v,xL,yL,zL,t)
19
20   if( nargin < 1 )
21     Demo
22     return
23   end
24
25   persistent h
26
27   switch lower(action)
28     case 'initialize'
29
30       NewFigure('3D Bar Animation');
31       h = bar3(v);
32
33       colorbar
34
35       xlabel(xL)
36       xlabel(yL)
37       xlabel(zL)
38       title(t);
39       view(3)
40       rotate3d on
41
42     case 'update'
43       nRows = length(h);
44       for i = 1:nRows
45         z = get(h(i),'zdata');
46         n = size(v,1);
47         j = 2;
48         for k = 1:n
49           z(j,   2) = v(k,i);
50           z(j,   3) = v(k,i);
51           z(j+1,2) = v(k,i);
52           z(j+1,3) = v(k,i);
53           j        = j + 6;
54         end
55         set(h(i),'zdata',z);
56       end
57   end
```

The built-in demo of the function animates the product of a sine and cosine, similar to the MATLAB logo. Bar3D is called first with 'initialize' and then in a loop for the 'update'.

Bar3D.m

```
59   function Demo
60   %% Bar3D>Demo
61   % Animate the MATLAB logo
62
```

```
63   x = linspace(0,4*pi,10);
64   v = [sin(x);sin(x);sin(x);sin(x)];
65   t = linspace(0,100,25);
66   t = cos(0.1*t);
67
68   Bar3D('initialize',v,'X','Y','Z','Bar3D Animation');
69   for k = 1:length(t)
70     Bar3D('update',v*t(k));
71     pause(1);
72   end
```

The figure at the beginning and end of the animation is shown in Figure 3.15.

3.9 Drawing a Robot

This section shows the elements of writing graphics code to draw a robot. If you are doing machine learning involving humans or robots, this is useful code to have. We'll show how to animate a robot arm.

3.9.1 Problem

We want to animate a robot arm.

3.9.2 Solution

We write code to create vertices and faces for use with the MATLAB `patch` function.

3.9.3 How It Works

DrawSCARA draws and animates a robot. The first part of the code just organizes the operation of the function using a `switch` statement.

DrawSCARA.m

```
45   switch( lower(action) )
46       case 'defaults'
47           m = Defaults;
48
49       case 'initialize'
50           if( nargin < 2 )
51               d   = Defaults;
52           else
53               d   = x;
54           end
55
56           p = Initialize( d );
57
58       case 'update'
59           if( nargout == 1 )
60               m = Update( p, x );
61           else
```

```
62                    Update( p, x );
63            end
64    end
```

Initialize creates the vertices and faces using functions Box, Frustrum, and UChannel. These are tedious to write and are geometry specific. You can apply them to a wide variety of problems, however. You should note that it stores the patches so that we just have to pass in new vertices when animating the arm. The "new" vertices are just the vertices of the arm rotated and translated to match the position of the arm. The arm itself does not deform. We do the computations in the right order so that transformations are passed up/down the chain to get everything moving correctly.

Update updates the arm positions by computing new vertices and passing them to the patches. drawnow draws the arm. We can also save the frames to animate them using MATLAB's movie functions.

DrawSCARA.m

```
161    function m = Update( p, x )
162
163    for k = 1:size(x,2)
164
165        % Link 1
166        c        = cos(x(1,k));
167        s        = sin(x(1,k));
168
169        b1       = [c -s 0;s c 0;0 0 1];
170        v        = (b1*p.v1')';
171
172        set(p.link1,'vertices',v);
173
174        % Link 2
175        r2       = b1*[p.a1;0;0];
176
177             c        = cos(x(2,k));
178        s        = sin(x(2,k));
179
180        b2       = [c -s 0;s c 0;0 0 1];
181        v        = (b2*b1*p.v2')';
182
183        v(:,1)   = v(:,1) + r2(1);
184        v(:,2)   = v(:,2) + r2(2);
185
186        set(p.link2,'vertices',v);
187
188        % Link 3
189        r3       = b2*b1*[p.r3;0;0] + r2;
190        v        = p.v3;
191
192        v(:,1)   = v(:,1) + r3(1);
```

```
193        v(:,2)    = v(:,2) + r3(2);
194        v(:,3)    = v(:,3) + x(3,k);
195
196        set(p.link3,'vertices',v);
197
198        % Link 4
199        c              = cos(x(4,k));
200        s              = sin(x(4,k));
201
202        b4             = [c -s 0;s c 0;0 0 1];
203        v              = (b4*b2*b1*p.v4')';
204        r4             = b2*b1*[p.r4;0;0] + r2;
205
206        v(:,1)    = v(:,1) + r4(1);
207        v(:,2)    = v(:,2) + r4(2);
208        v(:,3)    = v(:,3) + x(3,k);
209
210        set(p.link4,'vertices',v);
211
212        if( nargout > 0 )
213            m(k) = getframe;
214        else
215            drawnow;
216        end
217
218    end
```

The SCARA robot arm in the demo is shown at the end in Figure 3.16. The demo code could be replaced by a simulation of the arm dynamics. In this case, we pick angular rates and generate an array of angles. Note that this alternate demo function does not need to be a built-in demo function at all. This same block of code can be executed directly from the command line.

DrawSCARA.m

```
236    function Demo
237
238    DrawSCARA( 'initialize' );
239    t          = linspace(0,100);
240    omega1  = 0.1;
241    omega2  = 0.2;
242    omega3  = 0.3;
243    omega4  = 0.4;
244    x          = [sin(omega1*t);sin(omega2*t);0.01*sin(omega3*t);sin(omega4*t)
           ];
245    DrawSCARA( 'update', x );
```

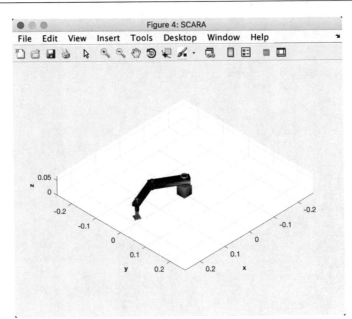

Figure 3.16: *Robot arm generated by* `DrawSCARA`

3.10 Importing a Model

3.10.1 Problem

This section shows how to import an external model and display it in MATLAB. We will import a model of the X-15 aircraft.

3.10.2 Solution

We write code to read in an OBJ file.

3.10.3 How It Works

You need to first put the OBJ file in the MATLAB path. We add the file "X15.obj".

`LoadOBJFile` reads in the file, parses the file, and then draws it. It creates a data structure that breaks the file into components if the file has components defined.

The main file loads in the data from the OBJ file. It tessellates it, converting quadrilaterals to triangles. It then groups the vertices and faces into components.

LoadOBJFile.m

```
24  function g = LoadOBJFile( file, kScale )
25
26  % Input processing
27  if nargin == 0
28    % Demo
29    file = 'X15.obj';
30    LoadOBJFile( file );
31    return
32  end
33
34  % Read in the file
35  g = GetDataOBJ( file );
36
37  % Convert to triangles
38  g = Tesselate( g );
39
40  if( isempty( g ) )
41    return;
42  end
43
44  if( nargin < 2 )
45    kScale = 1;
46  end
47
48  g.radius = 0;
49  if( isfield( g, 'component' ) )
50    for k = 1:length(g.component)
51      g.component(k).v = g.component(k).v*kScale;
52      g.radius          = max([Mag(g.component(k).v') g.radius]);
53    end
54
55    if( nargout == 0 )
56      DrawPicture( g );
57    end
58  end
```

The `DrawPicture` subfunction draws a picture of the object. It calls the subfunction `DrawMesh`. It uses `NewFigure` to create the figure window. The other functions are standard MATLAB plotting functions.

LoadOBJFile.m

```
65  %% Draw the picture
66  function DrawPicture( g )
67
68  NewFigure( g.name )
69  axes('DataAspectRatio',[1 1 1],'PlotBoxAspectRatio',[1 1 1] );
70
71  for k = 1:length(g.component)
72    DrawMesh( g.component(k) );
```

```
73   end
74
75   xlabel('X')
76   ylabel('Y')
77   zlabel('Z')
78
79   grid on
80   view(3)
81   rotate3d on
82   hold off
```

This subfunction parses the obj file. It starts by breaking each line into tokens. The first token gives the meaning of the line. It looks for "v" for vertices, "f" for faces, and "g" for components. It ignores all other data. If it finds a "g," it begins creating groups of faces. component is a data structure used to organize the faces and vertices.

LoadOBJFile.m

```
84   %% Get the polygon data
85   function g = GetDataOBJ( file )
86
87   % Initialize counters
88   kV       = 0;
89   kF       = 0;
90   nG       = 0;
91   hasG     = false;
92   g.name   = file;
93   lines    = readlines(file);
94   n        = size(lines);
95
96   % Read the file
97   for j = 1:n
98     t = split(lines(j));
99
100      if( ~isempty(t{1}) )
101
102        % The first token determines the action
103        switch t{1}
104          case '#'
105            % A Comment
106
107          case 'v'
108            kV       = kV + 1;
109            v(kV,:) = [str2double(t{2}) str2double(t{3}) str2double(t{4})
                     ];
110
111          case 'vn'
112            % Normals not used
113
114          case 'vt'
115            % Texture map coordinates not used
```

```
116
117        case 'f'
118          if( ~hasG )
119            kG        = 1;
120            group{1}  = 'Default';
121            nG        = 1;
122          end
123          if( isempty(t{end}))
124            t = t(1:end-1);
125          end
126
127          lT        = length(t) - 1;
128          vT        = zeros(1,lT);
129          for k = 1:lT
130            if( ~isempty(t{k+1}) )
131              gVO = GetVertexOBJ(t{k+1});
132
133              if( ~isempty(gVO) )
134                vT(k) = gVO;
135              end
136            end
137          end
138
139          % Assign the faces to all groups
140          for k = 1:length(kG)
141            i      = kG(k);
142            kF(i) = kF(i) + 1;
143            component(i).f(kF(i),1:lT) = vT;
144          end
145
146        case 'g'
147          hasG = true;
148          n = length(t) - 1;
149          if( n > 0 && ~isempty( t{2} ) )
150            kG = [];
151            for k = 1:n
152              isANewGroup = 1;
153              if( nG > 0 )
154                for i = 1:nG
155                  if( strcmp( group{i}, t{k+1} ) )
156                    isANewGroup = 0;
157                    break;
158                  end
159                end
160                if( isANewGroup )
161                  nG        = nG + 1;
162                  kF(nG)    = 0;
163                  group{nG} = t{k+1};
164                  i         = nG;
165                end
166              kG = [kG i];
```

79

```
167        else
168           nG      = 1;
169           kG      = 1;
170           group{1} = t{2};
171        end
172      end
173    end
174
175  case 'l'
176    fprintf(1,'%s: Line ''%s'' Lines not implemented\n',mfilename,t
           {1});
```

This code sorts the data into components.

LoadOBJFile.m

```
191  % Sort into groups
192  kG = 0;
193  for k = 1:nG
194    [n,m]  = size( component(k).f );
195    fC     = sort(reshape( component(k).f, n*m, 1 ));
196
197    % Delete duplicates
198    fC(fC == 0) = [];
199    kDelete = [];
200    for j = 2:length(fC)
201      if( fC(j) == fC(j-1) )
202        kDelete = [kDelete j];
203      end
204    end
205
206    fC(kDelete) = [];
207    if( ~isempty(fC) )
208      kG                = kG + 1;
209      g.component(kG)   = CreateComponent;
210      g.component(kG).nV = [];
211      g.component(kG).f  = component(kG).f;
212      g.component(kG).v  = v(fC,:);
213      [rF,cF]           = size( g.component(kG).f );
214
215      for i = 1:rF
216        for j = 1:cF
217          if( g.component(kG).f(i,j) == 0 )
218            break;
219          else
220            p = find( fC == g.component(kG).f(i,j) ); % Reindexing
221            g.component(kG).f(i,j) = p;
222          end
223        end
224        nM = find(g.component(kG).f(i,:) == 0);
225        if( isempty(nM) )
226          nM = length( g.component(kG).f(i,:) );
```

```
227        else
228          nM = min(nM) - 1;
229        end
230        g.component(kG).nV(i) = nM; % The number of vertices per face
231      end
232
233      g.component(kG).name   = group{k};
234      g.component(kG).color        = [0.6 0.6 0.6];
235    end
```

This block takes a line of the obj file and extracts vertices. A face line looks like

```
f 3/23/23 7/24/24 8/21/21
```

because a face can have normals and texture vertices too. We only want the regular vertices.

LoadOBJFile.m

```
248  %% Get the vertex from the face vertex list
249  function v = GetVertexOBJ( t )
250
251  k = strfind(char(t),'/');
252
253  if( isempty(k) )
254    v = str2num(t); %#ok<ST2NM>
255  else
256    k = k(1);
257    v = str2num(t(1:(k-1))); %#ok<ST2NM>
258  end
```

The following code draws the vertices and faces using `patch`. The color is hard-coded as is "phong," the lighting mode.

LoadOBJFile.m

```
260  %% Draw a mesh
261  function h = DrawMesh( m )
262
263  kMax = max(m.nV);
264  kMin = min(m.nV);
265  i    = 1;
266  for k = kMin:kMax
267    j = find( m.nV == k );
268    if( ~isempty(j) )
269      h(i) = patch( 'Vertices', m.v, 'Faces',   m.f(j,1:k),...
270        'facecolor',[0.7 0.7 0.7],...
271        'EdgeLighting', 'phong',...
272        'FaceLighting', 'phong');
273      i = i + 1;
274    end
275  end
```

The following code looks for faces with four vertices and breaks each into two triangles. This won't work if the polygon has more than four vertices. More general code is not too hard to write.

LoadOBJFile.m

```
277  %% Tesselate 4 corner polygons
278  function g = Tesselate( g )
279
280  for k = 1:length(g.component)
281
282    nD = size( g.component(k).f, 2 );
283
284    % Tesselate if necessary
285    if( nD == 4 )
286      f = g.component(k).f;
287
288      % Tesselate
289      i = 1;
290      while i <= size(f,1)
291        if f(i,4) ~= 0
292          % Tesselate into 2 triangles: 1/2/4 and 2/3/4
293          fv = f(i,:);
294          f(i,:) = [fv(1) fv(2) fv(4) 0];
295          f = [f(1:i,:); [fv(2) fv(3) fv(4) 0]; f(i+1:end,:)];
296          i = i+2;
297        else
298          i = i+1;
299        end
300      end
301      f(:,4) = [];
302      g.component(k).f = f;
303      g.component(k).nV = 3*ones(size(f,1),1);
304    end
305  end
```

Figure 3.17 shows the resulting X-15 aircraft. The triangle lines are black.

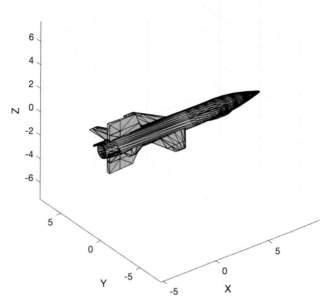

Figure 3.17: *X-15 model*

3.11   Summary

This chapter has demonstrated graphics that can help understand the results of machine learning software. Two- and three-dimensional graphics were demonstrated. The chapter also showed how to build a Graphical User Interface to help automate functions. Table 3.1 lists the functions and scripts included in the companion code.

Table 3.1: *Chapter code listing*

File	Description
Box	Draws a box
DrawSCARA	Draws a robot arm
DrawVertices	Draws a set of vertices and faces
Globe	Draws a texture-mapped globe
LoadOBJFile	Loads a Wavefront OBJ file
PlotSet	2D line plots
SecondOrderSystemSim	Simulates a second-order system
SimGUI	Code for the simulation GUI
SimGUI.fig	The figure file
TreeDiagram	Draws a tree diagram
TwoDDataDisplay	A script to display two-dimensional data in three-dimensional graphics

CHAPTER 4

■ ■ ■

Kalman Filters

Understanding or controlling a physical system often requires a model of the system, that is, knowledge of the characteristics and structure of the system. A model can be a predefined structure or can be determined solely through data. In the case of Kalman Filtering, we create a model and use the model as a framework for learning about the state of the system. This is part of the Control branch of our Autonomous Learning taxonomy from Chapter 1.

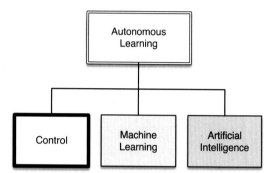

What is important about Kalman Filters is that they rigorously account for uncertainty in a system that you want to know more about. There is uncertainty in the model of the system, if you have a model, and uncertainty (i.e., noise) in measurements of a system.

A system can be defined by its dynamical states and its nominally constant parameters. For example, if you are studying an object sliding on a table, the states would be the position and velocity. The parameters would be the mass of the object and the friction coefficient. There might also be an external force on the object that we might want to estimate. The parameters and states compose the model. You need to know both to properly understand the system. Sometimes, it is hard to decide if something should be a state or a parameter. Mass is usually a parameter, but in a plane, car, or rocket where the mass changes as fuel is consumed, it is often modeled as a state.

Kalman Filters, invented by R. E. Kalman and others, are a mathematical framework for estimating or learning the states of a system. An estimator gives you statistically best estimates of the dynamical states of the system, such as the position and velocity of a moving point mass. Kalman Filters can also be written to identify the parameters of a system. Thus, the Kalman Filter provides a framework for both state and parameter identification.

© The Author(s), under exclusive license to APress Media, LLC, part of Springer Nature 2024
M. Paluszek, S. Thomas, *MATLAB Machine Learning Recipes*,
https://doi.org/10.1007/978-1-4842-9846-6_4

Another application of Kalman Filters is system identification. System identification is the process of identifying the structure and parameters of any system. For example, with a simple mass on a spring, it would be the identification or determination of the mass and spring constant values along with determining the differential equation for modeling the system. It is a form of machine learning that has its origins in control theory. There are many methods of system identification. In this chapter, we will only study the Kalman Filter. The term "learning" is not usually associated with estimation, but it is the same thing.

An important aspect of the system identification problem is determining what parameters and states can be estimated given the available measurements. This applies to all learning systems. The question is can we learn what we need to know about something through our observations? For this, we want to know if a parameter or state is observable and can be independently distinguished. For example, suppose we are using Newton's law

$$F = ma \tag{4.1}$$

where F is force, m is mass, and a is acceleration as our model, and our measurement is acceleration. Can we estimate both force and mass? The answer is no because we are measuring the *ratio* of force to mass:

$$a = \frac{F}{m} \tag{4.2}$$

We can't separate the two. If we had a force sensor or a mass sensor, we could determine each separately. You need to be aware of this issue in all learning systems including Kalman Filters.

4.1 Gaussian Distribution

In the limit, as you sum a large number of independent random variables, you get a Gaussian or Normal distribution. The Gaussian noise assumption is the underlying noise model for the conventional Kalman Filter.

The probability density function for the Gaussian distribution is

$$f(x) = \frac{1}{\sqrt{2\pi}} e^{-\frac{(x-\mu)^2}{2\sigma^2}} \tag{4.3}$$

With a Gaussian distribution, 68% of the values will fall within $\pm\sigma$, 95% of the values will fall within $\pm2\sigma$, and 99.7% of the values will fall within $\pm3\sigma$. The Gaussian Probability Density Function (PDF) and Cumulative Probability Density Function (CPDF) are computed in the following code. CPDF is the integral of PDF.

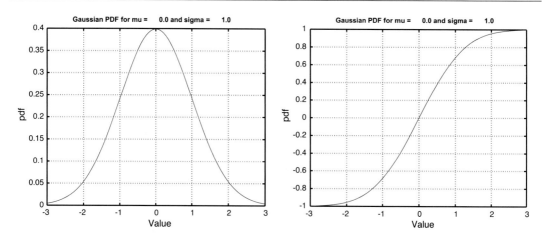

Figure 4.1: *Gaussian distribution*

GaussianExample.m

```
5  x      = linspace(-3,3);
6  mu     = 0;
7  sigma  = 1;
8  pDF    = GaussianPDF( x, mu, sigma );
9  cPDF   = GaussianCPDF( x, mu, sigma );
10 t      = sprintf('Gaussian PDF for mu = %8.1f and sigma = %8.1f',mu,
           sigma);
11 Plot2D( x, pDF, 'Value', 'pdf', t );
12 Plot2D( x, cPDF, 'Value', 'pdf', t );
```

The PDF and CPDF are shown in Figure 4.1.

Both plots are over the same 3 σ range. By the way, 6-σ is 99.9999998%. When we build a Kalman Filter, we assume that the measurement noise and the model noise are Gaussian. This is reasonable for most sensors but not necessarily for the model noise. Model noise is a combination of external inputs, model errors, and unmodeled dynamics.

4.2   A State Estimator Using a Linear Kalman Filter

4.2.1   Problem

You want to estimate the velocity and position of a mass attached through a spring and damper to a structure. The system is shown in Figure 4.2. m is the mass, k is the spring constant, c is the damping constant, and f is an external force. x is the position. The mass moves in only one direction.

Suppose we had a camera that was located near the mass. The camera would be pointed at the mass during its ascent. This would result in a measurement of the angle between the ground and the boresight of the camera. The angle measurement geometry is shown in Figure 4.3. The angle is measured from an offset baseline.

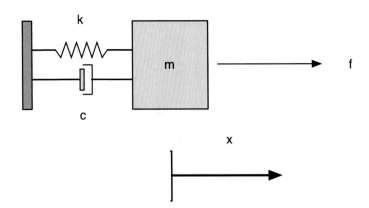

Figure 4.2: *Spring-mass-damper system. The mass is on the right. The spring is on the top to the left of the mass. The damper is below*

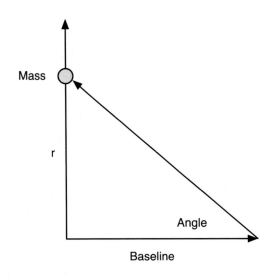

Figure 4.3: *The angle measurement geometry*

We want to use a conventional linear Kalman Filter to estimate the state of the system. This is suitable for a simple system that is modeled with linear equations.

4.2.2 Solution

First, we will need to define a mathematical model for the mass system and code it up. Then we will derive the Kalman Filter from the first principles, using the Bayes theorem. Finally, we present code implementing the Kalman Filter estimator for the spring-mass problem.

4.2.3 How It Works

Spring-Mass System Model

The continuous time differential equations modeling the system are

$$\frac{dr}{dt} = v \tag{4.4}$$

$$m\frac{dv}{dt} = f - cv - kx \tag{4.5}$$

This says the change in position r concerning time t is the velocity v. The change in velocity with respect to time (times mass) is an external force, minus the damping constant times velocity, minus the spring constant times the position. The second equation is just Newton's law where the total force is F and the total acceleration, a_T, is the total force divided by the mass, $\frac{F}{m}$:

$$F = f - cv - kx \tag{4.6}$$

$$\frac{dv}{dt} = a_T \tag{4.7}$$

To simplify the problem, we divide both sides of the second equation by mass and get

$$\frac{dr}{dt} = v \tag{4.8}$$

$$\frac{dv}{dt} = a - 2\zeta\omega v - \omega^2 x \tag{4.9}$$

where

$$\frac{c}{m} = 2\zeta\omega \tag{4.10}$$

$$\frac{k}{m} = \omega^2 \tag{4.11}$$

a is the acceleration due to external forces $\frac{f}{m}$, ζ is the damping ratio, and ω is the undamped natural frequency. The undamped natural frequency is the frequency at which the mass would oscillate if there was no damping. The damping ratio indicates how fast the system damps and what level of oscillations we observe. With a damping ratio of zero, the system never damps and the mass oscillates forever. With a damping ratio of one, you don't see any oscillation. This form makes it easier to understand what damping and oscillation to expect. You immediately know the frequency and the rate at which the oscillation should subside. m, c, and k, while they embody the same information, don't make this as obvious.

The following shows a simulation of the oscillator with damping (OscillatorDamping RatioSim). It shows different damping ratios. The loop that simulates different damping ratios is shown.

89

OscillatorDampingRatioSim.m

```
18  for j = 1:length(zeta)
19    % Initial state [position;velocity]
20    x = [0;1];
21    % Select damping ratio from array
22    d.zeta= zeta(j);
23
24    % Print a string for the legend
25    s{j} = sprintf('zeta = %6.4f',zeta(j));
26    for k = 1:nSim
27      % Plot storage
28      xPlot(j,k)  = x(1);
29
30      % Propagate (numerically integrate) the state equations
31      x  = RungeKutta( @RHSOscillator, 0, x, dT, d );
32    end
33  end
```

The results of the damping ratio demo are shown in Figure 4.4. The initial conditions are zero position and a velocity of one. The responses to different levels of damping ratios are seen. When zeta is zero, it is undamped and oscillates forever. Critical damping, which is desirable for minimizing actuator effort, is 0.7071. A damping ratio of 1 results in no overshoot to a step disturbance. In this case, we have "overshoot" since we are not at a rest initial condition.

The dynamical equations are in what is called state space form because the derivative of the state vector

$$x = \left[\begin{array}{c} r \\ v \end{array} \right] \tag{4.12}$$

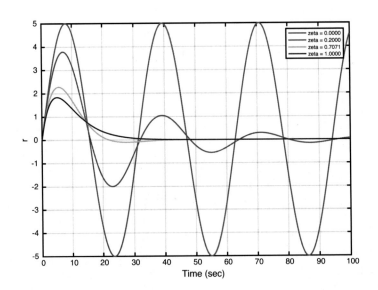

Figure 4.4: *Spring-mass-damper system simulation with different damping ratios* zeta

has nothing multiplying it, and there are only the first derivatives on the left-hand side. Sometimes, you see equations like

$$Q\dot{x} = Ax + Bu \tag{4.13}$$

If Q is not invertible, then you can't do

$$\dot{x} = Q^{-1}Ax + Q^{-1}Bu \tag{4.14}$$

to make state space equations. Conceptually, if Q is not invertible, that is the same thing as having fewer than N unique equations (where N is the length of x, the number of states).

All of our filter derivations work with dynamical equations in a state space form. Also, most numerical integration schemes are designed for sets of first-order differential equations.

The right-hand side for the state equations (first-order differential equations), RHSOscillator, is shown in the following listing. Notice that if no inputs are requested, it returns the default data structure. The code, if (margin < 1), tells the function to return the data structure if no inputs are given. This is a convenient way of making your functions self-documenting and keeping your data structures consistent. The actual working code is just one line.

RHSOscillator.m

```
40   xDot = [x(2);d.a-2*d.zeta*d.omega*x(2)-d.omega^2*x(1)];
```

The following listing gives the simulation script OscillatorSim. It causes the right-hand side, RHSOscillator, to be numerically integrated using the RungeKutta function. We start by getting the default data structure from the right-hand side. We fill it in with our desired parameters. Measurements y are created for each step including random noise. There are two measurements: position and angle.

The following code shows just the simulation loop of OscillatorSim. The angle measurement is just trigonometry. The first measurement line computes the angle, which is a nonlinear measurement. The second measures the vertical distance which is linear.

OscillatorSim.m

```
23   for k = 1:nSim
24       % Measurements
25       yTheta = atan(x(1)/baseline) + yTheta1Sigma*randn(1,1);
26       yR     = x(1) + yR1Sigma*randn(1,1);
27
28       % Plot storage
29       xPlot(:,k) = [x;yTheta;yR];
30
31       % Propagate (numerically integrate) the state equations
32       x = RungeKutta( @RHSOscillator, 0, x, dT, dRHS );
33   end
```

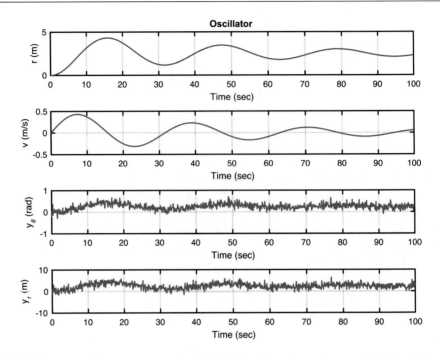

Figure 4.5: *Spring-mass-damper system simulation. The input is a step acceleration. The oscillation slowly damps out, that is, it goes to zero over time. The position r develops an offset due to the constant acceleration*

The results of the simulation are shown in Figure 4.5. The input is a disturbance acceleration that goes from zero to its value at time $t = 0$. It is constant for the duration of the simulation. This is known as a step disturbance. This causes the system to oscillate. The magnitude of the oscillation slowly goes to zero due to the damping. If the damping ratio were 1, we would not see any oscillation, as seen in Figure 4.4.

The offset seen in the plot of r can be found analytically by setting $v = 0$. Essentially, the spring force is balancing the external force.

$$0 = \frac{dv}{dt} = a - \omega^2 x \tag{4.15}$$

$$x = \frac{a}{\omega^2} \tag{4.16}$$

We have now completed the derivation of our model and can move on to building the Kalman Filters.

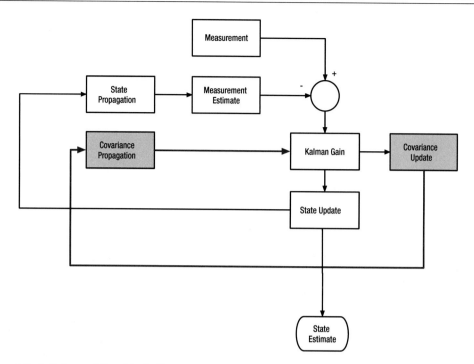

Figure 4.6: *A Kalman Filter block diagram*

Kalman Filter Derivation

The Kalman Filter is shown in Figure 4.6. The measurements are compared with the latest estimate of the measurements, based on the current state, and the error is multiplied by the Kalman gain to get a new estimate. The covariance is updated after every state and measurement update. The covariance gives the uncertainty of the states.

Kalman Filters can be derived from the Bayes theorem. What is Bayes' theorem? Bayes' theorem is

$$P(A_i|B) = \frac{P(B|A_i)P(A_i)}{\sum P(B|A_i)} \tag{4.17}$$

$$P(A_i|B) = \frac{P(B|A_i)P(A_i)}{P(B)} \tag{4.18}$$

which is just the probability of A_i given B. P means "probability." The vertical bar | means "given." This assumes that the probability of B is not zero, that is, $P(B) \neq 0$. In the Bayesian interpretation, the theorem introduces the effect of evidence on belief. This provides a rigorous framework for incorporating any data for which there is a degree of uncertainty. Put simply, given all evidence (or data) to date, Bayes' theorem allows you to determine how new evidence affects the belief. In the case of state estimation, this is the belief in the accuracy of the state estimate.

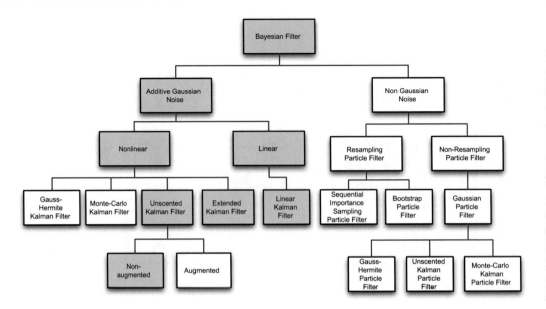

Figure 4.7: *The Kalman Filter family tree. All of the filter types are derived from a Bayesian Filter. This chapter covers the types in colored boxes*

Figure 4.7 shows the Kalman Filter family and how it relates to the Bayesian Filter. In this book, we are covering only the ones in the colored boxes. The complete derivation of the Kalman Filter is given as follows; this provides a coherent framework for all Kalman Filtering implementations. The different filters fall out of the Bayesian models based on assumptions about the model and sensor noise and the linearity or nonlinearity of the measurement and dynamics models. Let's look at the branch that is colored blue. Additive Gaussian noise filters can be linear or nonlinear depending on the type of dynamical and measurement models. In many cases, you can take a nonlinear system and linearize it about the normal operating conditions. You can then use a linear Kalman Filter. For example, a spacecraft dynamical model is nonlinear, and an Earth sensor that measures the Earth's chord width for roll and pitch information is nonlinear. However, if we are only concerned with Earth pointing, and small deviations from nominal pointing, we can linearize both the dynamical equations and measurement equations and use a linear Kalman Filter.

If nonlinearities are important, we have to use a nonlinear filter. The Extended Kalman Filter uses partial derivatives of the measurement and dynamical equations. These are computed each time step or with each measurement input. In effect, we are linearizing the system at each step and using the linear equations. We don't have to do a linear state propagation, that is, propagating the dynamical equations, and could propagate them using numerical integration. If we can get analytical derivatives of the measurement and dynamical equations, this is a reasonable approach. If there are singularities in any of the equations, this may not work.

The Unscented Kalman Filter uses nonlinear equations directly. There are two forms, augmented and non-augmented. In the former, we created an augmented state vector that includes both the states and the state and measurement noise variables. This may result in better results at the expense of more computation.

All of the filters in this chapter are Markov, that is, the current dynamical state is entirely determined from the previous state. Particle filters are not addressed in this book. They are a class of Monte-Carlo methods. Monte-Carlo (named after the famous casino) methods are computational algorithms that rely on random sampling to obtain results. For example, a Monte-Carlo approach to our oscillator simulation would be to use the MATLAB function `nrandn` to generate the accelerations. `randn` generates normally distributed random numbers. We'd run many tests to verify that our mass moved as expected.

Our derivation will use the notation $N(\mu, \sigma^2)$ to represent a normal variable. A normal variable is another word for a Gaussian variable. Gaussian means it is distributed as the normal distribution with mean μ (average) and variance σ^2. The following code from `Gaussian` computes a Gaussian or Normal distribution around a mean of 2 for a range of standard deviations. Figure 4.8 shows a plot. The height of the plot indicates how likely a given measurement of the variable is to have that value.

Gaussian.m

```
 6  %% Initialize
 7  mu            = 2;              % Mean
 8  sigma         = [1 2 3 4];     % Standard deviation
 9  n             = length(sigma);
10  x             = linspace(-7,10);
11
12  %% Simulation
13  xPlot = zeros(n,length(x));
```

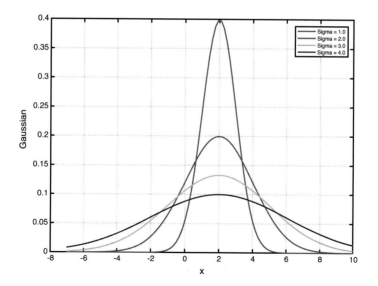

Figure 4.8: *Normal or Gaussian random variable about a mean of 2*

```
14   s        = cell(1,n);
15
16   for k = 1:length(sigma)
17     s{k}        = sprintf('Sigma = %3.1f',sigma(k));
18     f           = -(x-mu).^2/(2*sigma(k)^2);
19     xPlot(k,:) = exp(f)/sqrt(2*pi*sigma(k)^2);
20   end
```

Given the probabilistic state space model in discrete time [27]

$$x_k = f_k(x_{k-1}, w_{k-1}) \tag{4.19}$$

where x is the state vector and w is the noise vector, the measurement equation is

$$y_k = h_k(x_k, v_n) \tag{4.20}$$

where v_n is the measurement noise. This has the form of a hidden Markov model (HMM) because the state is hidden.

If the process is Markovian, then the future state x_k is dependent only on the current state x_{k-1} and is not dependent on the past states. This can be expressed in the equation

$$p(x_k|x_{1:k-1}, y_{1:k-1}) = p(x_k|x_{k-1}) \tag{4.21}$$

The | means given. In this case, the first term is read as "the probability of x_k given $x_{1:k-1}$ and $y_{1:k-1}$". This is the probability of the current state given all past states and all measurements up to the $k-1$ measurement. The past x_{k-1} is independent of the future given the present:

$$p(x_{k-1}|x_{k:T}, y_{k:T}) = p(x_{k-1}|x_k) \tag{4.22}$$

where T is the last sample and the measurements y_k are conditionally independent given x_k; that is, they can be determined using only x_k and are not dependent on $x_{1:k}$ or $y_{1:k-1}$. This can be expressed as

$$p(y_k|x_{1:k}, y_{1:k-1}) = p(y_k|x_k) \tag{4.23}$$

We can define the recursive Bayesian optimal filter that computes the distribution

$$p(x_k|y_{1:k}) \tag{4.24}$$

given

- The prior distribution $p(x_0)$, where x_0 is the state prior to the first measurement

- The state space model

$$x_k \sim p(x_k|x_{k-1}) \tag{4.25}$$
$$y_k \sim p(y_k|x_k) \tag{4.26}$$

96

- The measurement sequence $y_{1:k} = y_1, \ldots, y_k$

Computation is based on the recursion rule:

$$p(x_{k-1}|y_{1:k-1}) \rightarrow p(x_k|y_{1:k}) \tag{4.27}$$

This means we get the current state x_k from the prior state x_{k-1} and all the past measurements $y_{1:k-1}$. Assume we know the posterior distribution of the previous time step:

$$p(x_{k-1}|y_{1:k-1}) \tag{4.28}$$

The joint distribution of x_k, x_{k-1} given $y_{1:k-1}$ can be computed as

$$p(x_k, x_{k-1}|y_{1:k-1}) = p(x_k|x_{k-1}, y_{1:k-1})p(x_{k-1}|y_{1:k-1}) \tag{4.29}$$
$$= p(x_k|x_{k-1})p(x_{k-1}|y_{1:k-1}) \tag{4.30}$$

because this is a Markov process. Integrating over x_{k-1} gives the prediction step of the optimal filter which is the Chapman-Kolmogorov equation:

$$p(x_k|y_{1:k-1}) = \int p(x_k|x_{k-1}, y_{1:k-1})p(x_{k-1}|y_{1:k-1})dx_{k-1} \tag{4.31}$$

The Chapman-Kolmogorov equation is an identity relating the joint probability distributions of different sets of coordinates on a stochastic process. The measurement update state is found in Bayes' rule:

$$P(x_k|y_{1:k}) = \frac{1}{C_k}p(y_k|x_k)p(x_k|y_{k-1}) \tag{4.32}$$

$$C_k = p(y_k|y_{1:k-1}) = \int p(y_k|x_k)p(x_k|y_{1:k-1})dx_k \tag{4.33}$$

C_k is the probability of the current measurement, given all past measurements.

If the noise is additive and Gaussian with the state covariance Q_n and the measurement covariance R_n, the model and measurement noise have zero mean, we can write the state equation as

$$x_k = f_k(x_{k-1}) + w_{k-1} \tag{4.34}$$

where x is the state vector and w is the noise vector. The measurement equation becomes

$$y_k = h_k(x_k) + v_n \tag{4.35}$$

Given that Q is not time dependent, we can write

$$p(x_k|x_{k-1}, y_{1:k-1}) = N(x_k; f(x_{k-1}), Q) \tag{4.36}$$

where recall that N is a normal variable, in this case with mean x_k; $f(x_{k-1})$ which means (x_k given $f(x_{k-1})$ and variance Q. We can now write the prediction step Equation (4.31) as

$$p(x_k|y_{1:k-1}) = \int N(x_k; f(x_{k-1}), Q)p(x_{k-1}|y_{1:k-1})dx_{k-1} \tag{4.37}$$

We need to find the first two moments of x_k. A moment is the expected value (or mean) of the variable. The first moment is of the variable, the second is of the variable squared, and so forth. They are

$$E[x_k] = \int x_k p(x_k|y_{1:k-1})dx_k \tag{4.38}$$

$$E[x_k x_k^T] = \int x_k x_k^T p(x_k|y_{1:k-1})dx_k \tag{4.39}$$

E means expected value. $E[x_k]$ is the mean and $E[x_k x_k^T]$ is the covariance. Expanding the first moment and using the identity $E[x] = \int x N(x; f(s), \Sigma)dx = f(s)$ where s is any argument.

$$E[x_k] = \int x_k \left[\int d(x_k; f(x_{k-1}), Q)p(x_{k-1}|y_{1:k-1})dx_{k-1} \right] dx_k \tag{4.40}$$

$$= \int x_k \left[\int N(x_k; f(x_{k-1}), Q)dx_k \right] p(x_{k-1}|y_{1:k-1})dx_{k-1} \tag{4.41}$$

$$= \int f(x_{k-1})p(x_{k-1}|y_{1:k-1})dx_{k-1} \tag{4.42}$$

Assuming that $p(x_{k-1}|y_{1:k-1}) = N(x_{k-1}; \hat{x}_{k-1|k-1}, P^{xx}_{k-1|k-1})$ where P^{xx} is the covariance of x and noting that $x_k = f_k(x_{k-1}) + w_{k-1}$, we get

$$\hat{x}_{k|k-1} = \int f(x_{k-1})N(x_{k-1}; \hat{x}_{k-1|k-1}, P^{xx}_{k-1|k-1})dx_{k-1} \tag{4.43}$$

For the second moment

$$E[x_k x_k^T] = \int x_k x_k^T p(x_k|y_{1:k-1})dx_k \tag{4.44}$$

$$= \int \left[\int N(x_k; f(x_{k-1}), Q)x_k x_k^T dx_k \right] p(x_{k-1}|y_{1:k-1})dx_{k-1} \tag{4.45}$$

which results in

$$P^{xx}_{k|k-1} = Q + \int f(x_{k-1})f^T(x_{k-1})N(x_{k-1}; \hat{x}_{k-1|k-1}, P^{xx}_{k-1|k-1})dx_{k-1} - \hat{x}^T_{k|k-1}\hat{x}_{k|k-1} \tag{4.46}$$

The covariance for the initial state is Gaussian and is P_0^{xx}. The Kalman Filter can be written without further approximations as

$$\hat{x}_{k|k} = \hat{x}_{k|k-1} + K_n \left[y_k - \hat{y}_{k|k-1} \right] \tag{4.47}$$

$$P_{k|k}^{xx} = P_{k|k-1}^{xx} - K_n P_{k|k-1}^{yy} K_n^T \tag{4.48}$$

$$K_n = P_{k|k-1}^{xy} \left[P_{k|k-1}^{yy} \right]^{-1} \tag{4.49}$$

where K_n is the Kalman gain and P^{yy} is the measurement covariance. The Kalman gain is meant to provide an optimal balance between the model prediction and the measurements.

The solution of these equations requires the solution of five integrals of the form

$$I = \int g(x) N(x; \hat{x}, P^{xx}) dx \tag{4.50}$$

The three integrals needed by the filter are

$$P_{k|k-1}^{yy} = R + \int h(x_n) h^T(x_n) N(x_n; \hat{x}_{k|k-1}, P_{k|k-1}^{xx}) dx_k - \hat{x}_{k|k-1}^T \hat{y}_{k|k-1} \tag{4.51}$$

$$P_{k|k-1}^{xy} = \int x_n h^T(x_n) N(x_n; \hat{x}_{k|k-1}, P_{k|k-1}^{xx}) dx \tag{4.52}$$

$$\hat{y}_{k|k-1} = \int h(x_k) N(x_k; \hat{x}_{k|k-1}, P_{k|k-1}^{xx}) dx_k \tag{4.53}$$

Assume we have a model of the form

$$x_k = A_{k-1} x_{k-1} + B_{k-1} u_{k-1} + q_{k-1} \tag{4.54}$$

$$y_k = H_k x_k + r_k \tag{4.55}$$

where

- $x_k \in \Re^n$ is the state of system at time k.

- A_{k-1} is the state transition matrix at time $k - 1$.

- B_{k-1} is the input matrix at time $k - 1$.

- u_{k-1} is the input at time $k - 1$.

- $q_{k-1}, N(0, Q_k)$, is the Gaussian process noise at time $k - 1$.

- $y_k \in \Re^m$ is the measurement at time k.

- H_k is the measurement matrix at time k. This is found from the Jacobian (derivatives) of $h(x)$.

- $r_k = N(0, R_k)$ is the Gaussian measurement noise at time k.

- The prior distribution of the state is $x_0 = N(m_0, P_0)$ where parameters m_0 and P_0 contain all prior knowledge about the system. m_0 is the mean at time zero and P_0 is the covariance. Since our state is Gaussian, this completely describes the state.

- $\hat{x}_{k|k-1}$ is the mean of x at k given $\hat{x}$ at $k - 1$.

- $\hat{y}_{k|k-1}$ is the mean of y at k given $\hat{x}$ at $k - 1$.

$\Re^n$ means real numbers in a vector of order n, that is, the state has n quantities. In probabilistic terms, the model is

$$
\begin{aligned}
p(x_k|x_{k-1}) &= N(x_k; A_{k-1}x_{k-1}, Q_k) & (4.56) \\
p(y_k|x_k) &= N(y_k; H_k x_k, R_k) & (4.57)
\end{aligned}
$$

The integrals become simple matrix equations. In the following equations, P_k^- means the covariance before the measurement update:

$$
\begin{aligned}
P_{k|k-1}^{yy} &= H_k P_k^- H_k^T + R_k & (4.58) \\
P_{k|k-1}^{xy} &= P_k^- H_k^T & (4.59) \\
P_{k|k-1}^{xx} &= A_{k-1} P_{k-1} A_{k-1}^T + Q_{k-1} & (4.60) \\
\hat{x}_{k|k-1} &= m_k^- & (4.61) \\
\hat{y}_{k|k-1} &= H_k m_k^- & (4.62)
\end{aligned}
$$

The prediction step becomes

$$
\begin{aligned}
m_k^- &= A_{k-1} m_{k-1} & (4.63) \\
P_k^- &= A_{k-1} P_{k-1} A_{k-1}^T + Q_{k-1} & (4.64)
\end{aligned}
$$

The first term in the preceding covariance equation propagates the covariance based on the state transition matrix, A. Q_{k+1} adds to this to form the next covariance. Process noise Q_k is a measure of the accuracy of the mathematical model, A, in representing the system. For example, suppose A was a mathematical model that damped all states to zero. Without Q, P would go to zero. But if we weren't that certain about the model, the covariance would never be less than Q. Picking Q can be difficult. In a dynamical system with uncertain disturbances, you can compute the standard deviation of the disturbances to compute Q. If the model, A, is uncertain, then you might do a statistical analysis of the range of models. Or you can try different Q in simulation and see which ones work the best!

The update step is

$$
\begin{aligned}
v_k &= y_k - H_k m_k^- & (4.65) \\
S_k &= H_k P_k^- H_k^T + R_k & (4.66) \\
K_k &= P_k^- H_k^T S_k^{-1} & (4.67) \\
m_k &= m_k^- + K_k v_k & (4.68) \\
P_k &= P_k^- - K_k S_k K_k^T & (4.69)
\end{aligned}
$$

S_k is an intermediate quantity. v_k is the residual. The residual is the difference between the measurement and your estimate of the measurement given the estimated states. R is just the covariance matrix of the measurements. If the noise is not white, a different filter should be used. White noise has equal energy at all frequencies. Many types of noise, such as the noise from an imager, are not white noise but band limited, that is, it has noise in a limited range of frequencies. You can sometimes add additional states to A to model the noise better. For example, a low-pass filter can be added to band limit the noise. This makes A bigger but is generally not an issue.

Kalman Filter Implementation

Now we will implement a Kalman Filter estimator for the mass-spring oscillator. First, we need a method of converting the continuous time problem to discrete time. We only need to know the states at discrete times or fixed intervals, T. We use the continuous to discrete transform that uses the MATLAB expm function, which performs the matrix exponential. This transform is coded in CToDZOH, the body of which is shown in the following listing. T is the sampling period.

CToDZOH.m

```
45  [n,m]  = size(b);
46  q      = expm([a*T b*T;zeros(m,n+m)]);
47  f      = q(1:n,1:n);
48  g      = q(1:n,n+1:n+m);
```

CToDZOH includes a demo for a double integrator. A double integrator is a system in which the second derivative of the state is directly dependent upon an external input. In this example, x is the state, representing a position, and a is an external input of acceleration:

$$\frac{d^2r}{dt^2} = a \tag{4.70}$$

Written in state space form, it is

$$\frac{dr}{dt} = v \tag{4.71}$$

$$\frac{dv}{dt} = a \tag{4.72}$$

or in matrix form

$$\dot{x} = Ax + Bu \tag{4.73}$$

where

$$x = \begin{bmatrix} r \\ v \end{bmatrix} \tag{4.74}$$

$$u = \begin{bmatrix} 0 \\ a \end{bmatrix} \tag{4.75}$$

$$A = \begin{bmatrix} 0 & 1 \\ 0 & 0 \end{bmatrix} \tag{4.76}$$

$$B = \begin{bmatrix} 0 \\ 1 \end{bmatrix} \tag{4.77}$$

To run the demo, simply run CToDZOH from the command line without any inputs:

```
>> CToDZOH
Double integrator with a 0.5-second time step.
a =
        0        1
        0        0
b =
        0
        1
f =
     1.0000     0.5000
          0     1.0000
g =
     0.1250
     0.5000
```

The discrete plant matrix f transitions the state from step k to step $k + 1$. The position state at step $k + 1$ is the state at k plus the velocity at step k multiplied by the time step T of 0.5 seconds. The velocity at step $k + 1$ is the velocity at k plus the time step times the acceleration at step k. The acceleration at the time k multiplies $\frac{1}{2}T^2$ to get the contribution to position. This is just the standard solution to a particle under constant acceleration.

$$r_{k+1} = r_k + Tv_k + \frac{1}{2}T^2 a_k \tag{4.78}$$

$$v_{k+1} = v_k + Ta_k \tag{4.79}$$

In matrix form, this is

$$x_{k+1} = fx_k + bu_k \tag{4.80}$$

With the discrete-time approximation, we can change the acceleration every step k to get the time history. This assumes that the acceleration is constant over the period T. We need to pick T to be sufficiently small so that this is approximately true if we are to get good results.

The script for testing the Kalman Filter is KFSim.m. KFInitialize is used to initialize the filter (a Kalman Filter, 'kf', in this case). This function has been written to handle multiple types of Kalman Filters, and we will use it again in the recipes for Extended and Unscented Kalman Filters ('ekf' and 'ukf', respectively). We show it as follows. This function uses dynamic field names to assign the input values to each field.

The simulation starts by assigning values to all of the variables used in the simulation. We get the data structure from the function RHSOscillator and then modify its values. We write the continuous time model in matrix form and then convert it to discrete time. Note that the measurement equation matrix that multiplies the state, h, is [1 0], indicating we are measuring the position of the mass. MATLAB's randn random number function is used to add Gaussian noise to the simulation. The rest of the script is the simulation loop with plotting afterward.

The first part of the script creates continuous time state space matrices and converts them to discrete time using CToDZOH. You then use KFInitialize to initialize the Kalman Filter.

KFSim.m

```
11  tEnd        = 100.0;              % Simulation end time (sec)
12  dT          = 0.1;                % Time step (sec)
13  d           = RHSOscillator();    % Get the default data structure
14  d.a         = 0.1;                % Disturbance acceleration
15  d.omega     = 0.2;                % Oscillator frequency
16  d.zeta      = 0.1;                % Damping ratio
17  x           = [0;0];              % Initial state [position;velocity]
18  y1Sigma     = 1;                  % 1 sigma position measurement noise
19
20  % xdot = a*x + b*u
21  a = [0 1;-2*d.zeta*d.omega -d.omega^2]; % Continuous time model
22  b = [0;1];                             % Continuous time input matrix
23
24  % x[k+1] = f*x[k] + g*u[k]
25  [f,g]       = CToDZOH(a,b,dT);     % Discrete time model
26  xE          = [0.3; 0.1];          % Estimated initial state
27  q           = [1e-6 1e-6];         % Model noise covariance ;
28                                     % [1e-6 1e-6] is for low model noise test
29                                     % [1e-4 1e-4] is for high model noise test
30  dKF  = KFInitialize('kf','m',xE,'a',f,'b',g,'h',[1 0],...
31                      'r',y1Sigma^2,'q',diag(q),'p',diag(xE.^2));
```

The simulation loop cycles through measurements of the state and the Kalman Filter update and prediction state with the code KFPredict and KFUpdate. The integrator is between the two to get the phasing of the update and prediction correct. You have to be careful to put the predict and update steps in the right places in the script so that the estimator is synchronized with the simulation time.

KFSim.m

```
34  nSim  = floor(tEnd/dT) + 1;
35  xPlot = zeros(5,nSim);
36
37  for k = 1:nSim
38    % Position measurement with random noise
39    y = x(1) + y1Sigma*randn(1,1);
40
41    % Update the Kalman Filter
42    dKF.y = y;
43    dKF   = KFUpdate(dKF);
44
45    % Plot storage
46    xPlot(:,k) = [x;y;dKF.m-x];
47
48    % Propagate (numerically integrate) the state equations
49    x = RungeKutta( @RHSOscillator, 0, x, dT, d );
50
51    % Propagate the Kalman Filter
52    dKF.u = d.a;
53    dKF   = KFPredict(dKF);
54  end
```

The prediction Kalman Filter step, KFPredict, is shown in the following listing with an abbreviated header. The prediction propagates the state one-time step and propagates the covariance matrix with it. It is saying that when we propagate the state, there is uncertainty, so we must add that to the covariance matrix.

KFPredict.m

```
1   %% KFPREDICT Linear Kalman Filter prediction step.
27
28  function d = KFPredict( d )
29
30  % The first path is if there is no input matrix b
31  if( isempty(d.b) )
32    d.m = d.a*d.m;
33  else
34    d.m = d.a*d.m + d.b*d.u;
35  end
36
37  d.p = d.a*d.p*d.a' + d.q;
```

The update Kalman Filter step, KFUpdate, is shown in the following listing. This adds the measurements to the estimate and accounts for the uncertainty (noise) in the measurements.

KFUpdate.m

```
1   %% KFUPDATE Linear Kalman Filter measurement update step.
27
28  function d = KFUpdate( d )
29
30  s   = d.h*d.p*d.h' + d.r;        % Intermediate value
31  k   = d.p*d.h'/s;           % Kalman gain
32  v   = d.y - d.h*d.m;        % Residual
33  d.m = d.m + k*v;            % Mean update
34  d.p = d.p - k*s*k';         % Covariance update
```

You will note that the "memory" of the filter is stored in the data structure d. No persistent data storage is used. This makes it easier to use these functions in multiple places in your code. Note also that you don't have to call KFUpdate every time step. You need only call it when you have new data. However, the filter does assume uniform time steps.

The script gives two examples of the model noise covariance matrix. Figure 4.9 shows results when high numbers, [1e-4 1e-4], for the model covariance are used. Figure 4.10 when lower numbers, [1e-6 1e-6], are used. We don't change the measurement covariance because only the ratio between noise covariance and model covariance is important.

When the higher numbers are used, the errors are Gaussian but noisy. When the low numbers are used, the result is very smooth, with little noise seen. However, the errors are large in the low model covariance case. This is because the filter is essentially ignoring the measurements

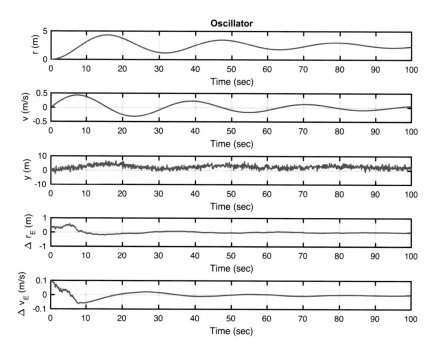

Figure 4.9: *The Kalman Filter results with the higher model noise matrix, [1e-4 1e-4]*

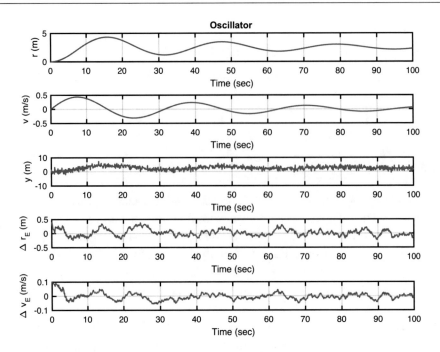

Figure 4.10: *The Kalman Filter results with the lower model noise matrix, [1e-6 1e-6]. Less noise is seen but the errors are large*

since it thinks the model is very accurate. You should try different options in the script and see how it performs. As you can see, the parameters make a huge difference in how well the filter learns about the states of the system.

4.3   Using the Extended Kalman Filter for State Estimation

4.3.1   Problem

We want to track the damped oscillator using an Extended Kalman Filter with the nonlinear angle measurement. The Extended Kalman Filter was developed to handle models with nonlinear dynamical models and/or nonlinear measurement models. The conventional, or linear, filter requires linear dynamical equations and linear measurement models, that is, the measurement is a linear function of the state. If the model is not linear, linear filters will not track the states very well.

Given a nonlinear model of the form

$$x_k = f(x_{k-1}, k-1) + q_{k-1} \tag{4.81}$$
$$y_k = h(x_k, k) + r_k \tag{4.82}$$

The prediction step is

$$m_k^- = f(m_{k-1}, k-1) \tag{4.83}$$
$$P_k^- = F_x(m_{k-1}, k-1)P_{k-1}F_x(m_{k-1}, k-1)^T + Q_{k-1} \tag{4.84}$$

F is the Jacobian of f. The update step is

$$v_k = y_k - h(m_k^-, k) \tag{4.85}$$
$$S_k = H_x(m_k^-, k)P_k^- H_x(m_k^-, k)^T + R_k \tag{4.86}$$
$$K_k = P_k^- H_x(m_k^-, k)^T S_k^{-1} \tag{4.87}$$
$$m_k = m_k^- + K_k v_k \tag{4.88}$$
$$P_k = P_k^- - K_k S_k K_k^T \tag{4.89}$$

$F_x(m, k-1)$ and $H_x(m, k)$ are the Jacobians of the nonlinear functions f and h. The Jacobians are just a matrix of partial derivatives of F and H. This results in matrices from the vectors F and H. For example, assume we have $f(x, y)$ which is

$$f = \begin{bmatrix} f_x(x, y) \\ f_y(x, y) \end{bmatrix} \tag{4.90}$$

The Jacobian is

$$F_k = \begin{bmatrix} \frac{\partial f_x(x_k, y_k)}{\partial x} & \frac{\partial f_x(x_k, y_k)}{\partial y} \\ \frac{\partial f_y(x_k, y_k)}{\partial x} & \frac{\partial f_y(x_k, y_k)}{\partial y} \end{bmatrix} \tag{4.91}$$

The matrix is computed at x_k, y_k.

The Jacobians can be found analytically or numerically. If done numerically, the Jacobian needs to be computed about the current value of m_k. In the Iterated Extended Kalman Filter, the update step is done in a loop using updated values of m_k after the first iteration. $H_x(m, k)$ needs to be updated on each step.

4.3.2 Solution

We will use the same KFInitialize function as created in the previous recipe, but now using the 'ekf' input. We will need functions for the derivative of the model dynamics, the measurement, and the measurement derivatives. These are implemented in RHSOscillatorPartial, AngleMeasurement, and AngleMeasurementPartial.

We will also need custom versions of the filter predict and update steps.

4.3.3 How It Works

The EKF requires a measurement function, a measurement derivative function, and a state derivative function. The state derivative function computes the a matrix:

$$x_{k+1} = a_k x_k \tag{4.92}$$

You would only use the EKF if a_k changed with time. In this problem, it does not. The function to compute a is RHSOscillatorPartial. It uses CToDZOH. We could have computed a once, but using CToDZOH makes the function more general.

RHSOscillatorPartial.m

```
24  function a = RHSOscillatorPartial( ~, ~, dT, d )
25
26  if( nargin < 1 )
27    a = struct('zeta',0.7071,'omega',0.1);
28    return
29  end
30
31  b = [0;1];
32  a = [0 1;d.omega^2 -2*d.zeta*d.omega];
33  a = CToDZOH( a, b, dT );
```

Our measurement is nonlinear (being an arctangent) and needs to be linearized about each value of position. AngleMeasurement computes the measurement which is nonlinear but smooth.

AngleMeasurement.m

```
26  y = atan(x(1)/d.baseline);
```

AngleMeasurementPartial computes the derivative. The following function computes the c matrix:

$$y_k = c_k x_k \tag{4.93}$$

The partial measurement is found by taking the derivative of the arctangent of the angle from the baseline. The comment reminds you of this fact.

AngleMeasurementPartial.m

```
24  % y = atan(x(1)/d.baseline);
25
26  u   = x(1)/d.baseline;
27  dH  = 1/(1+u^2);
28  h   = [dH 0]/d.baseline;
```

Conveniently, the measurement function is smooth. If there were discontinuities, the measurement partials would be difficult to compute. The EKF implementation can handle either function for the derivatives or matrices. In the case of the functions, we use `feval` to call them. This can be seen in the `EKFPredict` and `EKFUpdate` functions.

`EKFPredict` is the state propagation step for an Extended Kalman Filter. It numerically integrates the right-hand side using `RungeKutta`. `RungeKutta` might be overkill in some problems, and a simple Euler integration may be appropriate. Euler integration is just

$$x_{k+1} = x_k + \Delta T f(x, u, t) \tag{4.94}$$

where $f(x, u, t)$ is the right-hand side that can be a function of the state, x, time t, and the inputs u.

EKFPredict.m

```
26  function d = EKFPredict( d )
27
28  % Get the state transition matrix
29  if( isempty(d.a) )
30    a = feval( d.fX, d.m, d.t, d.dT, d.fData );
31  else
32    a = d.a;
33  end
34
35  % Propagate the mean
36  d.m = RungeKutta( d.f, d.t, d.m, d.dT, d.fData );
37
38  % Propagate the covariance
39  d.p = a*d.p*a' + d.q;
```

EKFUpdate.m

```
1   %% EKFUPDATE Extended Kalman Filter measurement update step.
2   %% Form
3   %    d = EKFUpdate( d )
4   %
5   %% Description
6   % All inputs are after the predict state (see EKFPredict). The h
7   % data field may contain either a function name for computing
8   % the estimated measurements or an m by n matrix. If h is a function
9   % name you must include hX which is a function to compute the m by n
10  % matrix as a linearized version of the function h.
11  %
12  %% Inputs
13  %    d    (.)    EKF data structure
14  %                .m      (n,1)  Mean
15  %                .p      (n,n)  Covariance
16  %                .h      (m,n)  Either a matrix or name/handle of
       function
17  %                .hX     (*)    Name or handle of Jacobian function for h
18  %                .y      (m,1)  Measurement vector
19  %                .r      (m,m)  Measurement covariance vector
```

```
20  %                    .hData   (.)    Data structure for the h and hX functions
21  %
22  %% Outputs
23  %    d    (.)    Updated EKF data structure
24  %                .m      (n,1)    Mean
25  %                .p      (n,n)    Covariance
26  %                .v      (m,1)    Residuals
27
28  function d = EKFUpdate( d )
29
30  % Residual
31  if( isnumeric( d.h ) )
32    h    = d.h;
33    yE   = h*d.m;
34  else
35    h    = feval( d.hX, d.m, d.hData );
36    yE   = feval( d.h,  d.m, d.hData );
37  end
38
39  % Residual
40  d.v      = d.y - yE;
41
42  % Update step
43  s    = h*d.p*h' + d.r;
44  k    = d.p*h'/s;
45  d.m = d.m + k*d.v;
46  d.p = d.p - k*s*k';
```

The EKFSim script implements the Extended Kalman Filter with all of the preceding functions as shown in the following listing. The functions are passed to the EKF in the data structure produced by KFInitialize. Note the use of function handles using @, that is, @RHSOscillator. Notice that KFInitialize requires hX and fX for computing partial derivatives of the dynamical equations and measurement equations.

```
28  %% Simulation
29  xPlot = zeros(5,nSim);
30
31  for k = 1:nSim
32    % Angle measurement with random noise
33    y = AngleMeasurement( x, dMeas ) + y1Sigma*randn;
34
35    % Update the Kalman Filter
36    dKF.y = y;
37    dKF   = EKFUpdate(dKF);
38
39    % Plot storage
40    xPlot(:,k) = [x;y;dKF.m-x];
41
42    % Propagate (numerically integrate) the state equations
43    x = RungeKutta( @RHSOscillator, 0, x, dT, d );
```

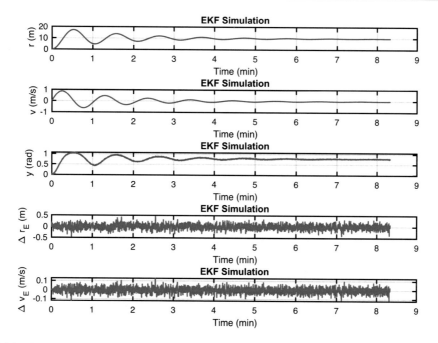

Figure 4.11: *The Extended Kalman Filter tracks the oscillator using the angle measurement*

```
44
45    % Propagate the Kalman Filter
46    dKF = EKFPredict(dKF);
47  end
```

Figure 4.11 shows the results. The errors are small. Since the problem dynamics are linear, we don't expect any differences from a conventional Kalman Filter.

4.4   Using the UKF for State Estimation

4.4.1   Problem

You want to learn the states of the spring, damper, and mass system given a nonlinear angle measurement. This time, we'll use an Unscented Kalman Filter. With the Unscented Kalman Filter, we work with the nonlinear dynamical and measurement equations directly. We don't have to linearize them as we did for the EKF with RHSOscillatorPartial and AngleMeasurementPartial. The Unscented Kalman Filter is also known as a sigma σ point filter because it simultaneously maintains models one sigma (standard deviation) from the mean.

4.4.2 Solution

We will create an Unscented Kalman Filter as a state estimator. The UKF has the advantage over the EKF that it does not have a bias in the error that cannot be removed. It also does not require taking derivatives of the dynamical model every time you want to update the covariance. This will absorb measurements and determine the state. It will autonomously learn about the state of the system based on a preexisting model.

In the following text, we develop the equations for the non-augmented Kalman Filter. This form only allows for additive Gaussian noise. Given a nonlinear model of the form

$$x_k = f(x_{k-1}, k-1) + q_{k-1} \tag{4.95}$$
$$y_k = h(x_k, k) + r_k \tag{4.96}$$

There are a set of weights used to combine the sigma point states and measurements. These are

$$W_m^0 = \frac{\lambda}{n+\lambda} \tag{4.97}$$

$$W_c^0 = \frac{\lambda}{n+\lambda} + 1 - \alpha^2 + \beta \tag{4.98}$$

$$W_m^i = \frac{\lambda}{2(n+\lambda)}, i = 1, \dots, 2n \tag{4.99}$$

$$W_c^i = \frac{\lambda}{2(n+\lambda)}, i = 1, \dots, 2n \tag{4.100}$$

m are weights on the mean state (m for mean) and c weights on the covariances. Note that $W_m^i = W_c^i$.

$$\lambda = \alpha^2(n+\kappa) - n \tag{4.101}$$
$$c = \lambda + n = \alpha^2(n+\kappa) \tag{4.102}$$

c scales the covariances to compute the sigma points, that is, the distribution of points around the mean for computing the additional states to propagate. α, β, and κ are scaling constants. The general rules for the scaling constants are

- α: 0 for state estimation, 3 minus the number of states for parameter estimation.

- β: Determines spread of sigma points. Smaller means more closely spaced sigma points.

- κ: Constant for prior knowledge. Set to 2 for Gaussian processes.

n is the order of the system. The weights can be put into matrix form

$$w_m = \left[W_m^0 \cdots W_m^{2n}\right]^T \tag{4.103}$$

$$W = (I - [w_m \cdots w_m]) \begin{bmatrix} W_c^0 & \cdots & 0 \\ \vdots & \ddots & \vdots \\ 0 & \cdots & W_c^{2n} \end{bmatrix} (I - [w_m \cdots w_m])^T \tag{4.104}$$

I is the $2n + 1$ by $2n + 1$ identity matrix. In the equation vector, w_m is replicated $2n + 1$ times. W is $2n + 1$ by $2n + 1$.

The prediction step is

$$X_{k-1} = \begin{bmatrix} m_{k-1} & \cdots & m_{k-1} \end{bmatrix} + \sqrt{c} \begin{bmatrix} 0 & \sqrt{P_{k-1}} & -\sqrt{P_{k-1}} \end{bmatrix} \quad (4.105)$$

$$\hat{X}_k = f(X_{k-1}, k-1) \quad (4.106)$$

$$m_k^- = \hat{X}_k w_m \quad (4.107)$$

$$P_k^- = \hat{X}_k W \hat{X}_k^T + Q_{k-1} \quad (4.108)$$

where X is a matrix where its column is the state vector possibly with an added sigma point vector. The update step is

$$X_k^- = \begin{bmatrix} m_k^- & \cdots & m_k^- \end{bmatrix} + \sqrt{c} \begin{bmatrix} 0 & \sqrt{P_k^-} & -\sqrt{P_k^-} \end{bmatrix} \quad (4.109)$$

$$Y_k^- = h(X_k^-, k) \quad (4.110)$$

$$\mu_k = Y_k^- w_m \quad (4.111)$$

$$S_k = Y_k^- W [Y_k^-]^T + R_k \quad (4.112)$$

$$C_k = X_k^- W [Y_k^-]^T \quad (4.113)$$

$$K_k = C_k S_k^{-1} \quad (4.114)$$

$$m_k = m_k^- + K_k(y_k - \mu_k) \quad (4.115)$$

$$P_k = P_k^- - K_k S_k K_k^T \quad (4.116)$$

μ_k is a matrix of the measurements in which each column is a copy modified by the sigma points. S_k and C_k are intermediate quantities. The brackets around Y_k^- are just for clarity.

4.4.3 How It Works

The weights are computed in UKFWeight.

UKFWeight.m

```
1  %% UKFWEIGHT Unscented Kalman Filter weight calculation
2  %% Form
3  %   d = UKFWeight( d )
4  %
5  %% Description
6  % Unscented Kalman Filter weights.
7  %
8  % The weight matrix is used by the matrix form of the Unscented
9  % Transform. Both UKFPredict and UKFUpdate use the data structure
10 % generated by this function.
11 %
12 % The constant alpha determines the spread of the sigma points around x
13 % and is usually set to between 10e-4 and 1. beta incorporates prior
14 % knowledge of the distribution of x and is 2 for a Gaussian
```

```
15  % distribution. kappa is set to 0 for state estimation and 3 - number
       of
16  % states for parameter estimation.
17  %
18  %% Inputs
19  %    d    (.)         Data structure with constants
20  %                     .kappa  (1,1)     0 for state estimation, 3-#states for
21  %                                       parameter estimation
22  %                     .m      (:,1)     Vector of mean states
23  %                     .alpha  (1,1)     Determines spread of sigma points
24  %                     .beta   (1,1)     Prior knowledge - 2 for Gaussian
25  %
26  %% Outputs
27  %    d    (.)         Data structure with constants
28  %                     .w      (2*n+1,2*n+1)  Weight matrix
29  %                     .wM     (1,2*n+1)      Weight array
30  %                     .wC     (2*n+1,1)      Weight array
31  %                     .c      (1,1)          Scaling constant
32  %                     .lambda (1,1)          Scaling constant
33  %
34
35  function d = UKFWeight( d )
36
37  % Compute the fundamental constants
38  n           = length(d.m);
39  a2          = d.alpha^2;
40  d.lambda    = a2*(n + d.kappa) - n;
41  nL          = n + d.lambda;
42  wMP         = 0.5*ones(1,2*n)/nL;
43  d.wM        = [d.lambda/nL                wMP]';
44  d.wC        = [d.lambda/nL+(1-a2+d.beta)  wMP];
45
46  d.c         = sqrt(nL);
47
48  % Build the matrix
49  f           = eye(2*n+1) - repmat(d.wM,1,2*n+1);
50  d.w         = f*diag(d.wC)*f';
```

The prediction Unscented Kalman Filter step is shown in the following excerpt from UKFPredict.

UKFPredict.m

```
 2  %% UKFPREDICT Unscented Kalman Filter measurement update step
34  function d = UKFPredict( d )
35
36  pS      = chol(d.p)';
37  nS      = length(d.m);
38  nSig    = 2*nS + 1;
39  mM      = repmat(d.m,1,nSig);
40  x       = mM + d.c*[zeros(nS,1) pS -pS];
```

```
41
42  xH        = Propagate( x, d );
43  d.m       = xH*d.wM;
44  d.p       = xH*d.w*xH' + d.q;
45  d.p       = 0.5*(d.p + d.p'); % Force symmetry
47
48  %% Propagate each sigma point state vector
49  function x = Propagate( x, d )
50
51  for j = 1:size(x,2)
52          x(:,j) = RungeKutta( d.f, d.t, x(:,j), d.dT, d.fData );
53  end
```

UKFPredict uses RungeKutta for prediction that is done by numerical integration. In effect, we are simulating the model and just correcting the results with the next function, UKFUpdate. This gets to the core of the Kalman Filter. It is just a simulation of your model with a measurement correction step. In the case of the conventional linear Kalman Filter, we use a linear discrete-time model.

The update Unscented Kalman Filter step is shown in the following listing. The update propagates the state one-time step.

UKFUpdate.m

```
2   %% UKFUPDATE Unscented Kalman Filter measurement update step.
26  function d = UKFUpdate( d )
27
28  % Get the sigma points
29  pS        = d.c*chol(d.p)';
30  nS        = length(d.m);
31  nSig      = 2*nS + 1;
32  mM        = repmat(d.m,1,nSig);
33  x         = mM + [zeros(nS,1) pS -pS];
34  [y, r]    = Measurement( x, d );
35  mu        = y*d.wM;
36  s         = y*d.w*y' + r;
37  c         = x*d.w*y';
38  k         = c/s;
39  d.v       = d.y - mu;
40  d.m       = d.m + k*d.v;
41  d.p       = d.p - k*s*k';
43
44  %%        Measurement estimates from the sigma points
45  function [y, r] = Measurement( x, d )
46
47  nSigma    = size(x,2);
48
49  % Create the arrays
50  lR  = length(d.r);
51  y   = zeros(lR,nSigma);
52  r   = d.r;
```

115

```
53
54   for j = 1:nSigma
55         f          = feval(d.hFun, x(:,j), d.hData );
56         iR         = 1:lR;
57         y(iR,j)    = f;
58   end
```

The sigma points are generated using `chol`. `chol` is Cholesky factorization and generates an approximate square root of a matrix. A true matrix square root is more computationally expensive, and the results don't justify the penalty. The idea is to distribute the sigma points around the mean and `chol` works well. Here is an example that compares the two approaches:

```
>> z = [1 0.2;0.2 2]
z =
     1.0000    0.2000
     0.2000    2.0000
>> b = chol(z)
b =
     1.0000    0.2000
          0    1.4000
>> b*b
ans =
     1.0000    0.4800
          0    1.9600
>> q = sqrtm(z)
q =
     0.9965    0.0830
     0.0830    1.4118
>> q*q
ans =
     1.0000    0.2000
     0.2000    2.0000
```

The square root produces a square root! The diagonal of `b*b` is close to `z` which is all that is important.

The script for testing the Unscented Kalman Filter, `UKFSim`, is shown in the following listing. As noted earlier, we don't need to convert the continuous time model into discrete time as we did for the KF and EKF. Instead, we pass the filter to the right-hand side of the differential equations. You must also pass it a measurement model which can be nonlinear. You add `UKFUpdate` and `UKFPredict` function calls to the simulation loop. We start by initializing all parameters. `KFInitialize` takes parameter pairs, after `'ukf'`, to initialize the filter. The remainder is the simulation loop and plotting. Initialization requires the computation of the weighting matrices after calling `KFInitialize`.

UKFSim.m

```
24  dKF   = KFInitialize( 'ukf','m',xE,'f',@RHSOscillator,'fData',d,...
25                        'r',y1Sigma^2,'q',q,'p',p,...
26                        'hFun',@AngleMeasurement,'hData',dMeas,'dT',dT);
27  dKF   = UKFWeight( dKF );
```

We show the simulation loop here:

UKFSim.m

```
29  %% Simulation
30  xPlot = zeros(5,nSim);
31
32  for k = 1:nSim
33    % Measurements
34    y = AngleMeasurement( x, dMeas ) + y1Sigma*randn;
35
36    % Update the Kalman Filter
37    dKF.y = y;
38    dKF   = UKFUpdate(dKF);
39
40    % Plot storage
41    xPlot(:,k)  = [x;y;dKF.m-x];
42
43    % Propagate (numerically integrate) the state equations
44    x = RungeKutta( @RHSOscillator, 0, x, dT, d );
45
46    % Propagate the Kalman Filter
47    dKF = UKFPredict(dKF);
48  end
```

The results are shown in Figure 4.12. The errors Δr_E and Δv_E are just noise. The measurement goes over a large angle range which would make a linear approximation problematic.

4.5   Using the UKF for Parameter Estimation

4.5.1   Problem

You want to learn the parameters of the spring, damper, and mass system given a nonlinear angle measurement. The UKF can be configured to do this.

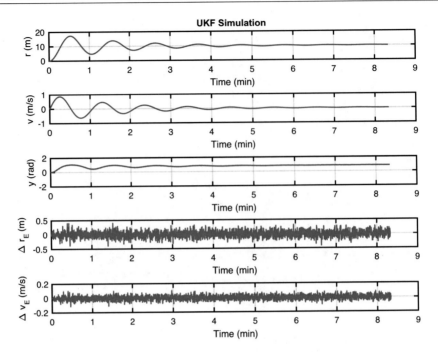

Figure 4.12: *The Unscented Kalman Filter results for state estimation*

4.5.2 Solution

The solution is to create an Unscented Kalman Filter configured as a parameter estimator. This will absorb measurements and determine the undamped natural frequency. It will autonomously learn about the system based on a preexisting model. We develop the version that requires an estimate of the state that could be generated with a UKF running in parallel, as in the previous recipe.

4.5.3 How It Works

Initialize the parameter filter with the expected value of the parameters, η [32]:

$$\hat{\eta}(t_0) = E\{\hat{\eta}_0\} \tag{4.117}$$

and the covariance for the parameters

$$P_{\eta_o} = E\{(\eta(t_0) - \hat{\eta}_0)(\eta(t_0) - \hat{\eta}_0)^T\} \tag{4.118}$$

The update sequence begins by adding the parameter model uncertainty, Q, to the covariance, P,

$$P = P + Q \tag{4.119}$$

Q is for the parameters, not the states. The sigma points are then calculated. These are points found by adding the square root of the covariance matrix to the current estimate of the parameters.

$$\eta_\sigma = \begin{bmatrix} \hat{\eta} & \hat{\eta} + \gamma\sqrt{P} & \hat{\eta} - \gamma\sqrt{P} \end{bmatrix} \tag{4.120}$$

γ is a factor that determines the spread of the sigma points. We use `chol` for the square root. If there are L parameters, the P matrix is $L \times L$, so this array will be $L \times (2L + 1)$.

The state equations are of the form

$$\dot{x} = f(x, u, t) \tag{4.121}$$

and the measurement equations are

$$y = h(x, u, t) \tag{4.122}$$

x is the previous state of the system, as identified by the state estimator or other processes. u is a structure with all other inputs to the system that are not being estimated. η is a vector of parameters that are being estimated and t is time. y is the vector of measurements. This is the dual estimation approach in that we are not estimating x and η simultaneously.

The script, `UKFPSim`, for testing the Unscented Kalman Filter parameter estimation is shown in the following listing. We are not doing the UKF state estimation to simplify the script. Normally, you would run the UKF in parallel. We start by initializing all parameters. `KFInitialize` takes parameter pairs to initialize the filters. The remainder is the simulation loop and plotting. Notice that there is only an update call since parameters, unlike states, do not propagate.

UKFPSim.m

```
30   %% Simulation
31   xPlot = zeros(5,nSim);
32
33   for k = 1:nSim
34     % Update the Kalman Filter parameter estimates
35     dKF.x = x;
36
37     % Plot storage
38     xPlot(:,k) = [y;x;dKF.eta;dKF.p];
39
40     % Propagate (numerically integrate) the state equations
41     x = RungeKutta( @RHSOscillator, 0, x, dT, d );
42
43     % Incorporate measurements
44     y      = LinearMeasurement( x ) + y1Sigma*randn;
45     dKF.y = y;
46     dKF    = UKFPUpdate(dKF);
47   end
```

The Unscented Kalman Filter parameter update function is shown in the following code. It uses the state estimate generated by the UKF. As noted, we are using the exact value of the state generated by the simulation. This function needs a specialized right-hand side that uses the parameter estimate, `d.eta`. We modified `RHSOscillator` for this purpose and wrote `RHSOscillatorUKF`.

UKFPUpdate.m

```
43  function d = UKFPUpdate( d )
44
45  d.wA    = zeros(d.L,d.n);
46  D       = zeros(d.lY,d.n);
47  yD      = zeros(d.lY,1);
48
49  % Update the covariance
50  d.p = d.p + d.q;
51
52  % Compute the sigma points
53  d = SigmaPoints( d );
54
55  % We are computing the states, then the measurements
56  % for the parameters +/- 1 sigma
57  for k = 1:d.n
58    d.fData.eta = d.wA(:,k);
59    x             = RungeKutta( d.f, d.t, d.x, d.dT, d.fData );
60    D(:,k)        = feval( d.hFun, x, d.hData );
61    yD            = yD + d.wM(k)*D(:,k);
62  end
63
64  pWD = zeros(d.L,d.lY);
65  pDD = d.r;
66  for k = 1:d.n
67    wD  = D(:,k) - yD;
68    pDD = pDD + d.wC(k)*(wD*wD');
69    pWD = pWD + d.wC(k)*(d.wA(:,k) - d.eta)*wD';
70  end
71
72  pDD = 0.5*(pDD + pDD');
73
74  % Incorporate the measurements
75  K       = pWD/pDD;
76  dY      = d.y - yD;
77  d.eta   = d.eta + K*dY;
78  d.p     = d.p - K*pDD*K';
79  d.p     = 0.5*(d.p + d.p'); % Force symmetry
80
81  %% Create the sigma points for the parameters
82  function d = SigmaPoints( d )
83
84  n       = 2:(d.L+1);
85  m       = (d.L+2):(2*d.L + 1);
```

```
86  etaM         = repmat(d.eta,length(d.eta));
87  sqrtP        = chol(d.p);
88  d.wA(:,1)    = d.eta;
89  d.wA(:,n)    = etaM + d.gamma*sqrtP;
90  d.wA(:,m)    = etaM - d.gamma*sqrtP;
```

It also has its weight initialization function UKFPWeight.m. The weight matrix is used by the matrix form of the Unscented Transform. The constant alpha determines the spread of the sigma points around the parameter vector and is usually set to between 10e-4 and 1. beta incorporates prior knowledge of the distribution of the parameter vector and is 2 for a Gaussian distribution. kappa is set to 0 for state estimation and 3 for the the number of states for parameter estimation.

UKFPWeight.m

```
34  function d = UKFPWeight( d )
35
36  d.L          = length(d.eta);
37  d.lambda     = d.alpha^2*(d.L + d.kappa) - d.L;
38  d.gamma      = sqrt(d.L + d.lambda);
39  d.wC(1)      = d.lambda/(d.L + d.lambda) + (1 - d.alpha^2 + d.beta);
40  d.wM(1)      = d.lambda/(d.L + d.lambda);
41  d.n          = 2*d.L + 1;
42  for k = 2:d.n
43     d.wC(k)  = 1/(2*(d.L + d.lambda));
44     d.wM(k)  = d.wC(k);
45  end
46
47  d.wA         = zeros(d.L,d.n);
48  y            = feval( d.hFun, d.x, d.hData );
49  d.lY         = length(y);
50  d.D          = zeros(d.lY,d.n);
```

RHSOscillatorUKF is the oscillator model used by the Unscented Kalman Filter. It has a different input format than RHSOscillator. There is only one line of code.

RHSOscillatorUKF.m

```
39  xDot = [x(2);d.a-2*d.zeta*d.eta*x(2)-d.eta^2*x(1)];
```

LinearMeasurement is a simple measurement function for demonstration purposes. The Unscented Kalman Filter can use arbitrarily complex measurement functions.

The results of a simulation of an undamped oscillator are shown in Figure 4.13 and Figure 4.14. The filter rapidly estimates the undamped natural frequency. The result is noisy, however. You can explore this script by varying the numbers in the script.

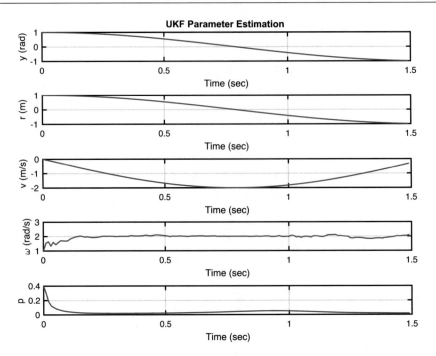

Figure 4.13: *The Unscented Kalman parameter estimation results. p is the covariance. It shows that our parameter estimate has converged*

4.6 Range to a Car

4.6.1 Problem

You want to compute the range of a car traveling in front of you.

4.6.2 Solution

The solution is to model the car in front of you with a random acceleration.

4.6.3 How It Works

The system model is shown in Figure 4.15.

Both cars are moving at a nearly steady speed. You are not using radar cruise control. You just want to know the range of the car in front of you. The dynamic model for the other car is

$$\dot{x} = v_2 \tag{4.123}$$

$$\dot{v} = \eta_{v_2} + a_w \tag{4.124}$$

$$\dot{a}_w = \eta_w \tag{4.125}$$

Since our only measurement is relative distance, we can't measure the position of our car, so the preceding equations are relative.

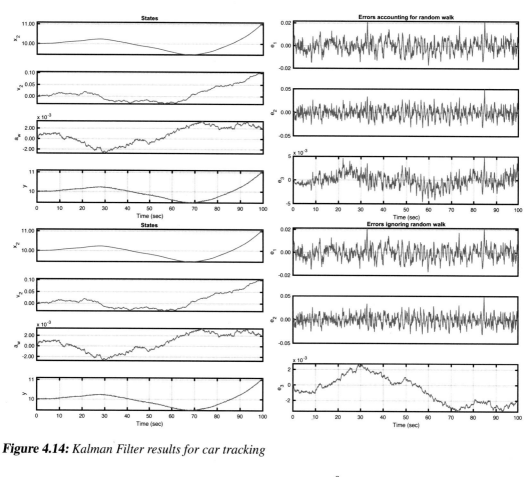

Figure 4.14: *Kalman Filter results for car tracking*

Figure 4.15: *The dynamical model for two cars*

The model is linear, so a conventional Kalman Filter can be used. The continuous state equations are

$$\dot{x} = \begin{bmatrix} 0 & 1 & 0 \\ 0 & 0 & 1 \\ 0 & 0 & 0 \end{bmatrix} x + \begin{bmatrix} 0 \\ a_2 \\ 0 \end{bmatrix} \tag{4.126}$$

123

The model noise covariance matrix is

$$
\begin{bmatrix}
0 & 0 & 0 \\
0 & \sigma_{v_2}^2 & 0 \\
0 & 0 & \sigma_w^2
\end{bmatrix}
\tag{4.127}
$$

There is no uncertainty in the position derivatives. We propagate the equations of motion using state propagation since they are linear. The following code shows the simulation. You have the option to set the noise value in the Kalman Filter for the random walk to zero.

KFAuto.m

```
10  rng('default');
11
12  %% Initialize
13  tEnd          = 100.0;            % Simulation end time (sec)
14  dT            = 0.1;              % Time step (sec)
15  mSigma        = 0.01;            % 1 sigma position measurement noise
16  ignoreRW      = true;
17
18  % xdot = a*x + b*u
19  a = [0 1 0 ;0 0 1 ;0 0 0 ]; % Continuous time model
20  b = [0 0;1 0;0 1]; % Continuous time input matrix
21
22  % x[k+1] = f*x[k] + g*u[k]
23  [f,g]  = CToDZOH(a,b,dT);  % Discrete time model
24  x      = [10;0;0];          % Initial state
25  sig    = [1e-2; 1e-3]; % Model noise
26  if( ignoreRW )
27      q      = [0; sig(1); 0].^2;      % Model noise covariance ;
28  else
29      q      = [0; sig].^2;       % Model noise covariance ;
30  end
31  dKF    = KFInitialize('kf','m',x,'a',f,'b',g,'h',[1 0 0],...
32                   'r',mSigma^2,'q',diag(q),'p',diag((0.01*x).^2),...
33                   'u',[0;0]);
34
35  %% Simulation
36  nSim = floor(tEnd/dT) + 1;
37  xPlot = zeros(7,nSim);
38
39  for k = 1:nSim
40      % Position measurement with random noise
41      y = x(1) + mSigma*randn(1,1);
42
43      % Update the Kalman Filter
44      dKF.y = y;
45      dKF   = KFUpdate(dKF);
46
47      % Plot storage
48      xPlot(:,k) = [x;y;dKF.m-x];
```

```
49
50    % Propagate the state equations
51    x = f*x + g*(sig.*randn(2,1));
52
53    % Propagate the Kalman Filter
54    dKF    = KFPredict(dKF);
55  end
56
57  %% Plot the results
58  yL     = {'x_2' 'v_2' 'a_w' 'y' 'e_1' 'e_2' 'e_3' };
59  t      = dT*(0:(nSim-1));
60
61  if( ignoreRW)
62    s = sprintf('Errors ignoring random walk');
63  else
64    s = sprintf('Errors accounting for random walk');
65  end
```

■ **TIP**   Use rng('default') to reset the random number generators so that you can get the same random numbers in each run.

The following plots show the results. We use `rng('default')` to set the random number generators to the same value each run.

As expected, the random walk estimate is not as good when the Kalman Filter thinks its plant covariance is zero. The overall estimate doesn't change much because the random walk is not that large.

4.7   Summary

This chapter has demonstrated learning using Kalman Filters. In this case, learning is the estimation of states and parameters for a damped oscillator. We looked at conventional Kalman Filters, Extended Kalman Filters, and Unscented Kalman Filters. We looked at the parameter learning version of the Unscented Kalman Filters. All examples were done using a damped oscillator. Table 4.1 lists the functions and scripts included in the companion code.

Table 4.1: *Chapter Code Listing*

File	Description
AngleMeasurement	Angle measurement of the mass
AngleMeasurementPartial	Angle measurement derivative
LinearMeasurement	Position measurement of the mass
OscillatorSim	Simulation of the damped oscillator
OscillatorDampingRatioSim	Simulation of the damped oscillator with different damping ratios
RHSOscillator	Dynamical model for the damped oscillator
RHSOscillatorPartial	Derivative model for the damped oscillator
RungeKutta	Fourth-order Runge-Kutta integrator
PlotSet	Creates two-dimensional plots from a data set
TimeLabel	Produces time labels and scaled time vectors
Gaussian	Plots a Gaussian distribution
GaussianExample	Plots the PDF and CPDF
KFAuto	Kalman Filters automobile tracking
KFInitialize	Initializes Kalman Filters
KFSim	Demonstration of a conventional Kalman Filter
KFPredict	Prediction step for a conventional Kalman Filter
KFUpdate	Update step for a conventional Kalman Filter
EKFPredict	Prediction step for an Extended Kalman Filter
EKFUpdate	Update step for an Extended Kalman Filter
UKFPredict	Prediction step for an Unscented Kalman Filter
UKFUpdate	Update step for an Unscented Kalman Filter
UKFPUpdate	Update step for an Unscented Kalman Filter parameter update
UKFSim	Demonstration of an Unscented Kalman Filter
UKFPSim	Demonstration of parameter estimation Unscented Kalman Filter
UKFWeights	Generates weights for the Unscented Kalman Filter
UKFPWeights	Generates weights for the Unscented Kalman Filter parameter estimator
RHSOscillatorUKF	Dynamical model for the damped oscillator for use in UKF parameter estimation

CHAPTER 5

■ ■ ■

Adaptive Control

Control systems need to react to the environment in a predictable and repeatable fashion. Control systems take measurements and use them to control the process. For example, a ship measures its heading and changes its rudder angle to attain a desired heading.

Typically, control systems are designed and implemented with all of the parameters hard-coded into the software. This works very well in most circumstances, particularly when the

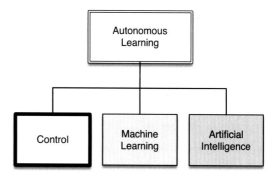

system is well known during the design process. When the system is not well defined or is expected to change significantly during operation, it may be necessary to implement learning control. For example, the batteries in an electric car degrade over time. This leads to less range. An autonomous driving system would need to learn that range was decreasing. This would be done by comparing the distance traveled with the battery's state of charge. More drastic, and sudden, changes can alter a system. For example, in an aircraft, the air data system might fail due to a sensor malfunction. If GPS were still operating, the plane would want to switch to a GPS-only system. In a multi-input-multi-output control system, a branch may fail, due to a failed actuator or sensor. The system might have to be modified to operate branches in that case.

Learning and adaptive control are often used interchangeably. In this chapter, you will learn a variety of techniques for adaptive control for different systems. Each technique is applied to a different system, but all are generally applicable to any control system.

Figure 5.1 provides a taxonomy of adaptive and learning control. The paths depend on the nature of the dynamical system. The rightmost branch is tuning. This is something a designer would do during testing, but it could also be done automatically as will be described in the self-tuning Recipe 5.1. The next path is for systems that will vary with time. Our first example of a system with time-varying parameters applies Model Reference Adaptive Control (MRAC) for a spinning wheel. This is discussed in Section 5.2.

M. Paluszek, S. Thomas, *MATLAB Machine Learning Recipes*,
https://doi.org/10.1007/978-1-4842-9846-6_5

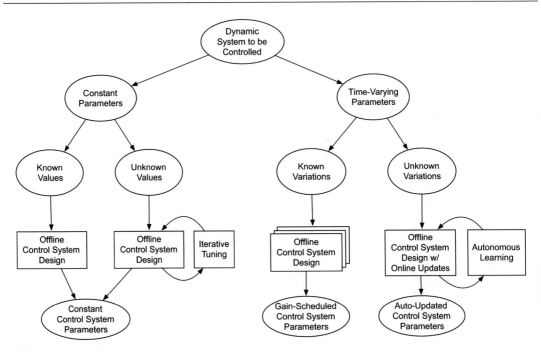

Figure 5.1: *Taxonomy of adaptive or learning control*

The next example is ship control. Your goal is to control the heading angle. The dynamics of the ship are a function of the forward speed. While it isn't learning from experience, it is adapting based on information about its environment.

The last example is a spacecraft with variable inertia. This shows very simple parameter estimation.

5.1   Self-Tuning: Tuning an Oscillator

We want to tune a damper so that we critically damp a spring system for which the spring constantly changes. Our system will work by perturbing the undamped spring with a step and measuring the frequency using a Fast Fourier Transform. We then compute the damping using the frequency and add a damper to the simulation. We then measure the undamped natural frequency again to see that it is the correct value. Finally, we set the damping ratio to 1 and observe the response. The frequency is measured during operation, so this is an example of online learning. The system is shown in Figure 5.2.

In Chapter 4, we introduced parameter identification in the context of Kalman Filters, which is another way of finding the frequency. The approach here is to collect a large sample of data and process it in batch to find the natural frequency. The equations for the system are

$$\dot{r} = v \tag{5.1}$$

$$m\dot{v} = -cv - kr \tag{5.2}$$

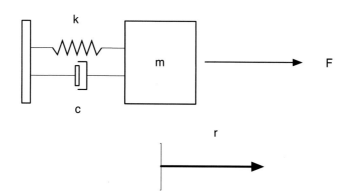

Figure 5.2: *Spring-mass-damper system. The mass is on the right. The spring is on the top to the left of the mass. The damper is below. F is the external force, m is the mass, k is the stiffness, and c is the damping*

c is the damping and k is the stiffness. The damping term causes the velocity to go to zero. The stiffness term bounds the range of motion (unless the damping is negative). The dot above the symbols means the first derivative with respect to time. That is

$$\dot{r} = \frac{dr}{dt} \tag{5.3}$$

The equations state that the change in position with respect to time is the velocity, and the mass times the change in velocity with respect to time is equal to a force proportional to its velocity and position. The second equation is Newton's law:

$$F = ma \tag{5.4}$$

where F is force, m is mass, and a is acceleration.

■ **TIP**   Weight is the mass times the acceleration of gravity.

$$F = -cv - kr \tag{5.5}$$
$$a = \frac{dv}{dt} \tag{5.6}$$

5.1.1 Problem

We want to identify the frequency of an oscillator and tune a control system to that frequency.

5.1.2 Solution

The solution is to have the control system measure the frequency of the spring. We will use an FFT to identify the frequency of the oscillation.

5.1.3 How It Works

The following script shows how an FFT identifies the oscillation frequency for a damped oscillator.

The function is shown in the following code. We use the RHSOscillator dynamical model for the system. We start with a small initial position to get it to oscillate. We also have a small damping ratio so it will damp out. The resolution of the spectrum is dependent on the number of samples:

$$r = \frac{2\pi}{nT} \tag{5.7}$$

where n is the number of samples and T is the sampling period. The maximum frequency is

$$\omega = \frac{nr}{2} \tag{5.8}$$

The following shows the simulation loop and FFTEnergy call.

FFTSim.m

```
7   nSim          = 2^16;              % Number of time steps
8   dT            = 0.1;               % Time step (sec)
9   dRHS          = RHSOscillator;     % Get the default data structure
10  dRHS.omega    = 0.1;                % Oscillator frequency
11  dRHS.zeta     = 0.1;               % Damping ratio
12  x             = [1;0];             % Initial state [position;velocity]
13  y1Sigma       = 0.001;             % 1 sigma position measurement noise
14
15  %% Simulation
16  xPlot = zeros(3,nSim);
17
18  for k = 1:nSim
19    % Measurements
20    y           = x(1) + y1Sigma*randn;
21    % Plot storage
22    xPlot(:,k)  = [x;y];
23    % Propagate (numerically integrate) the state equations
24    x           = RungeKutta( @RHSOscillator, 0, x, dT, dRHS );
25  end
```

FFTEnergy is shown as follows.

FFTEnergy.m

```
21  function [e, w, wP] = FFTEnergy( y, tSamp, aPeak )
35  n = size( y, 2 );
36
37  % If the input vector is odd drop one sample
38  if( 2*floor(n/2) ~= n )
39    n = n - 1;
40    y = y(1:n,:);
41  end
42
43  x   = fft(y);
44  e   = real(x.*conj(x))/n;
45
46  hN = n/2;
47  e   = e(1,1:hN);
48  r   = 2*pi/(n*tSamp);
49  w   = r*(0:(hN-1));
50
51  if( nargout > 2 )
52    k   = e > aPeak*max(e) ;
53    wP = w(k);
54  end
```

The Fast Fourier Transform takes the sampled time sequence and computes the frequency spectrum. We compute the FFT using MATLAB's fft function. We take the result and multiply it by its conjugate to get the energy. The first half of the result has the frequency information. aPeak is to indicate peaks for the output. It is just looking for values greater than a certain threshold.

Figure 5.3 shows the damped oscillation. Figure 5.4 shows the spectrum. We find the peak by searching for the maximum value. The noise in the signal is seen at the higher frequencies. A noise-free simulation is shown in Figure 5.5.

The tuning approach is to

1. Excite the oscillator with a pulse

2. Run it for 2^n steps

3. Do an FFT

4. If there is only one peak, compute the damping gain

131

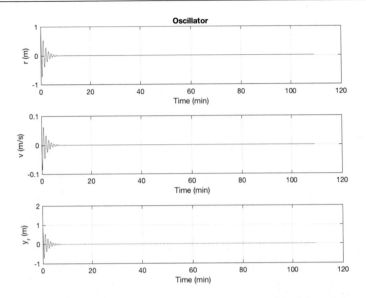

Figure 5.3: *Simulation of the damped oscillator. The damping ratio ζ is 0.5, and the undamped natural frequency ω is 0.1 rad/s*

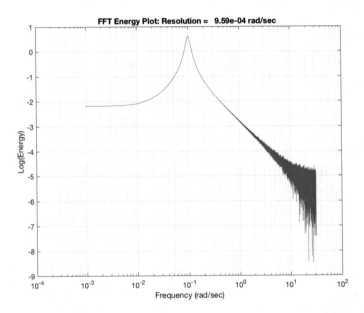

Figure 5.4: *The frequency spectrum. The peak is at the oscillation frequency of 0.1 rad/sec*

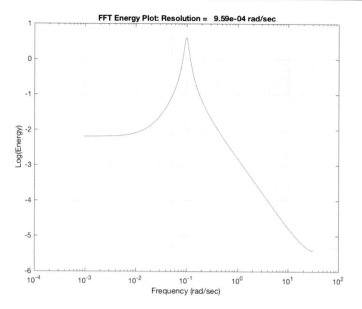

Figure 5.5: *The frequency spectrum without noise. The peak of the spectrum is at 0.1 rad/s in agreement with the simulation*

The script `TuningSim` calls `FFTEnergy.m` with `aPeak` set to 0.7. The value for `aPeak` is found by looking at a plot and picking a suitable number. The disturbances are Gaussian-distributed accelerations, and there is noise in the measurement. Note that this simulation uses a different right-hand-side function `RHSOscillatorControl`. The measurement with noise is implemented as

TuningSim.m

```
33      % Measurements
34      y            = x(1) + y1Sigma*randn;
```

The disturbances are implemented with a step perturbation, which ends at a given step, and random noise:

TuningSim.m

```
39      dRHS.a       = aJ + a1Sigma*randn;
40      if( k == kPulseStop )
41          aJ = 0;
42      end
```

133

The tuning code using `FFTEnergy` is shown in the following snippet.

TuningSim.m

```
47    FFTEnergy( yFFT, dT );
48    [ ~, ~, wP] = FFTEnergy( yFFT, dT );
49    if( length(wP) == 1 )
50      wOsc    = wP;
51      fprintf(1,'\tEstimated oscillator frequency %12.4f rad/s\n',wP);
52      dRHS.c    = 2*zeta*wOsc;
53    else
54      fprintf(1,'\tTuned\n');
55    end
```

The entire loop is run four times, with the first time undamped and the second, third, and fourth times updating the tuned gain. The results in the command window are

```
>> TuningSim
1:        Estimated oscillator frequency        0.0997 rad/s
2:        Tuned
3:        Tuned
4:        Tuned
```

If the random noise is large enough, the loop may tune more than once. Running it a few times or increasing the noise will show this behavior.

As you can see from the FFT plots in Figure 5.6, the spectra are "noisy" due to the sensor noise and Gaussian disturbance. The criteria for determining that the system is underdamped it is a distinctive peak. If the noise is large enough, we have to set lower thresholds to trigger the tuning. The top-left FFT plot shows the 0.1 rad/s peak. After tuning, we damp the oscillator sufficiently so that the peak is diminished. The time plot in Figure 5.6 (the bottom plot) shows that, initially, the system is lightly damped. After tuning, it oscillates very little. There is a slight transient every time the tuning is adjusted at 1.9, 3.6, and 5.5 seconds. The FFT plots (the top right and middle two) show the data used in the tuning.

An important point is that we must stimulate the system to identify the peak. All system identification, parameter estimation, and tuning algorithms have this requirement. An alternative to a pulse (which has a broad frequency spectrum) would be to use a sinusoidal sweep. That would excite any resonances and make it easier to identify the peak. However, care must be taken when exciting a physical system at different frequencies to ensure it does not have an unsafe or unstable response at natural frequencies.

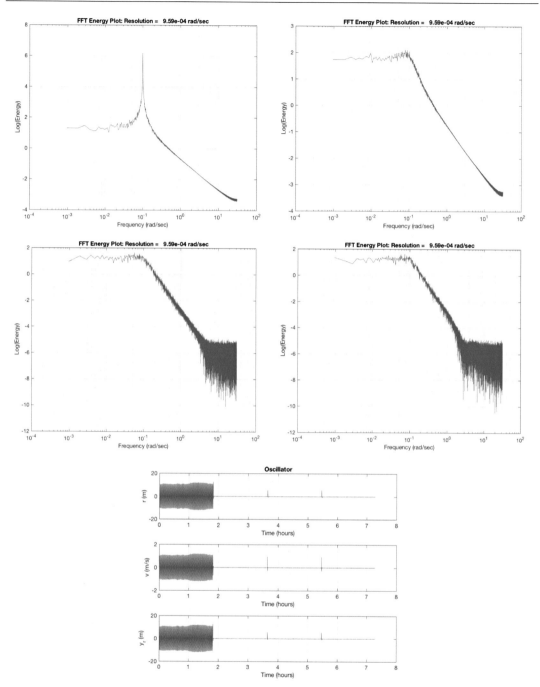

Figure 5.6: *Tuning simulation results. The first four plots are the frequency spectra taken at the end of each sampling interval; the last shows the results over time. Upper left, before tuning, the peak is seen*

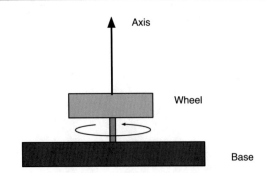

Figure 5.7: *Speed control of a rotor for the Model Reference Adaptive Control demo*

5.2   Implement MRAC

Our next example is to control a rotor with an unknown load so that it behaves in a desired manner. We will use Model Reference Adaptive Control (MRAC). The dynamical model of the rotary joint is [3] and is shown in Figure 5.7.

$$\frac{d\omega}{dt} = -a\omega + bu_c + u_d \tag{5.9}$$

where the damping a and/or input constants b are unknown. ω is the angular rate. u_c is the input voltage, and u_d is a disturbance angular acceleration. This is a first-order system that is modeled by one first-order differential equation. We would like the system to behave like the reference model:

$$\frac{d\omega}{dt} = -a_m\omega + b_mu_c + u_d \tag{5.10}$$

5.2.1   Problem

We want to control a system to behave like a particular model. Our example is a simple rotor.

5.2.2   Solution

The solution is to implement a Model Reference Adaptive Control (MRAC) function.

5.2.3   How It Works

The idea is to have a dynamic model that defines the behavior of your system. You want your system to have the same dynamics. This desired model is the reference, hence the name Model Reference Adaptive Control (MRAC). We will use the MIT rule [3] to design the adaptation system. The MIT rule was first developed at the MIT Instrumentation Laboratory (now Draper Laboratory), which developed the NASA Apollo and Space Shuttle guidance and control systems.

Consider a closed-loop system with one adjustable parameter, θ. θ is a parameter, not an angle. The desired output is y_m. The error is

$$e = y - y_m \tag{5.11}$$

Define a loss function (or cost) as

$$J(\theta) = \frac{1}{2}e^2 \tag{5.12}$$

The square removes the sign. If the error is zero, the cost is zero. We would like to minimize $J(\theta)$. To make J small, we change the parameters in the direction of the negative gradient of J or

$$\frac{d\theta}{dt} = -\gamma \frac{\partial J}{\partial \theta} = -\gamma e \frac{\partial e}{\partial \theta} \tag{5.13}$$

This is the MIT rule. If the system is changing slowly, then we can assume that θ is constant as the system adapts. γ is the adaptation gain. Our dynamic model is

$$\frac{d\omega}{dt} = a\omega + bu_c \tag{5.14}$$

We would like it to be the model:

$$\frac{d\omega_m}{dt} = a_m\omega_m + b_m u_c \tag{5.15}$$

a and b are the actual unknown parameters. a_m and b_m are the model parameters. We would like a and b to be a_m and b_m. Let the controller for our rotor be

$$u = \theta_1 u_c - \theta_2 \omega \tag{5.16}$$

The second term provides the damping. The controller has two adaptation parameters. If they are chosen to be

$$\theta_1 = \frac{b_m}{b} \tag{5.17}$$

$$\theta_2 = \frac{a_m - a}{b} \tag{5.18}$$

the input-output relations of the system and model are the same. This is called perfect model following. This is not required. To apply the MIT rule, write the error as

$$e = \omega - \omega_m \tag{5.19}$$

With the parameters θ_1 and θ_2, the system is

$$\frac{d\omega}{dt} = -(a + b\theta_2)\omega + b\theta_1 u_c \tag{5.20}$$

137

where γ is the adaptation gain. To continue with the implementation, we introduce the operator $p = \frac{d}{dt}$. We then write

$$p\omega = -(a + b\theta_2)\omega + b\theta_1 u_c \tag{5.21}$$

or

$$\omega = \frac{b\theta_1}{p + a + b\theta_2} u_c \tag{5.22}$$

We need to get the partial derivatives of the error with respect to θ_1 and θ_2. These are

$$\frac{\partial e}{\partial \theta_1} = \frac{b}{p + a + b\theta_2} u_c \tag{5.23}$$

$$\frac{\partial e}{\partial \theta_2} = -\frac{b^2\theta_1}{(p + a + b\theta_2)^2} u_c \tag{5.24}$$

from the chain rule for differentiation. Noting that

$$u_c = \frac{p + a + b\theta_2}{b\theta_1} \omega \tag{5.25}$$

the second equation becomes

$$\frac{\partial e}{\partial \theta_2} = \frac{b}{p + a + b\theta_2} y \tag{5.26}$$

Since we don't know a, let's assume that we are pretty close to it. Then let

$$p + a_m \approx p + a + b\theta_2 \tag{5.27}$$

Our adaptation laws are now

$$\frac{d\theta_1}{dt} = -\gamma \left(\frac{a_m}{p + a_m} u_c \right) e \tag{5.28}$$

$$\frac{d\theta_2}{dt} = \gamma \left(\frac{a_m}{p + a_m} \omega \right) e \tag{5.29}$$

Let

$$x_1 = \frac{a_m}{p + a_m} u_c \tag{5.30}$$

$$x_2 = \frac{a_m}{p + a_m} \omega \tag{5.31}$$

which are differential equations that must be integrated. The complete set is

$$\frac{dx_1}{dt} = -a_m x_1 + a_m u_c \tag{5.32}$$

$$\frac{dx_2}{dt} = -a_m x_2 + a_m \omega \tag{5.33}$$

$$\frac{d\theta_1}{dt} = -\gamma x_1 e \tag{5.34}$$

$$\frac{d\theta_2}{dt} = \gamma x_2 e \tag{5.35}$$

Our only measurement would be ω which would be measured with a tachometer. As noted before, the controller is

$$u = \theta_1 u_c - \theta_2 \omega \tag{5.36}$$

$$e = \omega - \omega_m \tag{5.37}$$

$$\frac{d\omega_m}{dt} = -a_m \omega_m + b_m u_c \tag{5.38}$$

The MRAC is implemented in the function MRAC shown in its entirety in the following listing. The controller has five differential equations that are propagated. The states are $[x_1, x_2, \theta_1, \theta_2, \omega_m]$. RungeKutta is used for the propagation, but a less computationally intensive lower-order integrator, such as Euler, could be used instead. The function returns the default data structure if no inputs and one output is specified. The default data structure has reasonable values. That makes it easier for a user to implement the function. It only propagates one step.

MRAC.m

```
23  function d = MRAC( omega, d )
24
25  if( nargin < 1 )
26    d = DataStructure;
27    return
28  end
29
30  d.x = RungeKutta( @RHS, 0, d.x, d.dT, d, omega );
31  d.u = d.x(3)*d.uC - d.x(4)*omega;
32
33  %% MRAC>DataStructure
34  function d = DataStructure
35  % Default data structure
36
37  d = struct('aM',2.0,'bM',2.0,'x',[0;0;0;0;0],'uC',0,'u',0,'gamma',1,'dT
       ',0.1);
39
40  %% MRAC>RHS
41  function xDot = RHS( ~, x, d, omega )
42  % RHS for MRAC
```

```
43
44  e      = omega - x(5);
45  xDot = [-d.aM*x(1) + d.aM*d.uC;...
46          -d.aM*x(2) + d.aM*omega;...
47          -d.gamma*x(1)*e;...
48           d.gamma*x(2)*e;...
49          -d.aM*x(5) + d.bM*d.uC];
```

Now that we have the MRAC controller done, we'll write some supporting functions and then test it all out in `RotorSim`.

5.3   Generating a Square Wave Input

5.3.1 Problem

We need to generate a square wave to stimulate the rotor in the previous recipe.

5.3.2 Solution

For simulation and testing our controller, we will generate a square wave with a function.

5.3.3 How It Works

`SquareWave` generates a square wave. The first few lines are our standard code for running a demo or returning the data structure.

SquareWave.m

```
26  function [v,d] = SquareWave( t, d )
27
28  if( nargin < 1 )
29    if( nargout == 0 )
30      Demo;
31    else
32      v = DataStructure;
33    end
34    return
35  end
36
37  if( d.state == 0 )
38    if( t - d.tSwitch >= d.tLow )
39      v          = 1;
40      d.tSwitch = t;
41      d.state   = 1;
42    else
43      v          = 0;
44    end
45  else
46    if( t - d.tSwitch >= d.tHigh )
47      v          = 0;
48      d.tSwitch = t;
```

```
49        d.state   = 0;
50    else
51        v         = 1;
52    end
53  end
```

This function uses `d.state` to determine if it is in the high or low part of a square wave. The width of the low part of the wave is set in `d.tLow`. The width of the high part of the square wave is set in `d.tHigh`. It stores the time of the last switch in `d.tSwitch`.

A square wave is shown in Figure 5.8. There are many ways to specify a square wave. This function produces a square wave with a minimum of zero and a maximum of one. You specify the time at zero and the time at one to create the square wave.

We adjusted the y-axis limit and line width using the following code.

SquareWave.m

```
76  PlotSet(t,v,'x label', 't (sec)', 'y label', 'v', 'plot title','Square
        Wave',...
77              'figure title', 'Square Wave');
78  set(gca,'ylim',[0 1.2])
79  h = get(gca,'children');
80  set(h,'linewidth',1);
```

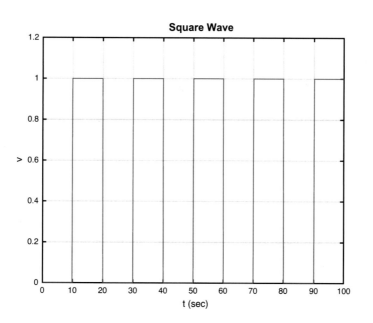

Figure 5.8: *Square wave*

141

■ **TIP**   h = get(gca,'children') gives you access to the line data structure in a plot for the most recent axes.

5.4   Demonstrate MRAC for a Rotor

5.4.1 Problem

We want to create a recipe to control our rotor using MRAC.

5.4.2 Solution

The solution is to implement our Model Reference Adaptive Control (MRAC) function in a MATLAB script from Recipe 5.2.

5.4.3 How It Works

MRAC is implemented in the script RotorSim. It calls MRAC to control the rotor. As in our other scripts, we use PlotSet for our 2D plots. Notice that we use two new options. One 'plot set' allows you to put more than one line on a subplot. The other 'legend' adds legends to each plot. The cell array argument to 'legend' has a cell array for each plot. In this case, we have two plots each with two lines, so the cell array is

```
{{'true', 'estimated'} ,{'Control' ,'Command'}}
```

Each plot legend is a cell entry within the overall cell array.

The rotor simulation script with MRAC is shown in the following listing. The square wave functions generate the command to the system that ω should track. RHSRotor, SquareWave, and MRAC all return default data structures. MRAC and SquareWave are called once per pass through the loop. The simulation right-hand-side, that is the dynamics of the rotor, in RHSRotor, are then propagated using RungeKutta. Note that we pass to pointer for RHSRotor to RungeKutta.

RotorSim.m

```
6   %% Initialize
7   nSim    = 4000;     % Number of time steps
8   dT    = 0.1;      % Time step (sec)
9   dRHS    = RHSRotor;     % Get the default data structure
10  dC    = MRAC;
11  dS    = SquareWave;
12  x       = 0.1;      % Initial state vector
13
14  %% Simulation
15  xPlot = zeros(4,nSim);
16  theta = zeros(2,nSim);
17  t     = 0;
18  for k = 1:nSim
```

```
19
20    % Plot storage
21    xPlot(:,k)    = [x;dC.x(5);dC.u;dC.uC];
22    theta(:,k)    = dC.x(3:4);
23    [uC, dS]      = SquareWave( t, dS );
24    dC.uC         = 2*(uC - 0.5);
25    dC            = MRAC( x, dC );
26    dRHS.u        = dC.u;
27
28    % Propagate (numerically integrate) the state equations
29    x             = RungeKutta( @RHSRotor, t, x, dT, dRHS );
30    t             = t + dT;
31  end
```

■ **TIP**   Pass pointers @fun instead of strings 'fun' to functions whenever possible.

RHSRotor is shown as follows.

RHSRotor.m

```
26  function xDot = RHSRotor( ˜, x, d )
27
28  if( nargin < 1 )
29    xDot = struct('a',1,'b',0.5,'u',0);
30    return
31  end
32
33  xDot    = -d.a*x + d.b*d.u;
```

The dynamics are just one line of code. The remaining returns the default data structure.

The results are shown in Figure 5.9. We set the adaptation gain, γ, to 1. a_m and b_m are set equal to 2. a is set equal to 1 and b to $\frac{1}{2}$.

The first plot shows the rotor's estimated and true angular rates on top and the control demand and actual control sent to the wheel on the bottom. The desired control is a square wave (generated by SquareWave). Notice the transient in the applied control at the transitions of the square wave. The control amplitude is greater than the commanded control. Notice also that the angular rate approaches the desired commanded square wave shape.

Figure 5.10 shows the convergence of the adaptive gains, θ_1 and θ_2. They have converged by the end of the simulation.

MRAC learns the gains of the system by observing the response to the control excitation. It requires excitation to converge. This is the nature of all learning systems. If there is insufficient stimulation, it isn't possible to observe the behavior of the system, so there is not enough information for learning. It is easy to find an excitation for a first-order system. For higher-order systems or nonlinear systems, this can be more difficult.

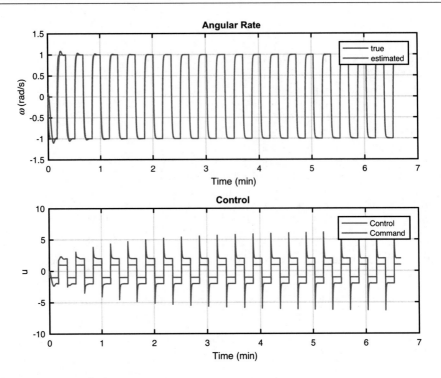

Figure 5.9: *MRAC control of a rotor*

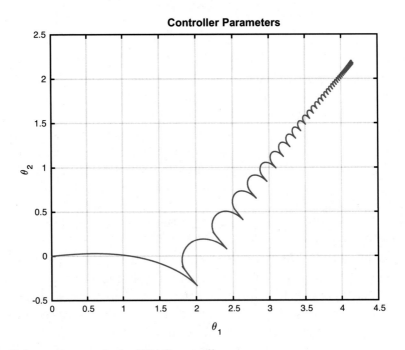

Figure 5.10: *Gain convergence in the MRAC controller*

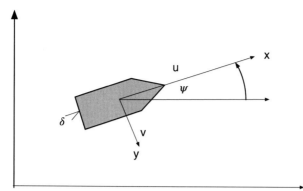

Figure 5.11: *Ship heading control for gain scheduling control*

5.5 Ship Steering: Implement Gain Scheduling for Steering Control of a Ship

5.5.1 Problem

We want to steer a ship at all speeds. The problem is that the dynamics are speed dependent, making this a nonlinear problem. The model is shown in Figure 5.11.

5.5.2 Solution

The solution is to use gain scheduling to set the gains based on speeds. The gain schedule is learned by automatically computing gains from the dynamical equations of the ship. This is similar to the self-tuning example except that we are seeking a set of gains for all speeds, not just one. In addition, we assume that we know the model of the system.

5.5.3 How It Works

The dynamical equations for the heading of a ship are in state space form [3]:

$$
\begin{bmatrix} \dot{v} \\ \dot{r} \\ \dot{\psi} \end{bmatrix} = \begin{bmatrix} \left(\frac{u}{l}\right) a_{11} & u a_{12} & 0 \\ \left(\frac{u}{l^2}\right) a_{21} & \left(\frac{u}{l}\right) a_{22} & 0 \\ 0 & 1 & 0 \end{bmatrix} \begin{bmatrix} v \\ r \\ \psi \end{bmatrix} + \begin{bmatrix} \left(\frac{u^2}{l}\right) b_1 \\ \left(\frac{u^2}{l^2}\right) b_2 \\ 0 \end{bmatrix} \delta + \begin{bmatrix} \alpha_v \\ \alpha_r \\ 0 \end{bmatrix} \tag{5.39}
$$

v is the transverse speed, u is the ship's speed, l is the ship length, r is the turning rate, and ψ is the heading angle. α_v and α_r are disturbances. The ship is assumed to be moving at speed u. This is achieved by the propeller that is not modeled. The control is rudder angle δ. Notice that if $u = 0$, the ship cannot be steered. All of the coefficients in the state matrix are functions of u, except for the heading angle. Our goal is to control the heading given the disturbance acceleration in the first equation and the disturbance angular rate in the second.

The disturbances only affect the dynamics states, r, and v. The last state, ψ, is a kinematic state and does not have a disturbance.

Table 5.1: *Ship parameters [3]*

Parameter	Minesweeper	Cargo	Tanker
l	55	161	350
a_{11}	−0.86	−0.77	−0.45
a_{12}	−0.48	−0.34	−0.44
a_{21}	−5.20	−3.39	−4.10
a_{22}	−2.40	−1.63	−0.81
b_1	0.18	0.17	0.10
b_2	1.40	−1.63	−0.81

The ship model is shown in the following code, RHSShip. The second and third outputs are for use in the controller. Notice that the differential equations are linear in the state and the control. Both matrices are a function of the forward velocity. We are not trying to control the forward velocity, it is an input to the system. The default parameters for the minesweeper are given in Table 5.1. These are the same numbers that are in the default data structure.

RHSShip.m

```
32  function [xDot, a, b] = RHSShip( ~, x, d )
33
34  if( nargin < 1 )
35    xDot = struct('l',100,'u',10,'a',[-0.86 -0.48;-5.2 -2.4],'b'
          ,[0.18;-1.4],'alpha',[0;0;0],'delta',0);
36    return
37  end
38
39  uOL   = d.u/d.l;
40  uOLSq = d.u/d.l^2;
41  uSqOl = d.u^2/d.l;
42  a     = [ uOL*d.a(1,1) d.u*d.a(1,2) 0;...
43            uOLSq*d.a(2,1) uOL*d.a(2,2) 0;...
44                    0               1 0];
45  b     = [uSqOl*d.b(1);...
46          uOL^2*d.b(2);...
47          0];
48
49  xDot  = a*x + b*d.delta + d.alpha;
```

In the ship simulation, ShipSim, we linearly increase the forward speed while commanding a series of heading psi changes. The controller takes the state space model at each time step and computes new gains which are used to steer the ship. The controller is a linear quadratic regulator. We can use full-state feedback because the states are easily modeled. Such controller will work perfectly in this case but are a bit harder to implement when you need to estimate some of the states or have unmodeled dynamics.

ShipSim.m

```
23   for k = 1:nSim
24     % Plot storage
25     xPlot(:,k)    = x;
26     dRHS.u        = u(k);
27     % Control
28     % Get the state space matrices
29     [~,a,b]       = RHSShip( 0, x, dRHS );
30     gain(k,:)     = QCR( a, b, qC, rC );
31     dRHS.delta    = -gain(k,:)*[x(1);x(2);x(3) - psi(k)]; % Rudder angle
32     delta(k)      = dRHS.delta;
33     % Propagate (numerically integrate) the state equations
34     x             = RungeKutta( @RHSShip, 0, x, dT, dRHS );
35   end
```

The quadratic regulator generator code is shown in the following listing. It generates the gain from the matrix Riccati equation. A Riccati equation is an ordinary differential equation that is quadratic in the unknown function. In steady state, this reduces to the algebraic Riccati equation that is solved in this function.

QCR.m

```
29   function k = QCR( a, b, q, r )
30
31   [sinf,rr] = Riccati( [a,-(b/r)*b';-q',-a'] );
32
33   if( rr == 1 )
34     disp('Repeated roots. Adjust q, r or n');
35   end
36
37   k = r\(b'*sinf);
38
39   function [sinf, rr] = Riccati( g )
40   %% Ricatti
41   %    Solves the matrix Riccati equation in the form
42   %
43   %    g = [a    r ]
44   %        [q   -a']
46
47   rg = size(g);
48
49   [w, e] = eig(g);
50
51   es = sort(diag(e));
52
53   % Look for repeated roots
54   j = 1:length(es)-1;
55
56   if ( any(abs(es(j)-es(j+1))<eps*abs(es(j)+es(j+1))) )
57     rr = 1;
```

```
58  else
59    rr = 0;
60  end
61
62  % Sort the columns of w
63  ws    = w(:,real(diag(e))) < 0);
64
65  sinf = real(ws(rg/2+1:rg,:)/ws(1:rg/2,:));
```

a is the state transition matrix, b is the input matrix, q is the state cost matrix, and r is the control cost matrix. The bigger the elements of q, the more cost we place on deviations of the states from zero. That leads to tight control at the expense of more control. The bigger the elements of b the more cost we place on control. Bigger b means less control. Quadratic regulators guarantee stability if all states are measured. They are a very handy controller to get something working. The results are given in Figure 5.12. Note how the gains evolve.

The gain on the angular rate r is nearly constant. Notice that the ψ range is very small! Normally, you would zoom out the plot. The other two gains increase with speed. This is an example of gain scheduling. The difference is that we autonomously compute the gains from perfect measurements of the ship's forward speed.

ShipSimDisturbance is a modified version of ShipSim that is a shorter duration, with only one-course change, and with disturbances in both angular rate and lateral velocity. The results are given in Figure 5.13.

5.6   Spacecraft Pointing

5.6.1 Problem

We want to control the orientation of a spacecraft with thrusters for control. We do not know the inertia, which has a major impact on control.

5.6.2 Solution

The solution is to use a parameter estimator to estimate the inertia and feed it into the control system.

5.6.3 How It Works

The spacecraft model is shown in Figure 5.14.

The dynamical equations are

$$I = I_0 + m_f r_f^2 \tag{5.40}$$

$$T_c + T_d = I\ddot{\theta} + \dot{m}_f r_f^2 \dot{\theta} \tag{5.41}$$

$$\dot{m}_f = -\frac{T_c}{r u_e} \tag{5.42}$$

148

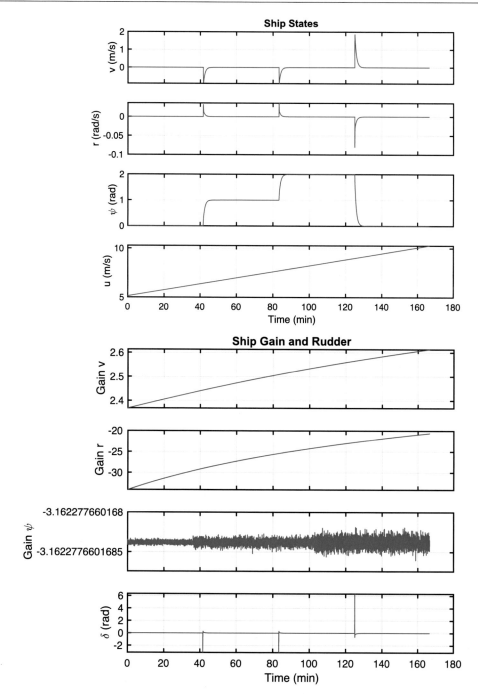

Figure 5.12: *Ship steering simulation. The states are shown on the top with the forward velocity. The gains and rudder angle are shown on the bottom. Notice the "pulses" in the rudder to make the maneuvers*

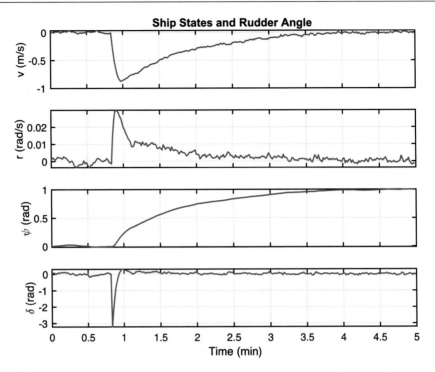

Figure 5.13: *Ship steering simulation. The states are shown on the left with the rudder angle. The disturbances are Gaussian white noise*

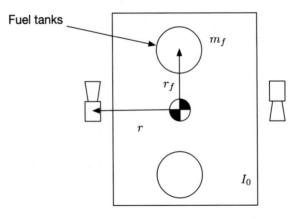

Figure 5.14: *Spacecraft model*

where I is the total inertia, I_0 is the constant inertia for everything except the fuel mass, T_c is the thruster control torque, T_d is the disturbance torque, m_f is the total fuel mass, r_f is the distance to the fuel tank center (moment arm), r is the vector to the thrusters, u_e is the thruster exhaust velocity, and θ is the angle of the spacecraft axis. Fuel consumption is balanced between the two tanks, so the center of mass remains at (0,0). The second term in the second equation is the inertia derivative term, which adds damping to the system.

Our controller is a PD (proportional derivative) controller of the form

$$T_c = Ia \qquad (5.43)$$

$$a = -K(\theta + \tau\dot{\theta}) \qquad (5.44)$$

K is the forward gain and τ the rate constant. We design the controller for unit inertia and then estimate the inertia so that our dynamic response is always the same. We will estimate the inertia using a very simple algorithm:

$$I_k = K_I I_{k-1} - (1 - K_I)\frac{T_{c_k}}{\ddot{\theta}_k} \qquad (5.45)$$

K_I is less than or equal to one. We will do this only when the control torque is not zero and the change in rate is not zero. This is a first difference approximation and should be good if we don't have a lot of noise. The following code snippet shows the simulation loop with the control system. The dynamics are in RHSSpacecraft.m.

SpacecraftSim.m

```
15  %% Controller
16  kForward   = 0.05;
17  tau        = 10;
18  tCThresh   = 0.00;
19  kI         = 0.9; % Inertia filter gain
20
21  %% Simulation
22  xPlot      = zeros(7,nSim);
23  inrEst     = 1.01*(dRHS.i0 + dRHS.rF^2*x(3)) + 0.05*randn(1)*dRHS.i0;
24  dRHS.tC    = 0;
25
26  for k = 1:nSim
27    % Control
28    dRHS.tC = -inrEst*kForward*(x(1) + tau*x(2));
29    % Collect plotting information
30    [xDot,inrTrue] = RHSSpacecraft(0,x,dRHS);
31    omegaDot = xDot(2); % from gyro
32    if( abs(dRHS.tC) > tCThresh  )
33      inrEst = kI*inrEst + (1-kI)*dRHS.tC/omegaDot;
34    end
35    xPlot(:,k) = [x;inrEst;dRHS.tD;dRHS.tC;inrTrue];
36        % Propagate (numerically integrate) the state equations
37        x = RungeKutta( @RHSSpacecraft, 0, x, dT, dRHS );
38  end
```

We only estimate inertia when the control torque is above a threshold. This prevents us from responding to noise. We also incorporate the inertia estimator in a simple low-pass filter. The results are shown in Figure 5.15. The threshold means the algorithm only estimates inertia at the very beginning of the simulation when it is reducing the attitude error.

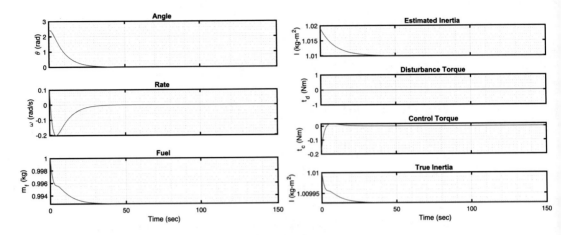

Figure 5.15: *States and control outputs from the spacecraft simulation*

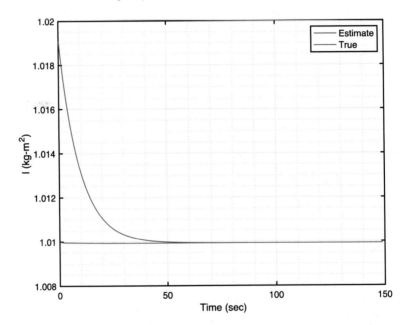

Figure 5.16: *Estimated and actual inertia from the spacecraft simulation*

The dynamics function computes the true inertia from the fuel mass state and the dry mass inertia. This allows the script to compare the estimate against the truth value in Figure 5.16.

This algorithm appears crude, but it is fundamentally all we can do in this situation given just angular rate measurements. Note that the inertia estimate happens while the control is operating, making this a nonlinear controller. More sophisticated filters or estimators could improve the performance.

5.7 Direct Adaptive Control

5.7.1 Problem

We want to control a system for which the plant is unknown. This is one in which the order and parameters for the model are unknown.

5.7.2 Solution

The solution is to use direct adaptation based on Lyapunov control.

5.7.3 How It Works

Assume the dynamics equation is

$$\dot{y} = ay + bu \tag{5.46}$$

u is the control. If a is < 0, the system will always converge. If we use feedback control of the form $u = -ky$, then

$$\dot{y} = (a - bk)y + bu_d \tag{5.47}$$

where u_d is an external disturbance. If $a - bk$ is positive, the system is unstable. If we don't know a or b, then we can't guarantee stability with a fixed gain control. We could try and estimate a and b and then design the controller in real time. A simple approach [18] is an adaptive controller. Assume that $b > 0$, then the gain is

$$\dot{k} = y^2 \tag{5.48}$$

This is known as a universal regulator. To show this is stable, pick the Lyapunov function:

$$V - \frac{y^2}{2} \tag{5.49}$$

Its derivative is

$$\dot{V} = (a - bk)y^2 = (-bk)\dot{k} \tag{5.50}$$

Integrating

$$\frac{y^2}{2} = ak - \frac{bk^2}{2} + C \tag{5.51}$$

Since $\dot{k} > 0$, k can only increase. k has to be bounded because, otherwise, the right-hand side could be negative, which is impossible because the left-hand side is always positive. The following script implements the controller with $a > 0$. Notice how the controller drives the error to zero.

153

DirectAdaptiveControl.m

```
1   %% Direct adaptive control demo
2   % Reference: ECE 517: Nonlinear and Adaptive Control Lecture Notes
3   % Daniel Liberzon November 3, 2021
4
5   n        = 1000;
6   dT       = 0.1;
7
8   % Plant
9   a        = 0.1;
10  b        = 1;
11  x        = 0.1;
12
13  % Initial gain
14  gain     = 0.1;
15
16  % Storage
17  xP       = zeros(3,n);
18  for k = 1:n
19    gain        = gain + dT*x^2;
20    u           = -gain*x;
21    xP(:,k)     = [x;u;gain];
22    x           = RungeKutta(@RHS,0,x,dT,a,b,u);
23  end
24
25  yL = {'x','u','K'};
26
27  t  = (0:n-1)*dT;
28
29  TimeHistory(t,xP,yL,'Direct Adaptive Control');
30
31  %% Right hand side
32  function xDot = RHS(~,x,a,b,u)
33
34  xDot = a*x + b*u;
35
36  end
```

The results are shown in Figure 5.17. Note the rapid convergence. No knowledge of a or b is required. a and b are never estimated.

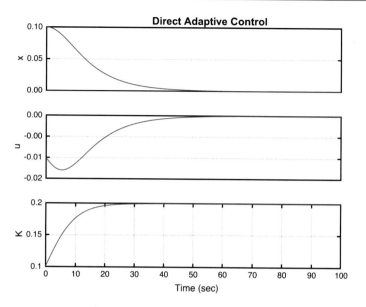

Figure 5.17: *Direct adaptive control*

Table 5.2: *Chapter Code Listing*

File	Description
DirectAdaptiveControl	Direct adaptive control simulations
FFTEnergy	Generates FFT energy
FFTSim	Demonstration of the Fast Fourier Transform
MRAC	Implements Model Reference Adaptive Control
QCR	Generates a full-state feedback controller
RHSOscillatorControl	Right-hand side of a damped oscillator with a velocity gain
RHSRotor	Right-hand side for a rotor
RHSShip	Right-hand side for a ship steering model
RHSSpacecraft	Right-hand side for a spacecraft model
RotorSim	Simulation of Model Reference Adaptive Control
ShipSim	Simulation of ship steering
ShipSimDisturbance	Simulation of ship steering with disturbances
SpacecraftSim	Spacecraft control with inertia estimation
SquareWave	Generates a square wave
TuningSim	Controller tuning demonstration
WrapPhase	Keeps angles between $-\pi$ and π

5.8   Summary

This chapter has demonstrated adaptive or learning control. You learned about model tuning, model reference adaptive control, adaptive control, and gain scheduling. Table 5.2 lists the functions and scripts included in the companion code.

CHAPTER 6

■ ■ ■

Fuzzy Logic

Fuzzy logic [30] is an alternative approach to control system design. Fuzzy logic works within the framework of set theory and is better at dealing with ambiguities. For example, three sets might be defined for a sensor: hard failure, soft failure, and no failure. The three sets might overlap, and at any given time, the sensor may have a degree of membership in each set. In effect, you would be applying a degree of fuzziness. The degree of membership in each set can

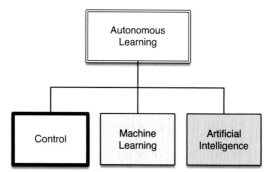

be used to determine what action to take. An algorithmic approach would have to assign a number to the state of the sensor. This could be problematic and not necessarily represent the actual state of the system.

When you go to a doctor with pain, the doctor will often try and get you to convert a subjective concept, pain, into a number from 0 to 10. As pain is personal and your impression is imprecise, you are giving a fuzzy concept or belief a hard number. As you may have experienced, this is not always productive or useful.

Surveys do the same thing. For example, you will be asked to rate the service in a restaurant from 0 to 5. You then rate a bunch of other things on the same scale. This allows the review to come up with a number for your overall impression of the restaurant. Does the resulting 4.8 mean anything? Netflix abandoned the numerical ratings of movies you have seen for thumbs up and down. It seems that they felt that a binary decision, really two sets, was a better data point than a number.

NASA and the US Department of Defense like to use technology readiness levels (TRLs) that go from 1 to 9 to determine where your work is in terms of readiness. Nine is a technology already operating in a target system. One is just an idea. All the other levels are fuzzy for anything moderately complicated. Even giving a technology a 9 is not informative. The M-16 rifle was deployed to Vietnam. It often jammed. In terms of TRL, it was 9, but a 9 doesn't say how well it is working. Again, the readiness of the rifle, when you read soldiers' and Marines' impressions, was best represented by fuzzy beliefs.

This chapter will show you how to build a simple fuzzy logic engine and implement a fuzzy logic control system for windshield wipers. Unlike the other chapters, we will be working with linguistic concepts, not hard numbers. Of course, when you set your wiper motor speed, you need to pick a number (defuzzify your output), but all the intermediate steps employ fuzzy logic. A second example shows control of an HVAC system in a home. Traditional thermostats must be manually switched from heating to cooling, while modern heat pumps can switch automatically. We will compare a traditional control option with two fuzzy examples.

6.1 Building Fuzzy Logic Systems

6.1.1 Problem

We want to have a tool to build a fuzzy logic controller.

6.1.2 Solution

Build a MATLAB function that takes parameter pairs that define everything needed for the fuzzy controller. This will be stored in a data structure.

6.1.3 How It Works

To create a fuzzy system, you must create inputs, outputs, and rules. You can also choose methods for some parts of the fuzzy inference. The fuzzy inference engine has three steps:

1. Fuzzify the inputs

2. Fire rules

3. Defuzzify the outputs

The fuzzy system data is stored in a MATLAB data structure. This structure has the following fields:

- input (:)

- output (:)

- rules (:)

- implication (@)

- aggregation (@)

- defuzzify (@)

The first three fields are arrays of struct arrays. There are separate structures for fuzzy sets and rules, described as follows. The last three fields are function handles for the implementation of these steps in the fuzzy process.

The fuzzy set structure, which is the same for inputs and outputs of the system, has the following fields:

- name

- range (2) (two-element array with minimum and maximum values)

- comp {:} (cell array of label strings)

- type {:} (cell array of membership function handles)

- params {:} (cell array of parameter vectors)

The fuzzy rule struct has the following fields:

- input (:) (vector of input component numbers)

- output (:) (vector of outputs)

- operator {:} (cell array of operator function handles)

Defuzzification requires three steps: implication, aggregation, and the defuzzification of the aggregate. These will be simply function handles. Implication applies the rule strength to the output membership functions, and aggregation combines this data from all the rules for each output across its range. The final defuzzification step produces a crisp value for each output.

This is a lot of data to organize. We do it with the function `BuildFuzzySystem`. The following code snippet shows how it assigns data to the data structure using parameter pairs. The `'id'` field increments the index used for either the input, output, or rule.

BuildFuzzySystem.m

```
53   d = struct;
54   j = 1;
55
56   for k = 1:2:length(varargin)
57     switch (lower(varargin{k}))
58       case 'id'
59         j = varargin{k+1};
60       case 'input comp'
61         d.input(j).comp = varargin{k+1};
62       case 'input type'
63         d.input(j).type = varargin{k+1};
64       case 'input name'
65          d.input(j).name = varargin{k+1};
66       case 'input params'
67         d.input(j).params = varargin{k+1};
68       case 'input range'
69         d.input(j).range = varargin{k+1};
70       case 'output comp'
71         d.output(j).comp = varargin{k+1};
```

This code continues with other cases. Since the fuzzy variables are by nature linguistic, a section of code will map any string names of the fuzzy variables in the rule definitions into their numerical indices using `contains`, which will save computation later.

BuildFuzzySystem.m

```
103   % match rules to sets if cell array
104   for k = 1:length(d.rules)
105     inputs = d.rules(k).input;
106     if iscell(inputs)
107       nIn = length(inputs);
108       input = zeros(1,nIn);
109       for j = 1:nIn
110         comp = d.input(j).comp;
111         val = find(contains(comp,inputs(j)));
112         if ~isempty(val)
113           input(j) = val;
114         end
115       end
116       d.rules(k).input = input;
117     end
118     outputs = d.rules(k).output;
119     if iscell(outputs)
120       nOut = length(outputs);
121       output = zeros(1,nOut);
122       for j = 1:nOut
123         comp = d.output(j).comp;
124         val = find(contains(comp,outputs(j)));
125         if ~isempty(val)
126           output(j) = val;
127         end
128       end
129       d.rules(k).output = output;
130     end
131   end % array of rules
```

The following is a snippet showing how to use `BuildFuzzySystem`, showing just the creation of the first input for the SmartWipers example. This example will be described fully in a later recipe.

```
>> SmartWipers = BuildFuzzySystem(...
            'id',1,...
            'input comp',{'Dry'  'Drizzle'  'Wet'},...
            'input type', {@TrapezoidMF  @TriangleMF  @TrapezoidMF}
              ,...
            'input params',{[0 0 10 50]  [40 50]  [50 90 101 101]},...
            'input range',[0 100],...
            'input name','Wetness')

SmartWipers =
    struct with fields:
```

```
            input: [1x1 struct]

>> SmartWipers.input(1)

ans =
   struct with fields:

      comp: {'Dry'  'Drizzle'  'Wet'}
      type: {@TrapezoidMF  @TriangleMF  @TrapezoidMF}
    params: {[0 0 10 50]  [40 50]  [50 90 101 101]}
     range: [0 100]
      name: 'Wetness'
```

Fuzzy sets in this context consist of a set of linguistic categories or components defining a variable. For instance, if the variable is "age," the components might be "young," "middle aged," and "old." Each fuzzy set has a range over which it is valid, for instance, a good range for "age" might be 0 to 100. Each component has a membership function that describes the degree to which a value in the set's range belongs to each component. For instance, a person who is 50 would rarely be described as "young," but might be described as "middle aged" or "old," depending on the person asked.

To build a fuzzy set, you must divide the variable into components. The simplest are triangles and trapezoids. The following membership functions are provided with this recipe: triangular, trapezoidal, Gaussian, general bell, and sigmoidal. Membership functions are limited in value to between zero and one. The membership functions are shown in Figure 6.1 and described further as follows:

Triangle: The triangular membership function requires two parameters: the center of the triangle and the half-width of the desired triangle base. Triangular membership functions are limited to symmetric triangles.

Trapezoid: The trapezoid membership function requires four parameters: the leftmost point, the start of the plateau, the end of the plateau, and the rightmost points.

Gaussian: A Gaussian membership function is a continuous function with two parameters: the center of the bell and the width (standard deviation) of the bell. Gaussian membership functions are symmetric.

Bell: A general bell function is also continuous and symmetric, but it has three parameters to allow for a flattened top, making it similar to a smoothed trapezoid. It requires three parameters: the center of the bell, the width of the bell at points $y = 0.5$, and the slope of the function at points $y = 0.5$.

Sigmoid: Just as a bell function is similar to a smoothed trapezoid, a sigmoidal membership function is similar to a smoothed step function. It takes two parameters: the point at which $y = 0.5$ and the slope of the function. As the slope approaches infinity, the sigmoidal function approaches the step function.

161

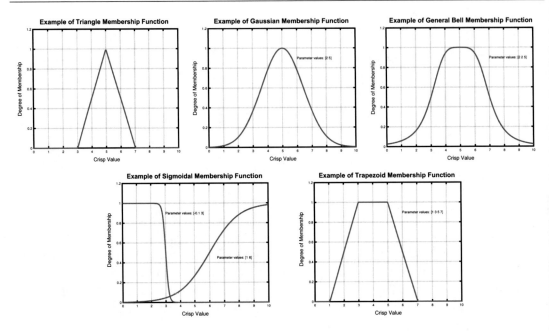

Figure 6.1: *Membership functions*

Fuzzy rules are if-then statements. For example, an air conditioner rule might say IF the room temperature IS high, THEN the blower level IS high. In this case, "room temperature" is the input fuzzy set, "high" is its component for this rule, "blower level" is the output fuzzy set, and "high" is its chosen component. Rules may combine inputs with either an AND or an OR operator. The AND operator is the minimum of the membership values, while the OR operator returns the maximum of the values. In our structure, the rules use numeric indices for the components of each input and output for computational efficiency. An example is

```
>> d.rules(1)

ans =

  struct with fields:

       input: [1 1]
      output: [1 3]
    operator: @FuzzyAND
```

This structure for a fuzzy system is supported by a set of helper functions for the fuzzy operations. This includes membership functions, with an MF suffix; operators, namely AND and OR; implication functions with an IMP suffix; and defuzzification. The following list gives all the support functions provided with this chapter. This is not an exhaustive list of algorithms, and other commercial or open source tools may provide additional methods:

- Membership functions

  - `TriangleMF.m`, `GaussianMF.m`, `GeneralBellMF.m`, `SigmoidalMF.m`, `TrapezoidMF.m`

- Fuzzy operators, for rules

  - `FuzzyAND.m`, `FuzzyOR.m`

- Implication

  - `ScaleIMP.m`, `ClipIMP.m`

- Aggregation

  - `max`

- Defuzzification

  - `CentroidDF.m`

6.2 Implement Fuzzy Logic

6.2.1 Problem

We want to implement fuzzy logic.

6.2.2 Solution

Build a fuzzy inference engine. This will be a function that calls the steps in fuzzy inference given a fuzzy system as defined in the previous recipe, using function handles to specify options within the algorithm.

6.2.3 How It Works

Let's repeat the three steps in fuzzy inference, adding the substeps within Defuzzify:

1. Fuzzify

2. Fire

3. Defuzzify

   (a) Implication

   (b) Aggregation

   (c) Defuzzify the aggregate

The control flow is in the main function, called `FuzzyInference`. It just calls subfunctions `Fuzzify`, `Fire`, and `Defuzzify` in order. It calls `warndlg` if the inputs are not sensible.

FuzzyInference.m

```
29  function [y,data] = FuzzyInference( x, system, verbosity )
39  if length(x) == length( system.input )
40    fuzzyX     = Fuzzify( x, system.input );
41    strength   = Fire( fuzzyX, system.rules );
42    y          = Defuzzify( strength, system, x );
43  else
44    warndlg({'The length of x must be equal to the',...
45      'number of input sets in the system.'})
46  end
```

Since this function is written for educational purposes, we added an informational output struct. This includes the extra step of fuzzifying the outputs after the crisp value is computed from the rules. Therefore, we can examine both `fuzzyX` and `fuzzyY` as well as the strength of the rules firing.

FuzzyInference.m

```
48  if (nargout>1)
49    data.x = x;
50    data.fuzzyX = fuzzyX;
51    data.strength = strength;
52    data.fuzzyY = Fuzzify( y, system.output );
53    data.y = y;
54  end
```

You will notice, in the body of functions, the use of `feval` to evaluate function handles as the input. Earlier versions of this tool used strings for the function names with `eval`, but using handles is now a much faster option than evaluating strings. You pass in the inputs after the handle which can be any expression or variable. For example, for the function

```
function y = MyFun(x)
y = x;
```

You can evaluate it with a number or a variable or an expression, such as

```
>> feval(@MyFun,2)

ans =
     2

>> feval(@MyFun,sin(2))

ans =
     0.9093
```

■ **TIP**   Use `feval` instead of `eval` whenever possible.

The `Fuzzify` subfunction code is shown as follows. It evaluates the degree of membership of the inputs in each membership set.

FuzzyInference.m

```
56    function fuzzyX = Fuzzify( x, sets )
57       %% Fuzzify the inputs with the type function
58       % fuzzyX = Fuzzify( x, sets )
65       n = length(sets);
66       fuzzyX = cell(1,n);
67       for i = 1:n
68         nC = length(sets(i).comp);
69         range = sets(i).range(:);
70         if (range(1) <= x(i)) && (x(i) <= range(2))
71           for j = 1:nC
72             fuzzyX{i}(j) = feval(sets(i).type{j},x(i),sets(i).params{j});
73           end
74         else
75           fuzzyX{i}(1:nC) = zeros(1,nC);
76         end
77       end
```

The fuzzy rule logic is shown in the following code. The code applies "Fuzzy AND" or "Fuzzy OR." "Fuzzy AND" is the minimum of a set of membership values. "Fuzzy OR" is the maximum of a set of membership values. Suppose we have a vector `[1 0 1 0]`. The maximum value is 1 and the minimum is 0.

```
>> 1 && 0 &&  1 && 0

ans =

  logical
    0

>> 1 || 0 ||  1 || 0

ans =

  logical
    1
```

This corresponds to the fuzzy logic AND and OR.

The next code snippet shows the `Fire` subfunction in `FuzzyInference`. "Firing" a rule is the process of applying the rule operators to the fuzzified inputs. This determines the numerical strength of each rule using the specific membership values of the inputs.

FuzzyInference.m

```
81   function strength = Fire( FuzzyX, rules )
82     %% Fire a rule using the specified rules.operator function
83     % strength = Fire( FuzzyX, rules )
90     p = length( rules );
91     n = length( FuzzyX );
92
93     strength = zeros(1,p);
94
95     for i = 1:p
96       method = rules(i).operator;
97       dom = zeros(1,n);
98       for j = 1:n
99         comp = rules(i).input(j);
100        if comp ~= 0
101          dom(j) = FuzzyX{j}(comp);
102        else
103          dom(j) = inf;
104        end
105      end
106      strength(i) = feval(method,dom(dom<=1));
107    end
```

Finally, we defuzzify the results. This function first uses the implication function to determine membership. It aggregates the output using the aggregate function which, in this case, is max. The final step to computing the crisp values is computing the centroid of the aggregate. For explanatory purposes, this function is annotated with a plot capability of the defuzzification if "verbose" output is requested.

FuzzyInference.m

```
111  function [result,aggregate] = Defuzzify( strength, system, xIn )
112    %% Defuzzify the rule output
113    % result = Defuzzify( strength, system )
120    rules   = system.rules;
121    output  = system.output;
122
123    m       = length( output );
124    p       = length( rules );
125    impfun  = system.implicate;
126    aggfun  = system.aggregate;
127    defuzz  = system.defuzzify;
128
129    nPts    = 200;
130    result  = zeros(1,m);
131
132    if verbose
133      figure('name','Fuzzy Inference')
134      subplot(m,1,1); hold on;
135      xstr = num2str(xIn);
```

```
136        title(sprintf('Fuzzy output for [%s]',xstr))
137    end
138
139    for i = 1:m
140        if verbose
141            subplot(m,1,i); hold on; grid on;
142        end
143        range = output(i).range(:);
144        xO = linspace( range(1),range(2),nPts );
145        mem = zeros(p,nPts);
146        % precompute membership for the output set
147        ls = [];
148        label = {};
149        nC = length(output(i).type);
150        ymf = zeros(nC,nPts);
151        for k = 1:nC
152            mfun   = output(i).type{k};
153            params = output(i).params{k};
154            ymf(k,:) = feval(mfun,xO,params);
155            if verbose
156                plot(xO,ymf(k,:),'-.','linewidth',1);
157            end
158        end
159        % compute the membership for each fired rule
160        for j = 1:p
161            comp = rules(j).output(i);
162            if( comp ~= 0 ) && strength(j)>0
163                mem(j,:)  = feval(impfun, ymf(comp,:),strength(j));
164                if verbose
165                    ls(end+1) = plot(xO,mem(j,:),'linewidth',1);
166                    label{end+1} = [num2str(j) ' (' num2str(strength(j),3) ')'
                        ];
167                end
168            else
169                mem(j,:)  = zeros(size(xO));
170            end
171        end % rules
172        aggregate = feval(aggfun,mem);
173        result(i) = feval(defuzz,aggregate,xO);
174        if verbose
175            plot(xO,aggregate,'k--','linewidth',2);
176            yy = axis;
177            plot(result(i)*[1 1],yy(3:4),'r','linewidth',3)
178            text(result(i),yy(3) + 0.75*(yy(4)-yy(3)),sprintf('  %g',result
                (i)))
179            xlabel(output(i).name)
180            if i == 1
181                ll = legend(ls,label,'location','best');
182                ll.Title.String = 'Rules';
183            end
184        end
```

The plots in Figure 6.2 show the total defuzzification process. First, the membership sets of each variable are drawn in dash-dot lines in the background of the plot. Each rule designates a fuzzy output. The implication function combines the strength of the rule with the membership function of that fuzzy output. Clip implication takes the minimum at each point, so the strength limits the membership value. Scale implication uses the product of the strength and the membership. Rules with nonzero strength are plotted as shown with the solid lines, and those rules with nonzero firing strength are shown in the legend. Aggregation then combines the output from each rule into a single vector of membership for the output across its range. The final step is defuzzification of this array, in our case with centroiding via `CentroidDF`. The final crisp value is designated by the thick red line and labeled with the crisp value.

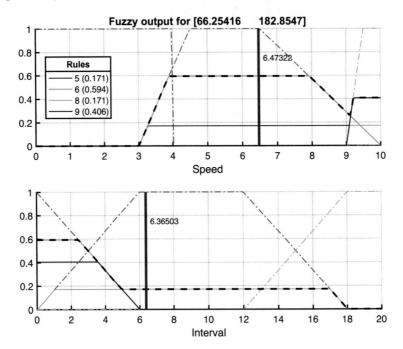

Figure 6.2: *Fuzzy rule plot for smart wipers*

6.3   Window Wiper Fuzzy Controller

6.3.1 Problem

We want a control system to select window wiper speed and interval based on rainfall. This is an implementation of the SmartWipers automatic windshield wiper control system from Cheok [8]. The inputs to the control system are the rain wetness and intensity, and the outputs are the wiper speed and interval.

6.3.2 Solution

Build a fuzzy logic control system using the tools we've developed. First, we will write a function to create the fuzzy system data structure, then a demo script to use it.

6.3.3 How It Works

To call a fuzzy system, use the function `y = FuzzyInference( x, system )`.

The script `SmartWipersDemo` implements the rainfall demo. The demo loads the fuzzy system from the function `SmartWipersSystem`, which uses `BuildFuzzySystem` from Recipe 6.1. The following code performs the fuzzy inference on a full range of the two inputs.

SmartWipersDemo.m

```
21  % Generate regularly space arrays in the 2 inputs
22  n = 30; % Number of samples
23  x = linspace(SmartWipers.input(1).range(1),SmartWipers.input(1).range
        (2),n);
24  y = linspace(SmartWipers.input(2).range(1),SmartWipers.input(2).range
        (2),n);
25
26  % Perform fuzzy inference over the input range
27  z1 = zeros(n,n);
28  z2 = zeros(n,n);
29  for k = 1:n
30    for j = 1:n
31      temp = FuzzyInference([x(k),y(j)], SmartWipers);
32      z1(k,j) = temp(1);
33      z2(k,j) = temp(2);
34    end
35  end
```

First, the demo will plot the input and output fuzzy variables using `FuzzyPlot`. Fuzzy inference is performed on each set of crisp inputs plotted. Figure 6.3 shows the inputs to the fuzzy logic system. Figure 6.4 shows the outputs. The rule base is displayed using `PrintFuzzyRules` and plotted using `surf`.

The inputs that are tested in the fuzzy logic system demo are given in Figure 6.5. This is just the full range of each input.

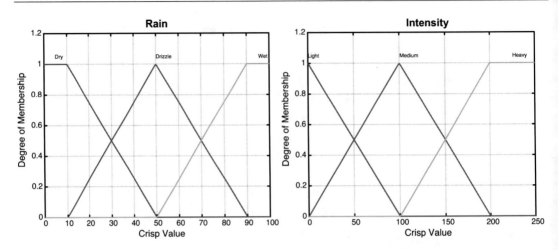

Figure 6.3: *Rain wetness and intensity are the inputs for the smart wiper control system*

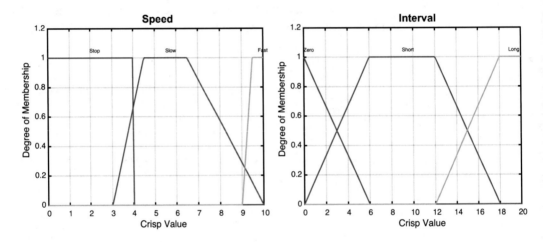

Figure 6.4: *Wiper speed and interval are the outputs for the smart wiper control system*

The printed rules are shown as follows:

```
>> SmartWipersDemo

1. if Wetness is Dry FuzzyAND Intensity is Light then    Speed is Stop
      Interval is Long
2. if Wetness is Dry FuzzyAND Intensity is Medium then   Speed is Slow
      Interval is Long
3. if Wetness is Dry FuzzyAND Intensity is Heavy then    Speed is Slow
      Interval is Short
4. if Wetness is Drizzle FuzzyAND Intensity is Light then        Speed is
      Slow    Interval is Long
5. if Wetness is Drizzle FuzzyAND Intensity is Medium then       Speed is
      Slow    Interval is Short
```

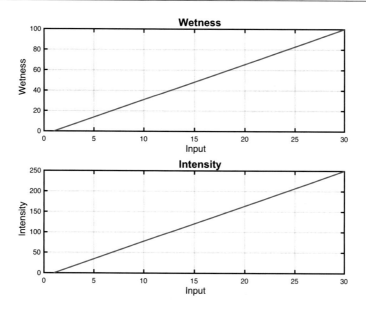

Figure 6.5: *Rain wetness and intensity input numbers*

6. if Wetness is Drizzle FuzzyAND Intensity is Heavy then ‸ Speed is
   Slow   Interval is Zero
7. if Wetness is Wet FuzzyAND Intensity is Light then   Speed is Slow
   Interval is Short
8. if Wetness is Wet FuzzyAND Intensity is Medium then Speed is Fast
   Interval is Short
9. if Wetness is Wet FuzzyAND Intensity is Heavy then   Speed is Fast
   Interval is Zero

Figure 6.6 gives surface plots to show how the outputs relate to the inputs via the rules. The surface plots are generated by the following code. We add a `colorbar` to make the plot more readable. The color is related to z value. We use `view` in the second plot to make it easier to read the figure. You can use `rotate3d on` to allow you to rotate the figure with the mouse.

SmartWipersDemo.m

```
41  % Plot the outputs as surfaces
42  NewFigure('Wiper Speed from Fuzzy Logic');
43  surf(x,y,z1)
44  xlabel('Raindrop Wetness')
45  ylabel('Droplet Frequency')
46  zlabel('Wiper Speed')
47  colorbar
48
49  NewFigure('Wiper Interval from Fuzzy Logic');
50  surf(x,y,z2)
51  xlabel('Raindrop Wetness')
52  ylabel('Droplet Frequency')
```

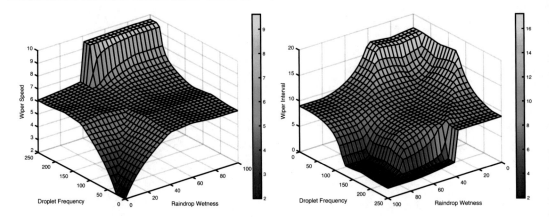

Figure 6.6: *Wiper speed and interval vs. droplet frequency and wetness*

```
53   zlabel('Wiper Interval')
54   view([142.5 30])
55   colorbar
```

■ **TIP**   Use `rotate3d` on to rotate a figure with the mouse.

The `SmartWipersTest` script tests the fuzzy inference using random inputs generated over the input range. This is done using the `FuzzyRand` function as follows.

FuzzyRand.m

```
1    %% FUZZYRAND Compute random inputs within range of the fuzzy input sets
2    %% Inputs
3    % system   (.)  Fuzzy system from BuildFuzzySystem
4    %% Outputs
5    % y        (n)  Random crisp values of the inputs
6
7    function y = FuzzyRand(system)
8
9    if nargin==0
10     system = SmartWipersSystem;
11     y = FuzzyRand(system)
12     return;
13   end
14
15   nIn = length(system.input);
16   y   = ones(1,nIn);
17
18   for k = 1:nIn
19     range = system.input(k).range;
20     y(k)  = range(1) + (range(2)-range(1))*rand(1);
21   end
```

The demo then prints out the crisp and fuzzy values of the inputs and outputs including the strength of the rules. This can provide useful insight when you are developing a new fuzzy system. In the random inputs captured as follows, the rain wetness is both drizzle and wet, the intensity is evenly split between medium and heavy, and the output is a slow speed with a short interval:

```
>> SmartWipersTest
------
Inputs
------
Wetness
----
Crisp: 64.4673
Range: 0 to 100
        Set         Value


    {'Dry'      }         0
    {'Drizzle'}     0.63832
    {'Wet'      }     0.36168


Intensity
----
Crisp: 152.816
Range: 0 to 250
        Set         Value


    {'Light'  }         0
    {'Medium'}     0.47184
    {'Heavy'  }     0.52816

Strength of rule firings:
----
    Input       Output      Fire Strength


    {[1 1]}     {[1 3]}             0
    {[1 2]}     {[2 3]}             0
    {[1 3]}     {[2 2]}             0
    {[2 1]}     {[2 3]}             0
    {[2 2]}     {[2 2]}       0.47184
    {[2 3]}     {[2 1]}       0.52816
    {[3 1]}     {[2 2]}             0
    {[3 2]}     {[3 2]}       0.36168
    {[3 3]}     {[3 1]}       0.36168


-------
Outputs
-------
Speed
```

```
----
Crisp: 6.48238
Range: 0 to 10
      Set          Value

      _____     _____

      {'Stop'}       0
      {'Slow'}       1
      {'Fast'}       0

Interval
----
Crisp: 8.14129
Range: 0 to 20
      Set          Value

      _____     _____

      {'Zero' }      0
      {'Short'}      1
      {'Long' }      0
```

6.4   Simple Discrete HVAC Fuzzy Controller

6.4.1 Problem

We want a control system to automatically switch between air conditioning and heating.

6.4.2 Solution

Build a fuzzy logic control system that can turn on the heating system and air conditioning based on the air temperature.

6.4.3 How It Works

Most older heating, ventilation, and air conditioning systems require the user to pick "AC" and "heat" modes. This doesn't work very well when the temperature is varying a lot from day to day such as during the fall or spring of a region, like New England in the United States, where the temperature varies significantly over the year.

The first step is fuzzifying the input. In the simplest implementation of the control system, there are two input variables: the measured internal temperature of the house and the target or setpoint temperature. The fuzzy categories are shown in Figure 6.7. These are overlapping trapezoids with the temperature in Celsius.

A simple fuzzy control matrix using these variables is shown in Table 6.1. This is the set of rules for the fuzzy controller in HVACSimplestFuzzyController. The rules are combined based on the degree of membership of the internal and target temperature in the different categories.

The dynamical model we will use to simulate the house temperature as a result of the control system is illustrated in Figure 6.8.

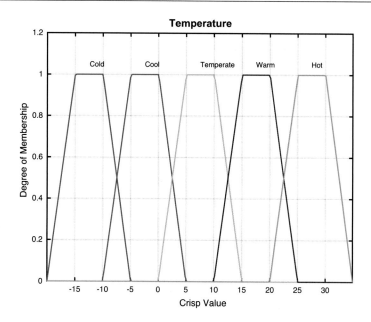

Figure 6.7: *Temperature categories (C) for HVAC*

Table 6.1: *The set of rules for the fuzzy HVAC system. The current value is in the top row; the target is in the first column*

	Cold	**Cool**	**Temperate**	**Warm**	**Hot**
Cold	No change	AC	AC	AC	AC
Cool	Heat	No change	AC	AC	AC
Temperate	Heat	Heat	No change	AC	AC
Warm	Heat	Heat	Heat	No change	AC
Hot	Heat	Heat	Heat	Heat	No change

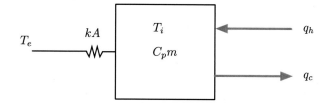

Figure 6.8: *House model*

The dynamical equations are

$$mC_p\frac{dT_i}{dt} = kA(T_e - T_i) + q_h - q_c \tag{6.1}$$

175

m is the thermal mass of the air in the house. A is the surface area of the house. C_p is the specific heat of air. k is the average thermal conductivity of the walls. q_h is the heater flux, and q_c is the air conditioning flux. Both are positive. The dynamic model is shown in the following code.

HVACSim.m

```
109  function [dT,qL] = RHS(~,tI,d)
110  %% Simulation right hand side (dynamics)
111  % Model the changing internal temperature of the house given the HVAC
112  % output and the external temperature.
113  %
114  % [dT,qL] = RHS(~,tI,d)
115  %
116  % dT: change in internal temperature
117  % qL: heat load on the house from outside
118
119  qL = d.k*d.A*(d.tE-tI);
120  dT = (qL + d.qH - d.qC)/(d.cP*d.m);
```

The simulation has two sets of initial conditions at the top: one in which the AC mode will be triggered, that is, a warm day, and one in which the heat will be triggered, a cold day.

HVACSim.m

```
24  % A/C example
25  %%{
26  tSet = 297; % Set point temperature (deg-K)
27  tI   = tSet+3; % Initial internal temperature (deg-K)
28  delT = 10; % Celsius
29  tE   = [ones(1,iS)*(tI + delT) ones(1,n-iS)*(tI - delT) ];
30  %}
31
32  % Heat example
33  %%{
34  tSet = 294; % Set point temperature (deg-K)
35  tI   = tSet-10; % Initial internal temperature (deg-K)
36  tE   = [ones(1,iS)*(tI - 20) ones(1,n-iS)*(tI - 10) ];
37  %}
```

A standard bang-bang controller has a deadband and hysteresis. The following code show the controller. The controller makes its decision to switch from heating to cooling based on the previous heating/cooling command and the demand. Note that it continues heating/cooling through 90% of the deadband. This prevents limit cycling.

HVACSim.m

```
124  function q = Controller(t,tSet,tDB,q,qMax)
125  %% Non-fuzzy Controller with hysteresis
126  % Typical crisp controller with a deadband and hysteresis
```

```
127  %
128  % q = Controller(t,tSet,tDB,q,qMax)
129
130  if( q < 0 )
131    if( t < tSet - 0.9*tDB)
132      q = 0;
133    end
134  elseif( q > 0 )
135    if( t > tSet + 0.9*tDB)
136      q = 0;
137    end
138  else
139    if( abs(t - tSet) > tDB )
140      if( t > tSet)
141        q = -qMax;
142      elseif ( t < tSet )
143        q = qMax;
144      end
145    end
146  end
```

The performance is shown in Figure 6.9. In this case, the external temperature drops 15 degrees Celsius in the middle of the simulation, and the heating system switches from heating to cooling. Hysteresis keeps the HVAC from shifting between heat and cool when the temperature crosses the setpoint.

The fuzzy controller has two modes, initialize and update. The initialize mode creates the fuzzy controller data structure. The following code shows the initialization through the first two rules.

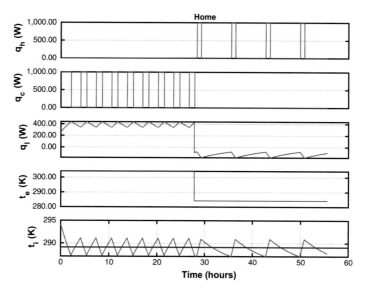

Figure 6.9: *Non-fuzzy hysteresis controller performance*

177

HVACSimplestFuzzyController.m

```
9   function [q,cat] = HVACSimplestFuzzyController(mode,tI,tSet,d)
21    case 'initialize'
22      if nargin<2
23        qMax = 1000;
24      else
25        qMax = tI;
26      end
27
28      % External and set point temps
29      bT = [0 4 6 10]; % 4 vertices of each input trapezoid
30      oT = [10 15 20 25 30];
31      iP = cell(1,5);
32      for k  = 1:5
33        iP{k} =  bT + oT(k);
34      end
35
36      % Define an arbitrary output range, 0 to 6 for the mode
37      oP = {[0 0 1.5 2.5] [1.5 2 4 4.5] [3.5 4.5 6 6]};
38
39      d = BuildFuzzySystem(...
40        'id',1,...
41        'input comp',{'Cold'  'Cool' 'Temperate' 'Warm'  'Hot'} ,...
42        'input type', {@TrapezoidMF @TrapezoidMF @TrapezoidMF
                  @TrapezoidMF @TrapezoidMF} ,...
43        'input params',iP,...
44        'input range',[bT(1) + oT(1) + eps bT(4)+ oT(5) - eps],...
45        'input name','Temperature',...
46        'id',2,...
47        'input comp',{'Cold'  'Cool' 'Temperate' 'Warm'  'Hot'} ,...
48        'input type', {@TrapezoidMF @TrapezoidMF @TrapezoidMF
                  @TrapezoidMF @TrapezoidMF} ,...
49        'input params',iP,...
50        'input range',[bT(1) + oT(1) bT(4)+ oT(5)],...
51        'input name','Target',...
52        'id',1,...
53        'output comp',{'AC'  'None'  'Heat'},...
54        'output type',{@TrapezoidMF  @TrapezoidMF  @TrapezoidMF},...
55        'output params',oP,...
56        'output name','Setting',...
57        'output range',[0 6],...
58        'implicate',@ClipIMP,...
59        'aggregate',@max,...
60        'defuzzify',@CentroidDF,...
61        'id',1,...
62        'rule input',[1 1],...
63        'rule output',2,...
64        'rule operator',@FuzzyAND,...
65        'id',2,...
66        'rule input',[2 2],...
67        'rule output',2,...
```

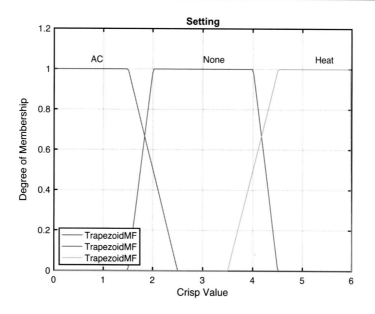

Figure 6.10: *Fuzzy controller output set (arbitrary mode setting)*

68   'rule operator',@FuzzyAND,...

The logic uses fuzzy AND only.

The following script plots the inputs to the simple fuzzy HVAC controller, which are the current temperature and the desired temperature, and the outputs. The output categories are shown in Figure 6.10.

HVACFuzzyPlot.m

```
1   %% Plot the HVAC fuzzy controller
2
3   h = waitbar(0,'HVAC Demo: plotting the rule base');
4
5   dFuzzy = HVACSimplestFuzzyController('initialize');
6   n = 30; % Number of samples
7
8   x = linspace(dFuzzy.input(1).range(1),dFuzzy.input(1).range(2),n);
9   y = linspace(dFuzzy.input(2).range(1),dFuzzy.input(2).range(2),n+2);
10
11  z = zeros(n,n+2);
12  for k = 1:n
13      for j = 1:n+2
14          z(k,j) = FuzzyInference([x(k),y(j)], dFuzzy);
15      end
16      waitbar(k/n)
17  end
18  close(h);
```

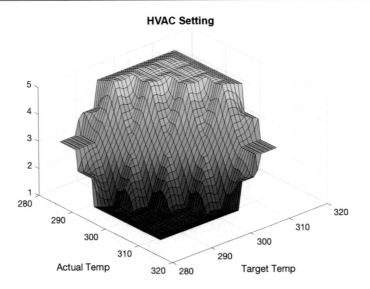

Figure 6.11: *Fuzzy controller inputs and outputs*

```
19
20   NewFigure('State from Fuzzy Logic');
21   surf(x,y,z');
22   xlabel(dFuzzy.input(1).name)
23   ylabel(dFuzzy.input(2).name)
24   zlabel('State')
25   colorbar
```

The results of the rule base are shown in Figure 6.11.

The simulation run with the fuzzy controller is shown in Figure 6.12. This is achieved by setting both useFuzzy and useSimple flags at the top of HVACSim to true. The simulation is much slower than the one using hysteresis. Note the deadband issue with the output to the HVAC; the temperature of the house is held constant in the face of the large external load q_l, but the system is constantly switching on and off to do so.

6.5 Variable HVAC Fuzzy Controller

6.5.1 Problem

The discrete fuzzy controller has a deadband issue. Modern HVAC, such as heat pumps, may have a variable setting, which will produce a smoother result.

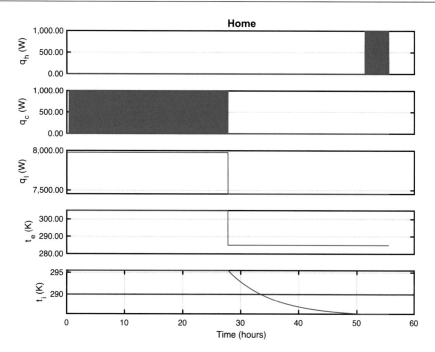

Figure 6.12: *Fuzzy simulation results*

6.5.2 Solution

In this version of the controller, we will have different inputs, rules, and outputs. The inputs will be the external temperature and the delta temperature from the setpoint. There will be a mode output and a value output which can be anywhere in the range, either negative from AC or positive for heat. There are less rules: if the delta is large, set the system on high; if it's smaller, set the system on low; and if it's close to the setpoint, keep the system off.

6.5.3 How It Works

As before, we build the system in an initialize section of the function. We then run the inference and compute the control output in an update section. A demo function will run if the function is called with no inputs, which plots the rule base. In this case, we used the string values of the fuzzy inputs and outputs to define the rules, which will be converted to indices by `BuildFuzzySystem`.

HVACFuzzyController.m

```
1   %% Fuzzy logic control system for HVAC
7   %% Form:
8   %   d      = HVACFuzzyController('initialize',qMax)
9   %   [d,q] = HVACFuzzyController('update',tE,t,tSet,d)
10  %
25  function [q,cat] = HVACFuzzyController(mode,tE,tI,tSet,d)
```

```
26
27  if( nargin < 1)
28    Demo;
29    return
30  end
31
32  %% Initialize
33  % Create the system
34  switch mode
35    case 'initialize'
36      if nargin<2
37        qMax = 1000;
38      else
39        qMax = tE;
40      end
41      oP = {qMax*[0 0 0.2 0.2] qMax*[0.2 0.3 0.5 0.6] qMax*[0.5 0.7 1.01
            1.01]};
42
43      d = BuildFuzzySystem(...
44              'id',1,...
45              'input comp',{'Cold'  'Temperate' 'Hot'} ,...
46              'input type', {@TrapezoidMF  @TrapezoidMF @TrapezoidMF}
                  ,...
47              'input params',{[-11 -11 50 70] [60 70 75 80] [70 80 111
                  111]},...
48              'input range',[-10 110],...
49              'input name','Ext Temp (F)',...
50              'id',2,...
51              'input comp',{'Chilly'  'OK'  'Warm'} ,...
52              'input type', {@TrapezoidMF  @TrapezoidMF @TrapezoidMF}
                  ,...
53              'input params',{[-21 -21 -8 0] [-5 -1 1 5] [0 8 21
                  21]},...
54              'input range',[-20 20],...
55              'input name','Delta-Temp (F)',...
56              'id',1,...
57              'output comp',{'AC' 'Off' 'Heat'},...
58              'output type',{@TrapezoidMF  @TrapezoidMF @TrapezoidMF
                  },...
59              'output params',{[-1.1 -1.1 -0.5 0] [-0.5 0 0 0.5] [0 0.5
                  1.1 1.1]},...
60              'output name','Mode',...
61              'output range',[-1 1],...
62              'id',2,...
63              'output comp',{'Zero'  'Low'  'High'},...
64              'output type',{@TrapezoidMF  @TrapezoidMF  @TrapezoidMF
                  },...
65              'output params',oP,...
66              'output name','Output',...
67              'output range',[0 qMax],...
68              'id',1,... % Cold and Too cold, Heat/high
```

```
69                      'rule input',{'Cold','Chilly'},...
70                      'rule output',{'Heat','High'},...
71                      'rule operator',@FuzzyAND,...
72                      'id',2,... % temperate and too cold, Heat/low
73                      'rule input',{'Temperate','Chilly'},...
74                      'rule output',{'Heat','Low'},...
75                      'rule operator',@FuzzyAND,...
76                      'id',3,... % Hot and too cold, AC/off
77                      'rule input',{'Hot','Chilly'},...
78                      'rule output',{'Off','Zero'},...
79                      'rule operator',@FuzzyAND,...
80                      'id',4,... % Cold and OK, Heat/zero
81                      'rule input',{'Cold','OK'},...
82                      'rule output',{'Off','Zero'},...
83                      'rule operator',@FuzzyAND,...
84                      'id',5,... % temperate and OK, off/off
85                      'rule input',{'Temperate','OK'},...
86                      'rule output',{'Off','Zero'},...
87                      'rule operator',@FuzzyAND,...
88                      'id',6,... % Hot and OK, AC/off
89                      'rule input',{'Hot','OK'},...
90                      'rule output',{'Off','Zero'},...
91                      'rule operator',@FuzzyAND,...
92                      'id',7,... % Cold and too hot, Heat/off
93                      'rule input',[1 3],...
94                      'rule output',[2 1],...
95                      'rule operator',@FuzzyAND,...
96                      'id',8,... % temperate and too hot, AC/low
97                      'rule input',[2 3],...
98                      'rule output',[1 2],...
99                      'rule operator',@FuzzyAND,...
100                     'id',9,... % Hot and too hot, AC/high
101                     'rule input',[3 3],...
102                     'rule output',[1 3],...
103                     'rule operator',@FuzzyAND,...
104                     'implicate',@ScaleIMP,...
105                     'aggregate','sum',...
106                     'defuzzify',@CentroidDF);
107         q = d;
108         cat = [];
109     case 'update'
110         kToC  = 273;
111         delta = (tI - tSet)*9/5;    % in K
112         tF    = (tE-kToC)*9/5 + 32; % in F
113         % expect internal temperature to be within specified range
114         if tF>d.input(1).range(2)
115             tF = d.input(1).range(2) - eps;
116         elseif tF<d.input(1).range(1)
117             tF = d.input(1).range(1) + eps;
118         end
119         % limit delta temperature range
```

```
120    if delta>d.input(2).range(2)
121       delta = d.input(2).range(2) - eps;
122    elseif tF<d.input(2).range(1)
123       delta = d.input(2).range(1) + eps;
124    end
125    [cat,data] = FuzzyInference([tF;delta], d);
126
127    mode = sign(cat(1));
128    if abs(cat(1))<0.01
129       mode = 0;
130    end
131    q = mode*cat(2);
132
133  end
```

In the update case, we check the inputs against the range and limit them if needed. This helps avoid numerical issues after the conversion from Celsius to Kelvin and allows us to have a smaller range for the delta variable without being concerned with large excursions in internal temperature. The mode output is computed using the `sign` function on the mode variable. If the mode value is very small, less than 0.01, we set the mode to 0. The final output setting requested of the HVAC, the q, is the product of the two variables.

The plots which follow are produced by the demo. Figure 6.13 and Figure 6.14 show the system in- puts and outputs. Since there are only two inputs, we can again produce surface plots of the outputs in Figure 6.15, Figure 6.16, and Figure 6.17.

HVACFuzzyController.m

```
135  %% Demonstrate the controller
136  function Demo
137
138  d = HVACFuzzyController('initialize');
139
140  % Plot the fuzzy variables
141  FuzzyPlot( d.input(1) );
142  FuzzyPlot( d.input(2) );
143  FuzzyPlot( d.output(1) );
144  FuzzyPlot( d.output(2) );
145
146  PrintFuzzyRules( d )
147
148  % differentiate btwn internal and external temp
149  t = linspace(d.input(1).range(1)+1e-12,d.input(1).range(2)-1e-12,51);
150  t_K = 5/9*(t-32)+273;   % convert input from C to K
151  tSet = 297;   % example setpoint (K)
152  delta = linspace(d.input(2).range(1)+1e-12,d.input(2).range(2)-1e
         -12,31);
153  q = zeros(length(delta),length(t_K));
154  mode = zeros(length(delta),length(t_K));
155  val = zeros(length(delta),length(t_K));
156  for k = 1:length(t)
```

```
157    for j = 1:length(delta)
158      [q(j,k),cat] = HVACFuzzyController('update',t_K(k),tSet+5/9*delta(j
             ),tSet,d);
159      mode(j,k) = cat(1);
160      val(j,k) = cat(2);
161    end
162  end
163
164  NewFigure('HVAC Output from Fuzzy Logic');
165  surf(t,delta,q)
166  xlabel('Outside Temperature (F)')
167  ylabel('Delta (F)')
168  zlabel('HVAC Output (W)')
169  colorbar
170  set(gca,'ydir','reverse')
171
172  NewFigure('HVAC Mode from Fuzzy Logic');
173  surf(t,delta,mode)
174  xlabel('Temperature (F)')
175  ylabel('Delta (F)')
176  zlabel('HVAC Mode')
177  colorbar
178  set(gca,'ydir','reverse')
179
180  NewFigure('HVAC Value from Fuzzy Logic');
181  surf(t,delta,val)
182  xlabel('Temperature (F)')
183  ylabel('Delta (F)')
184  zlabel('HVAC Value (W)')
185  colorbar
186  set(gca,'ydir','reverse')
187
188  tSet = 297;
189  tI   = tSet+5;
190  Q = zeros(size(t));
191  for k = 1:length(t)
192    Q(k) = HVACFuzzyController('update',t_K(k),tSet+5,tSet,d);
193  end
194
195  PlotSet(t_K,Q,'x label','T_e (K)','y label','Q (W)','plot title','Fuzzy
         HVAC');
196  y = get(gca,'ylim');
197  line(tSet*[1;1],y,'color','r')
198  line(tI*[1;1],y,'color','g')
199
200  tE = 280;
201  tSet = 294;
202  Q = zeros(size(delta));
203  for k = 1:length(delta)
204    Q(k) = HVACFuzzyController('update',tE,tSet+delta(k)*5/9,tSet,d);
205  end
```

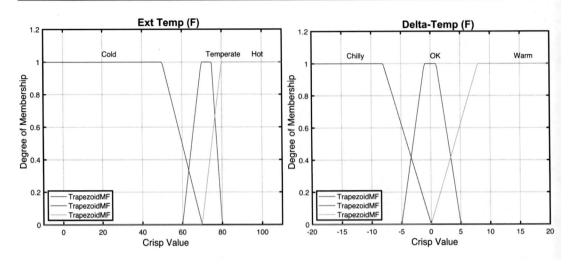

Figure 6.13: *The fuzzy inputs of the variable controller*

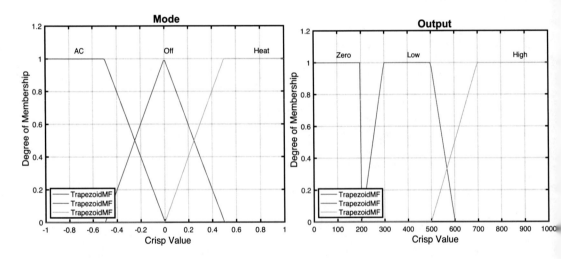

Figure 6.14: *The fuzzy outputs of the variable controller*

```
206   PlotSet(delta,Q,'x label','\Delta T (F)','y label','Q (W)',...
207      'figure title','Fuzzy HVAC','plot title',sprintf('Te = %g F',(tE-273)
             *9/5+32));
```

Here is the rule base:

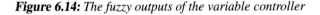

1. if Ext Temp (F) is Cold FuzzyAND Delta-Temp (F) is Chilly then   Mode is Heat   Output is High
2. if Ext Temp (F) is Temperate FuzzyAND Delta-Temp (F) is Chilly then Mode is Heat   Output is Low
3. if Ext Temp (F) is Hot FuzzyAND Delta-Temp (F) is Chilly then   Mode is Off   Output is Zero
4. if Ext Temp (F) is Cold FuzzyAND Delta-Temp (F) is OK then   Mode is Off   Output is Zero
5. if Ext Temp (F) is Temperate FuzzyAND Delta-Temp (F) is OK then   Mode is Off   Output is Zero
6. if Ext Temp (F) is Hot FuzzyAND Delta-Temp (F) is OK then   Mode is Off   Output is Zero

186

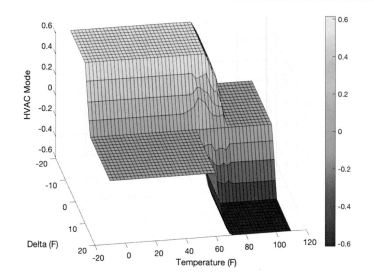

Figure 6.15: *The mode output of the variable controller following the reults*

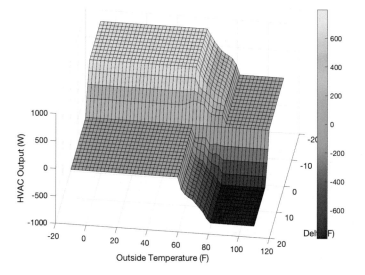

Figure 6.16: *The HVAC output of the variable controller*

```
7. if Ext Temp (F) is Cold FuzzyAND Delta-Temp (F) is Warm then        Mode is Off     Output is Zero

8. if Ext Temp (F) is Temperate FuzzyAND Delta-Temp (F) is Warm then   Mode is AC      Output is Low
9. if Ext Temp (F) is Hot FuzzyAND Delta-Temp (F) is Warm then  Mode is AC      Output is High
```

Finally, we try this version of the controller in the simulation. The plots in Figures 6.18 and 6.19 show results for both AC and heat. This updated fuzzy controller produces smooth results. Compare this to Figure 6.9 for the non-fuzzy bang-bang controller which produced limit cycling of the internal temperature.

Additional work for this system could include adding another input for the humidity. The system would need to be matched with the capabilities of the actual HVAC system.

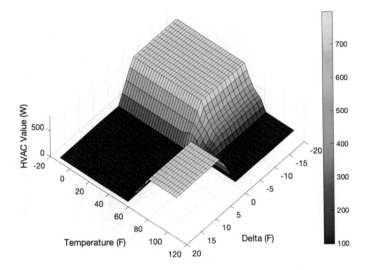

Figure 6.17: *The resulting combined AC or heat setting*

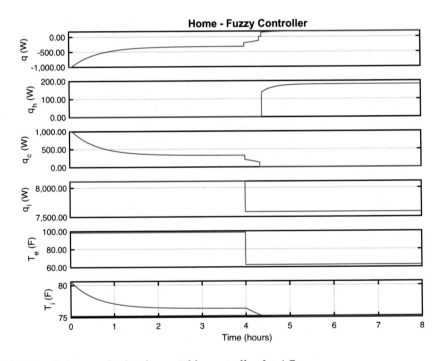

Figure 6.18: *Simulation results for the variable controller for AC*

188

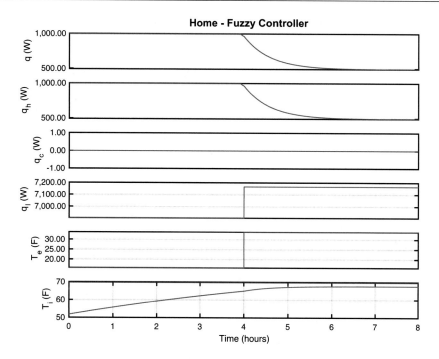

Figure 6.19: *Simulation results for the variable controller for heat*

6.6   Summary

This chapter demonstrated fuzzy logic. A windshield wiper demonstration gives an example of how it is used. The smart wiper system automatically adjusts wiper speed and wiper interval. A second system demonstrates a fuzzy HVAC system. Table 6.2 lists the functions and scripts included in the companion code. Fuzzy helper functions are grouped in Table 6.3.

Table 6.2: *Chapter code listing*

File	Description
BuildFuzzySystem	Builds a fuzzy logic system (data structure) using parameter pairs
SmartWipersSystem	Creates and returns the smart wipers data structure
SmartWipersDemo	Demonstrates a fuzzy logic control system for windshield wipers
FuzzyPlot	Plots a fuzzy set
FuzzyInference	Performs fuzzy inference given a fuzzy system and crisp data x
FuzzyRand	Creates a random set of inputs from a fuzzy system
HVACSim	Heating ventilation and air conditioning simulation
HVACSimplestFuzzyController	Discrete output simplest rule controller
HVACFuzzyController	Multi-input and output system fuzzy logic control system for HVAC
HVACFuzzyPlot	Plots the HVAC fuzzy controller rule base
PrintFuzzyRules	Prints fuzzy rules in a system struct to the command line

Table 6.3: *Fuzzy helper function listing*

CentroidDF	Centroid defuzzification
GeneralBellMF	General Bell membership function
GaussianMF	Gaussian membership function
TriangleMF	Triangle membership function
TrapezoidMF	Trapezoid membership function
SigmoidalMF	Displays a neural net with multiple layers
ClipIMP	Clip implication function
ScaleIMP	Scale implication function
FuzzyOR	Fuzzy OR (maximum of membership values)
FuzzAND	Fuzzy AND (minimum of membership values)

CHAPTER 7

■ ■ ■

Neural Aircraft Control

Longitudinal control is the control about the pitch axis of an aircraft, it needs to work at all altitudes and speeds. In this chapter, we will implement a neural net to produce the critical parameters for a nonlinear aircraft control system. This is an example of online learning and applies techniques from multiple previous chapters.

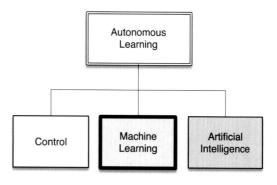

The longitudinal dynamics of an aircraft are also known as the pitch dynamics. The dynamics are entirely in the plane of symmetry of the aircraft. The plane of symmetry is defined as a plane that cuts the aircraft in half vertically. Most airplanes are symmetric about this plane. These dynamics include the forward and vertical motion of the aircraft and the pitching of the aircraft about the axis perpendicular to the plane of symmetry. Figure 7.1 shows an aircraft in flight. α is the angle of attack, the angle between the wing and the velocity vector. We assume that the wind direction is opposite that of the velocity vector, that is, the aircraft produces all of its wind. Drag is along the wind direction, and lift is perpendicular to drag. The pitch moment is around the center of mass. The model we will derive uses a small set of parameters, yet reproduces the longitudinal dynamics reasonably well. It is also easy for you to modify to simulate any aircraft of interest.

7.1   Longitudinal Motion

The next few recipes will involve the longitudinal control of an aircraft with a neural net to provide learning. We will

1. Model the aircraft dynamics

2. Find an equilibrium solution about which we will control the aircraft

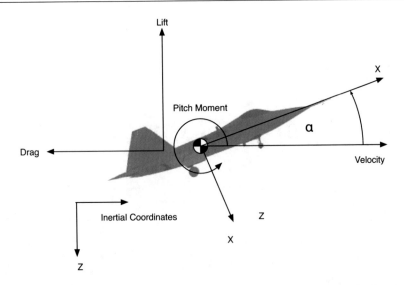

Figure 7.1: *Diagram of an aircraft in flight showing all the important quantities for longitudinal dynamics simulation*

3. Learn how to write a sigma-pi neural net

4. Implement the Proportional Integral Differential (PID) control

5. Implement the neural net

6. Simulate the system

In this recipe, we will model the longitudinal dynamics of an aircraft for use in learning control. We will derive a simple longitudinal dynamics model with a "small" number of parameters. Our control will use nonlinear dynamic inversion with a Proportional Integral Differential (PID) controller to control the pitch dynamics [19, 20]. Learning will be done using a sigma-pi neural network.

We will use the learning approach developed at NASA's Dryden Flight Research Center [35]. The baseline controller is a dynamic inversion type controller with a PID control law. A neutral net [17] provides learning while the aircraft is operating. The neutral network is a sigma-pi type network, meaning that the network sums the products of the inputs with their associated weights. The weights of the neural network are determined by a training algorithm that uses

1. Commanded aircraft rates from the reference model

2. PID errors

3. Adaptive control rates fed back from the neural network

7.1.1 Problem

We want to model the longitudinal dynamics of an aircraft.

7.1.2 Solution

The solution is to write the right-hand-side function for the aircraft longitudinal dynamics differential equations.

7.1.3 How It Works

We summarized the symbols for the dynamical model in Table 7.1. Our aerodynamic model is very simple. The lift and drag are

$$L = pSC_L \tag{7.1}$$
$$D = pSC_D \tag{7.2}$$

where S is the wetted area, the area that interacts with the airflow, and is the area that is counted in computing the aerodynamic forces, and p is the dynamic pressure, the pressure on the aircraft caused by its velocity:

$$p = \frac{1}{2}\rho v^2 \tag{7.3}$$

where ρ is the atmospheric density and v is the magnitude of the velocity. Atmospheric density is a function of altitude. For low-speed flight, this is mostly the wings. Most books use q for dynamic pressure. We use q for pitch angular rate (also a convention), so we use p for pressure here to avoid confusion.

The lift coefficient, C_L, is

$$C_L = C_{L_\alpha}\alpha \tag{7.4}$$

and the drag coefficient, C_D, is

$$C_D = C_{D_0} + kC_L^2 \tag{7.5}$$

The drag equation is called the drag polar. Increasing the angle of attack increases the aircraft's lift but also increases the aircraft's drag. The coefficient k is

$$k = \frac{1}{\pi \epsilon_0 AR} \tag{7.6}$$

where ϵ_0 is the Oswald Efficiency Factor that is typically between 0.75 and 0.85. AR is the wing aspect ratio. The aspect ratio is the ratio of the span of the wing to its chord. For complex shapes, it is approximately given by the formula

$$AR = \frac{b^2}{S} \tag{7.7}$$

Table 7.1: *Aircraft Dynamics Symbols*

Symbol	Description	Units
g	Acceleration of gravity at sea level	$9.806 \, \text{m/s}^2$
h	Altitude	m
k	Coefficient of lift-induced drag	
m	Mass	kg
p	Dynamic pressure	N/m^2
q	Pitch angular rate	rad/s
u	x-velocity	m/s
w	z-velocity	m/s
C_L	Lift coefficient	
C_D	Drag coefficient	
D	Drag	N
I_y	Pitch moment of inertia	kg-m^2
L	Lift	N
M	Pitch moment (torque)	Nm
M_e	Pitch moment due to elevator	Nm
r_e	Elevator moment arm	m
S	Wetted area of wings (the area that contributes to lift and drag)	m^2
S_e	Wetted area of elevator	m^2
T	Thrust	N
X	X force in the aircraft frame	N
Z	Z force in the aircraft frame	N
α	Angle of attack	rad
δ	Elevator angle	rad
γ	Flight path angle	rad
ρ	Air density	kg/m^3
θ	Pitch angle	rad

where b is the span and S is the wing area. Span is measured from wingtip to wingtip. Gliders have very high aspect ratios, and delta-wing aircraft have low aspect ratios.

The aerodynamic coefficients are nondimensional coefficients that when multiplied by the wetted area of the aircraft, and the dynamic pressure, produce the aerodynamic forces.

The dynamical equations, the differential equations of motion, are [1]

$$m(\dot{u} + qw) = X - mg\sin\theta + T\cos\epsilon \tag{7.8}$$

$$m(\dot{w} - qu) = Z + mg\cos\theta - T\sin\epsilon \tag{7.9}$$

$$I_y\dot{q} = M \tag{7.10}$$

$$\dot{\theta} = q \tag{7.11}$$

m is the mass, u is the x-velocity, w is the z-velocity, q is the pitch angular rate, θ is the pitch angle, T is the engine thrust, ϵ is the angle between the thrust vector and the x-axis, I_y is the pitch inertia, X is the x-force, Z is the z-force, and M is the torque about the pitch axis. The coupling between x and z velocities is due to writing the force equations in the rotating frame. The pitch equation is about the center of mass. These are a function of u, w, q, and altitude, h, which is found from

$$\dot{h} = u \sin \theta - w \cos \theta \tag{7.12}$$

The angle of attack, α, is the angle between the u and w velocities and is

$$\tan \alpha = \frac{w}{u} \tag{7.13}$$

The flight path angle γ is the angle between the vector velocity direction and the horizontal. It is related to θ and α by the relationship

$$\gamma = \theta - \alpha \tag{7.14}$$

The flight path angle does not appear in the equations, but it is useful to compute when studying aircraft motion. The forces are

$$X = L \sin \alpha - D \cos \alpha \tag{7.15}$$
$$Z = -L \cos \alpha - D \sin \alpha \tag{7.16}$$

The moment, or torque, is assumed due to the offset of the center of pressure and center of mass which is assumed to be along the x-axis:

$$M = (c_p - c)Z \tag{7.17}$$

where c_p is the location of the center of pressure. The moment due to the elevator is

$$M_e = q r_e S_e \sin(\delta) \tag{7.18}$$

S_e is the wetted area of the elevator, and r_E is the distance from the center of mass to the elevator. The dynamical model is in RHSAircraft. The atmospheric density model is an exponential model and is included as a subfunction in this function. RHSAircraft returns the default data structure if no inputs are given.

RHSAircraft.m

```
1   %% RHSAIRCRAFT Right hand side of an aircraft dynamical model
38  function [xDot, lift, drag, pD] = RHSAircraft( ˜, x, d )
39
40  if( nargin < 1 )
41    xDot = DataStructure;
42    return
43  end
```

195

```
44
45   g       = 9.806;  % m/s^2
46
47   u       = x(1);
48   w       = x(2);
49   q       = x(3);
50   theta   = x(4);
51   h       = x(5);
52
53   rho     = AtmDensity( h );
54
55   alpha   = atan(w/u);
56   cA      = cos(alpha);
57   sA      = sin(alpha);
58
59   v       = sqrt(u^2 + w^2);
60   pD      = 0.5*rho*v^2; % Dynamic pressure
61
62   cL      = d.cLAlpha*alpha;
63   cD      = d.cD0 + d.k*cL^2;
64
65   drag    = pD*d.s*cD;
66   lift    = pD*d.s*cL;
67
68   x       =  lift*sA - drag*cA;
69   z       = -lift*cA - drag*sA;
70   m       =  d.c*z + pD*d.sE*d.rE*sin(d.delta);
71
72   sT      = sin(theta);
73   cT      = cos(theta);
74
75   tEng    = d.thrust*d.throttle;
76   cE      = cos(d.epsilon);
77   sE      = sin(d.epsilon);
78
79   uDot    = (x + tEng*cE)/d.mass - q*w - g*sT + d.externalAccel(1);
80   wDot    = (z - tEng*sE)/d.mass + q*u + g*cT + d.externalAccel(2);
81   qDot    = m/d.inertia                        + d.externalAccel(3);
82   hDot    = u*sT - w*cT;
83
84   xDot    = [uDot;wDot;qDot;q;hDot];
```

We will use a model of the F-16 aircraft for our simulation. The F-16 is a single-engine supersonic multi-role combat aircraft used by many countries. The F-16 is shown in Figure 7.2.

196

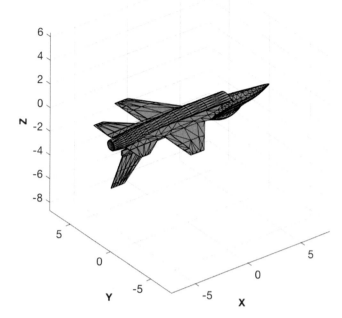

Figure 7.2: *F-16 model*

The inertia matrix is found by taking this model, distributing the mass among all the vertices, and computing the inertia from the formulas:

$$m_k = \frac{m}{N} \tag{7.19}$$

$$c = \sum_k m_k r_k \tag{7.20}$$

$$I = \sum_k m_k (r_k - c)^2 \tag{7.21}$$

where N is the number of nodes and r_k is the vector from the origin (which is arbitrary) to node k.

```
inr =

   1.0e+05 *

    0.3672     0.0002    -0.0604
    0.0002     1.4778     0.0000
   -0.0604     0.0000     1.7295
```

Table 7.2: *F-16 Data*

Symbol	Field	Value	Description	Units
C_{L_α}	cLAlpha	6.28	Lift coefficient	
C_{D_0}	cD0	0.0175	Zero lift drag coefficient	
k	k	0.1288	Lift coupling coefficient	
ϵ	epsilon	0	Thrust angle from the x-axis	rad
T	thrust	76.3e3	Engine thrust	N
S	s	27.87	Wing area	m^2
m	mass	12000	Aircraft mass	kg
I_y	inertia	1.7295e5	z-axis inertia	kg-m^2
$c - c_p$	c	1	Offset of center of mass from the center of pressure	m
S_e	sE	3.5	Elevator area	m^2
r_e	rE	4.0	Elevator moment arm	m

The F-16 data is given in Table 7.2.

There are many limitations to this model. First of all, the thrust is applied immediately with 100% accuracy. The thrust is also not a function of airspeed or altitude. Real engines take some time to achieve the commanded thrust, and the thrust levels change with airspeed and altitude. In the model, the elevator also responds instantaneously. Elevators are driven by motors, usually hydraulic but sometimes pure electric, and they take time to reach a commanded angle. In our model, the aerodynamics are very simple. In reality, lift and drag are complex functions of airspeed and angle of attack and are usually modeled with large tables of coefficients. We also model the pitching moment by a moment arm. Usually, the torque is modeled by a table. No aerodynamic damping is modeled, though this appears in most complete aerodynamic models for aircraft. You can easily add these features by creating functions:

```
C_L = CL(v,h,alpha,delta)
C_D = CD(v,h,alpha,delta)
C_M =CL(v,h,vdot,alpha,delta)
```

7.2   Numerically Finding Equilibrium

7.2.1 Problem

We want to determine the equilibrium state for the aircraft. This is the orientation at which all forces and torques balance.

7.2.2 Solution

The solution is to compute the Jacobian for the dynamics. The Jacobian is a matrix of all first-order partial derivatives of a vector-valued function, in this case, the dynamics of the aircraft.

7.2.3 How It Works

We want to start every simulation from an equilibrium state. This is done using the function EquilibriumState. It uses fminsearch to minimize

$$\dot{u}^2 + \dot{w}^2 \tag{7.22}$$

given the flight speed, altitude, and flight path angle. It then computes the elevator angle needed to zero the pitch angular acceleration. It has a built-in demo for equilibrium-level flight at 10 km.

EquilibriumState.m

```
39  function [x, thrust, delta, cost] = EquilibriumState( gamma, v, h, d )
40
41  if( nargin < 1 )
42     Demo;
43     return
44  end
45
46  x           = [v;0;0;0;h];
47  [~,~,drag]  = RHSAircraft( 0, x, d );
48  y0          = [0;drag];
49  cost(1)     = RHS( y0, d, gamma, v, h );
50  y           = fminsearch( @RHS, y0, [], d, gamma, v, h );
51  w           = y(1);
52  thrust      = y(2);
53  u           = sqrt(v^2-w^2);
54  alpha       = atan(w/u);
55  theta       = gamma + alpha;
56  cost(2)     = RHS( y, d, gamma, v, h );
57  x           = [u;w;0;theta;h];
58  d.thrust    = thrust;
59  d.delta     = 0;
60  [xDot,~,~,p] = RHSAircraft( 0, x, d );
61  delta       = -asin(d.inertia*xDot(3)/(d.rE*d.sE*p));
62  d.delta     = delta;
```

CostFun is the cost functional given as follows:

EquilibriumState.m

```
75  function cost = RHS( y, d, gamma, v, h )
76  %% EquilibriumState>RHS
77  % Cost function for fminsearch. The cost is the square of the velocity
78  % derivatives (the first two terms of xDot from RHSAircraft).
79  %
80  % See also RHSAircraft.
```

199

```
81
82   w            = y(1);
83   d.thrust     = y(2);
84   d.delta      = 0;
85   u            = sqrt(v^2-w^2);
86   alpha        = atan(w/u);
87   theta        = gamma + alpha;
88   x            = [u;w;0;theta;h];
89   xDot         = RHSAircraft( 0, x, d );
90   cost         = xDot(1:2)'*xDot(1:2);
```

The vector of values is the first input. Our first guess is that thrust equals drag. The vertical velocity and thrust are solved by fminsearch. fminsearch searches over thrust and vertical velocity to find an equilibrium state.

The results of the demo are

```
>> EquilibriumState
Velocity            250.00 m/s
Altitude          10000.00 m
Flight path angle     0.00 deg
Z speed              13.84 m/s
Thrust            11148.95 N
The angle of attack     3.17 deg
Elevator            -11.22 deg
Initial cost      9.62e+01
Final cost        1.17e-17
```

The initial and final costs show how successful fminsearch was in achieving the objective of minimizing the w and u accelerations.

7.3   Numerical Simulation of the Aircraft

7.3.1 Problem

We want to simulate the aircraft.

7.3.2 Solution

The solution is to create a script that calls the right-hand side of the dynamical equations, RHSAircraft, in a loop and plots the results.

7.3.3 How It Works

The simulation script is shown as follows. It computes the equilibrium state and then simulates the dynamics in a loop by calling `RungeKutta`. It applies a disturbance to the aircraft. It then uses `PlotSet` to plot the results.

AircraftSimOpenLoop.m

```
6   %% Initialize
7   nSim     = 2000;      % Number of time steps
8   dT       = 0.1;       % Time step (sec)
9   dRHS     = RHSAircraft;  % Get the default data structure has F-16 data
10  h        = 10000;
11  gamma    = 0.01;
12  v        = 250;
13  nPulse   = 10;
14  [x,  dRHS.thrust, dRHS.delta, cost] = EquilibriumState( gamma, v, h,
        dRHS );
15  fprintf(1,'Finding Equilibrium: Starting Cost %12.4e Final Cost %12.4e\
        n',cost);
16
17  accel = [0.0;0.1;0.0];
18
19  %% Simulation
20  xPlot = zeros(length(x)+2,nSim);
21  for k = 1:nSim
22          % Plot storage
23    [~,L,D]      = RHSAircraft( 0, x, dRHS );
24          xPlot(:,k)   = [x;L;D];
25    % Propagate (numerically integrate) the state equations
26    if( k > nPulse )
27      dRHS.externalAccel = [0;0;0];
28    else
29      dRHS.externalAccel = accel;
30    end
31    x = RungeKutta( @RHSAircraft, 0, x, dT, dRHS );
32    if( x(5) <= 0 )
33      break;
34    end
35  end
```

The applied external acceleration puts the aircraft into a slight climb with some noticeable oscillations:

```
>> AircraftSimOpenLoop
Velocity              250.00 m/s
Altitude            10000.00 m
Flight path angle       0.57 deg
Z speed                13.83 m/s
Thrust              12321.13 N
The angle of attack     3.17 deg
```

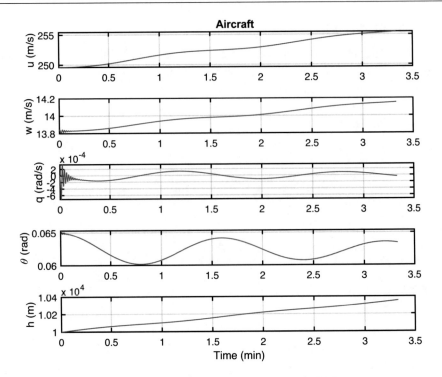

Figure 7.3: *Open-loop response to a pulse for the F-16 in a shallow climb*

```
Elevator              11.22 deg
Initial cost      9.62e+01
Final cost        5.66e-17
Finding Equilibrium: Starting Cost   9.6158e+01 Final Cost   5.6645e-17
```

The simulation results are shown in Figure 7.3. The aircraft climbs steadily. Two oscillations are seen. A high-frequency one is primarily associated with pitch and a low-frequency one with the velocity of the aircraft.

7.4   Activation Function

7.4.1 Problem

We are going to implement a neural net so that our aircraft control system can learn. We need an activation function to scale and limit measurements.

7.4.2 Solution

Use a sigmoid function as our activation function.

7.4.3 How It Works

The neural net uses the following sigmoid function:

$$g(x) = \frac{1 - e^{-kx}}{1 + e^{-kx}} \tag{7.23}$$

The sigmoid function with $k = 1$ is plotted in the following script.

Sigmoid.m

```
5  %% Initialize
6  x = linspace(-7,7);
7
8  %% Sigmoid
9  s = (1-exp(-x))./(1+exp(-x));
10
11  PlotSet( x, s, 'x label', 'x', 'y label', 's',...
12    'plot title', 'Sigmoid', 'figure title', 'Sigmoid' );
```

Results are shown in Figure 7.4.

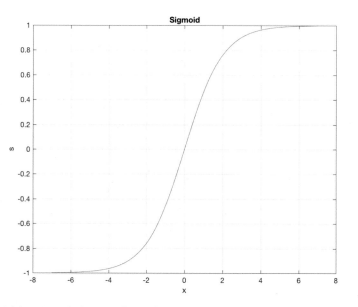

Figure 7.4: *Sigmoid function. At large values of x, the sigmoid function returns ± 1*

7.5   Neural Net for Learning Control

7.5.1 Problem

We want to use a neural net to add learning to the aircraft control system.

7.5.2 Solution

Use a sigma-pi neural net function. A sigma-pi neural net sums the inputs and products of the inputs to produce a model.

7.5.3 How It Works

The adaptive neural network for the pitch axis has seven inputs. The output of the neural network is a pitch angular acceleration that augments the control signal coming from the dynamic inversion controller. The control system is shown in Figure 7.5. The leftmost box produces the reference model given the pilot input. The output of the reference model is a vector of the desired states that are differenced from the true states and fed to the PID controller and the neural network. The output of the PID is differenced with the output of the neural network. This is fed into the model inversion block that drives the aircraft dynamics.

The sigma-pi neutral net is shown in Figure 7.6 for a two-input system. The output is

$$y = w_1 c + w_2 x_1 + w_3 x_2 + w_4 x_1 x_2 \tag{7.24}$$

The weights are selected to represent a nonlinear function. For example, suppose we want to represent the dynamic pressure:

$$y = \frac{1}{2} \rho v^2 \tag{7.25}$$

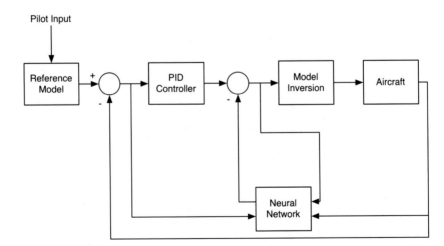

Figure 7.5: *Aircraft control system. It combines a PID controller with dynamic inversion to handle nonlinearities. A neural net provides learning*

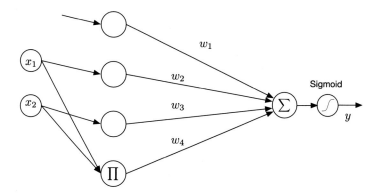

Figure 7.6: *Sigma-pi neural net.* Π *stands for product and* Σ *stands for sum*

We let $x_1 = \rho$ and $x_2 = v^2$. Set $w_4 = \frac{1}{2}$ and all other weights to zero. Suppose we didn't know the constant $\frac{1}{2}$. We would like our neural net to determine the weight through measurements. Learning for a neural net means determining the weights so that our net replicates the function it is modeling. Define the vector z which is the result of the product operations. In our two input cases, this would be

$$z = \begin{bmatrix} c \\ x_1 \\ x_2 \\ x_1 x_2 \end{bmatrix} \qquad (7.26)$$

c is a constant. The output is

$$y = w^T z \qquad (7.27)$$

We could assemble multiple inputs and outputs:

$$\begin{bmatrix} y_1 & y_2 & \cdots \end{bmatrix} = w^T \begin{bmatrix} z_1 & z_2 & \cdots \end{bmatrix} \qquad (7.28)$$

where z_k is a column array. We can solve for the weights w using least squares given the outputs y and inputs x. Define the vector of y to be Y and the matrix of z to be Z. The solution for w is

$$Y = Z^T w \qquad (7.29)$$

The least squares solution is

$$w = \left(Z Z^T \right)^{-1} Z Y^T \qquad (7.30)$$

This gives the best fit to w for the measurements Y and inputs Z. Suppose we take another measurement. We would then repeat this with bigger matrices. As a side note, you would compute this using an inverse. There are better numerical methods for doing least squares. MATLAB has the `pinv` function. For example:

```
>> z = rand(4,4);
>> w = rand(4,1);
>> y = w'*z;
>> wL = inv(z*z')*z*y'
wL =
     0.8308
     0.5853
     0.5497
     0.9172
>> w
w =
     0.8308
     0.5853
     0.5497
     0.9172

>> pinv(z')*y'
ans =
     0.8308
     0.5853
     0.5497
     0.9172
```

As you can see, they all agree! This is a good way to initially train your neural net. Collect as many measurements as you have values of z and compute the weights. Your net is then ready to go.

The recursive approach is to initialize the recursive trainer with n values of z and y:

$$p = \left(ZZ^T\right)^{-1} \tag{7.31}$$

$$w = pZY \tag{7.32}$$

The recursive learning algorithm is

$$p = p - \frac{pzz^Tp}{1+z^Tpz} \tag{7.33}$$

$$k = pz \tag{7.34}$$

$$w = w + k\left(y - z^Tw\right) \tag{7.35}$$

`RecursiveLearning` demonstrates recursive learning or training. It starts with an initial estimate based on a four-element training set. It then recursively learns based on new data.

RecursiveLearning.m

```
1   %% Test a recursive learning system
2
3   w    = rand(4,1); % Initial guess
4   Z    = randn(4,4);
5   Y    = Z'*w;
6
7   wN   = w + 0.1*randn(4,1); % True weights are a little different
8   n    = 100;         % Number of measurements
9   zA   = randn(4,n); % Random inputs
10  y    = wN'*zA;      % New measurements
11
12  % Batch training
13  p    = inv(Z*Z'); % Initial value
14  w    = p*Z*Y; % Initial value
15
16  %% Recursive learning
17  dW = zeros(4,n);
18  for j = 1:n
19      z        = zA(:,j);
20      p        = p - p*(z*z')*p/(1+z'*p*z);
21      w        = w + p*z*(y(j) - z'*w);
22      dW(:,j) = w - wN; % Store for plotting
23  end
24
25  %% Plot the results
26  yL = cell(1,4);
27  for j = 1:4
28      yL{j} = sprintf('\\Delta W_%d',j);
29  end
30
31  PlotSet(1:n,dW,'x label','Sample','y label',yL,...
32          'plot title','Recursive Training',...
33          'figure title','Recursive Training');
```

Figure 7.7 shows the results. After an initial transient, the learning converges. Every time you run this, you will get different answers because we initialize with random values. However, the error magnitude after the n inputs are processed will be similar because the standard deviation of the random offset from the truth values is the same.

You will notice that the recursive learning algorithm is identical in form to the Kalman Filter given in Section 4.2.3. Our learning algorithm was derived from batch least squares, which is an alternative derivation for the Kalman Filter.

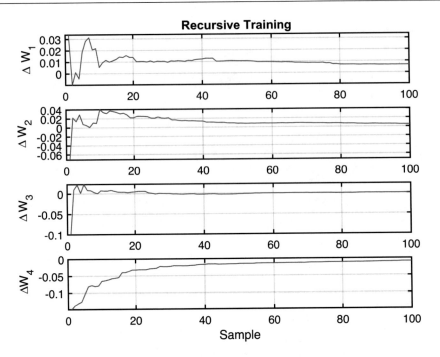

Figure 7.7: *Recursive training or learning. After an initial transient, the weights converge quickly*

7.6   Enumeration of All Sets of Inputs

7.6.1 Problem

One issue with a sigma-pi neural network is the number of possible nodes. For design purposes, we need a function to enumerate all possible sets of combinations of inputs. This reduces the complexity of a sigma-pi neural network.

7.6.2 Solution

Write a combination function that computes the number of sets.

7.6.3 How It Works

In our sigma-pi network, we hand-coded the products of the inputs. For more general code, we want to enumerate all combinations of inputs. If we have n inputs and want to take them k at a time, the number of sets is

$$\frac{n!}{(n-k)!k!} \tag{7.36}$$

The code to enumerate all sets is in the function `Combinations`.

Combinations.m

```matlab
14  function c = Combinations( r, k )
15
16  % Demo
17  if( nargin < 1 )
18    Combinations(1:4,3)
19    return
20  end
21
22  % Special cases
23  if( k == 1 )
24    c = r';
25    return
26  elseif( k == length(r) )
27    c = r;
28    return
29  end
30
31  % Recursion
32  rJ      = r(2:end);
33  c     = [];
34  if( length(rJ) > 1 )
35    for j = 2:length(r)-k+1
36      rJ              = r(j:end);
37      nC              = NumberOfCombinations(length(rJ),k-1);
38      cJ              = zeros(nC,k);
39      cJ(:,2:end)     = Combinations(rJ,k-1);
40      cJ(:,1)         = r(j-1);
41      if( ~isempty(c) )
42        c = [c;cJ];
43      else
44        c = cJ;
45      end
46    end
47  else
48    c = rJ;
49  end
50  c = [c;r(end-k+1:end)];
51
52  %% Combinations>NumberOfCombinations
53  function j = NumberOfCombinations(n,k)
54  % Compute the number of combinations
55  j = factorial(n)/(factorial(n-k)*factorial(k));
```

This handles two special cases on input and then calls itself recursively for all other cases. Here are some examples:

```
>> Combinations(1:4,3)
ans =
     1     2     3
     1     2     4
```

```
           1       3       4
           2       3       4
   >> Combinations(1:4,2)
  ans =
           1       2
           1       3
           1       4
           2       3
           2       4
           3       4
```

You can see that if we have 4 inputs and want all possible combinations, we end up with 14 total! This indicates a practical limit to a sigma-pi neural network as the number of weights will grow fast as the number of inputs increases.

7.7   Write a Sigma-Pi Neural Net Function

7.7.1 Problem

We need a sigma-pi net function for general problems.

7.7.2 Solution

Create an action-based sigma-pi function. This will use a generalized data structure format.

7.7.3 How It Works

The following code shows how we implement the sigma-pi neural net. `SigmaPiNeuralNet` has `action` as its first input. You use this to access the functionality of the function. Actions are

1. "Initialize": Initialize the function

2. "Set constant": Set the constant term

3. "Batch learning": Perform batch learning

4. "Recursive learning": Perform recursive learning

5. "Output": Generate outputs without training

You usually go in this order when running the function. Setting the constant is not needed if the default of one is fine. The functionality is distributed among subfunctions called from a `switch` statement.

SigmaPiNeuralNet.m

```
46  function [y, d] = SigmaPiNeuralNet( action, x, d )
47
48  % Demo or default data structure
```

210

```
49  if( nargin < 1 )
50    if( nargout == 1)
51      y = DefaultDataStructure;
52    else
53      Demo;
54    end
55    return
56  end
57
58  switch lower(action)
59    case 'initialize'
60      d   = CreateZIndices( x, d );
61      d.w = zeros(size(d.zI,1)+1,1);
62      y   = [];
63
64    case 'set constant'
65      d.c = x;
66      y   = [];
67
68    case 'batch learning'
69      [y, d] = BatchLearning( x, d );
70
71    case 'recursive learning'
72      [y, d] = RecursiveLearning( x, d );
73
74    case 'output'
75      [y, d] = NNOutput( x, d );
76
77    otherwise
78      error('%s is not an available action',action );
79  end
```

The demo shows an example of using the function to model dynamic pressure. Our inputs are the altitude and the square of the velocity. The neutral net will try to fit

$$y = w_1 c + w_2 h + w_3 v^2 + w_4 h v^2 \tag{7.37}$$

to

$$y = 0.6125 e^{-0.0817 h.^{1.15}} v^2 \tag{7.38}$$

We first get the default data structure. Then we initialize the filter with an empty x. We then get the initial weights by using batch learning. The number of columns of x should be at least twice the number of inputs. This gives a starting p matrix and an initial estimate of weights. We then perform recursive learning. The field kSigmoid must be small enough so that valid inputs are in the linear region of the sigmoid function. Note that this can be an array so that you can use different scalings on different inputs.

211

SigmaPiNeuralNet.m

```
176  function Demo
177  %% SigmaPiNeuralNet>Demo
178  % Demonstrate a sigma-pi neural net for dynamic pressure
179  x         = zeros(2,1);
180
181  d         = SigmaPiNeuralNet;
182  [~, d]    = SigmaPiNeuralNet( 'initialize', x, d );
183
184  h         = linspace(10,10000);
185  v         = linspace(10,400);
186  v2        = v.^2;
187  q         = 0.5*AtmDensity(h).*v2;
188
189  n         = 5;
190  x         = [h(1:n);v2(1:n)];
191  d.y       = q(1:n)';
192  [y, d]    = SigmaPiNeuralNet( 'batch learning', x, d );
193
194  fprintf(1,'Batch Results\n#        Truth    Neural Net\n');
195  for k = 1:length(y)
196    fprintf(1,'%d: %12.2f %12.2f\n',k,q(k),y(k));
197  end
198
199  n = length(h);
200  y = zeros(1,n);
201  x = [h;v2];
202  for k = 1:n
203    d.y = q(k);
204    [y(k), d]  = SigmaPiNeuralNet( 'recursive learning', x(:,k), d );
205  end
```

The batch results are as follows for five examples of dynamic pressures at low altitudes. As you can see, the truth model and neural net outputs are quite close:

```
>> SigmaPiNeuralNet
Batch Results
#         Truth    Neural Net
1:        61.22        61.17
2:       118.24       118.42
3:       193.12       192.88
4:       285.38       285.52
5:       394.51       394.48
```

The recursive learning results are shown in Figure 7.8. The results are pretty good over a wide range of altitudes. You could then just use the "update" action during aircraft operation.

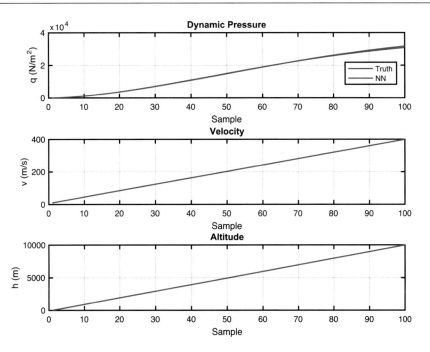

Figure 7.8: *Recursive training for the dynamic pressure example*

7.8 Implement PID Control

7.8.1 Problem

We want a Proportional Integral Differential controller to control the aircraft.

7.8.2 Solution

Write a function to implement PID control. The input will be the pitch angle error.

7.8.3 How It Works

Assume we have a double integrator driven by a constant input:

$$\ddot{x} = u \tag{7.39}$$

where $u = u_d + u_c$.

$$\ddot{x} = \frac{dx}{dt} \tag{7.40}$$

The result is

$$x = \frac{1}{2}ut^2 + x(0) + \dot{x}(0)t \tag{7.41}$$

213

The simplest control is to add a feedback controller:

$$u_c = -K\left(\tau_d \dot{x} + x\right) \tag{7.42}$$

where K is the forward gain and τ is the damping time constant. Our dynamical equation is now

$$\ddot{x} + K\left(\tau_d \dot{x} + x\right) = u_d \tag{7.43}$$

The damping term will cause the transients to die out. When that happens, the second and first derivatives of x are zero, and we end up with an offset:

$$x = \frac{u}{K} \tag{7.44}$$

This is generally not desirable. You could increase K until the offset was small, but that would mean your actuator would need to produce higher forces or torques. What we have at the moment is a PD controller or proportional derivative. Let's add another term to the controller:

$$u_c = -K\left(\tau_d \dot{x} + x + \frac{1}{\tau_i}\int x\right) \tag{7.45}$$

This is now a PID controller, or Proportional Integral Derivative controller. There is now a gain proportional to the integral of x. We add the new controller and then take another derivative to get

$$\dddot{x} + K\left(\tau_d \ddot{x} + \dot{x} + \frac{1}{\tau_i}x\right) = \dot{u}_d \tag{7.46}$$

Now in steady state

$$x = \frac{\tau_i}{K}\dot{u}_d \tag{7.47}$$

If u is constant, the offset is zero. Define s as the derivative operator:

$$s = \frac{d}{dt} \tag{7.48}$$

Then

$$s^3 x(s) + K\left(\tau_d s^2 x(s) + s x(s) + \frac{1}{\tau_i}x(s)\right) = s u_d(s) \tag{7.49}$$

Note that

$$\frac{u_c(s)}{x(s)} = K\left(1 + \tau_d s + \frac{1}{\tau_i s}\right) \tag{7.50}$$

where τ_d is the rate time constant which is how long the system will take to damp and τ_i is how fast the system will integrate out a steady disturbance.

The closed-loop transfer function is

$$\frac{x(s)}{u_d(s)} = \frac{s}{s^3 + K\tau_d s^2 + Ks + K/\tau_i} \tag{7.51}$$

where $s = j\omega$ and $j = \sqrt{-1}$.

The desired closed-loop transfer function is

$$\frac{x(s)}{u_d(s)} = \frac{s}{(s+\gamma)(s^2 + 2\zeta\sigma s + \sigma^2)} \tag{7.52}$$

or

$$\frac{x(s)}{u_d(s)} = \frac{s}{s^3 + (\gamma + 2\zeta\sigma)s^2 + \sigma(\sigma + 2\zeta\gamma)s + \gamma\sigma^2} \tag{7.53}$$

The parameters are

$$K = \sigma(\sigma + 2\zeta\gamma) \tag{7.54}$$

$$\tau_i = \frac{\sigma + 2\zeta\gamma}{\gamma\sigma} \tag{7.55}$$

$$\tau_d = \frac{\gamma + 2\zeta\sigma}{\sigma(\sigma + 2\zeta\gamma)} \tag{7.56}$$

This is a design for a PID. However, it is not possible to write this in the desired state space form:

$$\dot{x} = Ax + Au \tag{7.57}$$

$$y = Cx + Du \tag{7.58}$$

because it has a pure differentiator. We need to add a filter to the rate term so that it looks like

$$\frac{s}{\tau_r s + 1} \tag{7.59}$$

instead of s. We aren't going to derive the constants and will leave it as an exercise for the reader. The code for the PID is in PID.

PID.m

```matlab
47   function [a, b, c, d] = PID(  zeta, omega, tauInt, omegaR, tSamp )
48
49   % Demo
50   if( nargin < 1 )
51     Demo;
52     return
53   end
54
55   % Input processing
56   if( nargin < 4 )
57     omegaR = [];
58   end
59
60   % Default roll-off
61   if( isempty(omegaR) )
62     omegaR = 5*omega;
63   end
64
65   % Compute the PID gains
66   omegaI  = 2*pi/tauInt;
67
68   c2   = omegaI*omegaR;
69   c1   = omegaI+omegaR;
70   b1   = 2*zeta*omega;
71   b2   = omega^2;
72   g    = c1 + b1;
73   kI   = c2*b2/g;
74   kP   = (c1*b2 + b1.*c2  - kI)/g;
75   kR   = (c1*b1 + c2 + b2 - kP)/g;
76
77   % Compute the state space model
78   a    = [0 0;0 -g];
79   b    = [1;g];
80   c    = [kI -kR*g];
81   d    = kP + kR*g;
82
83   % Convert to discrete time
84   if( nargin > 4 )
85     [a,b] = CToDZOH(a,b,tSamp);
86   end
```

216

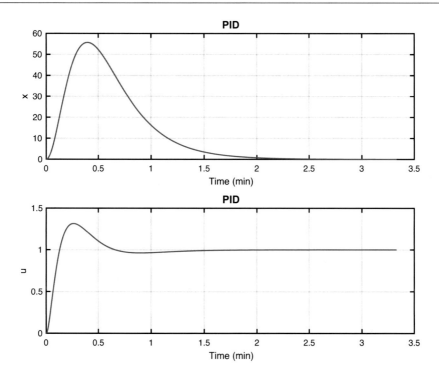

Figure 7.9: *PID control given a unit input*

It is interesting to evaluate the effect of the integrator. This is shown in Figure 7.9. The code is the demo in PID. Instead of numerically integrating the differential equations, we convert them into sampled time and propagate them. This is handy for linear equations. The double integrator equations are in the form

$$x_{k+1} = ax_k + bu_k \tag{7.60}$$

$$y = cx_k + du_k \tag{7.61}$$

This is the same form as the PID controller.

PID.m

```
96   % The double integrator plant
97   dT            = 0.1; % s
98   aP            = [0 1;0 0];
99   bP            = [0;1];
100  [aP, bP]      = CToDZOH( aP, bP, dT );
101
102  % Design the controller
103  [a, b, c, d]  = PID( 1, 0.1, 100, 0.5, dT );
104
105  % Run the simulation
106  n    = 2000;
107  p    = zeros(2,n);
```

```
108  x    = [0;0];
109  xC   = [0;0];
110
111  for k = 1:n
112    % PID Controller
113    y          = x(1);
114    xC         = a*xC + b*y;
115    uC         = c*xC + d*y;
116    p(:,k)     = [y;uC];
117    x          = aP*x + bP*(1-uC); % Unit step response
118  end
```

It takes about 2 minutes to drive x to zero, which is close to the 100 seconds specified for the integrator.

7.9   PID Control of Pitch

7.9.1 Problem

We want to control the pitch angle of an aircraft with a PID control.

7.9.2 Solution

Write a script to implement the controller with the PID controller and pitch dynamic inversion compensation.

7.9.3 How It Works

The PID controller changes the elevator angle to produce a pitch acceleration to rotate the aircraft. The elevator is the movable horizontal surface that is usually on the tail wing of an aircraft. In addition, additional elevator movement is needed to compensate for changes in the accelerations due to lift and drag as the aircraft changes its pitch orientation. This is done using the pitch dynamic inversion function. This returns the pitch acceleration that must be compensated for when applying the pitch control.

PitchDynamicInversion.m

```
30  function qDot = PitchDynamicInversion( x, d )
31
32  if( nargin < 1 )
33    qDot = DataStructure;
34    return
35  end
36
37  u    = x(1);
38  w    = x(2);
39  h    = x(5);
40
41  rho  = AtmDensity( h );
42
```

218

```
43   alpha = atan(w/u);
44   cA    = cos(alpha);
45   sA    = sin(alpha);
46
47   v     = sqrt(u^2 + w^2);
48   pD    = 0.5*rho*v^2; % Dynamic pressure
49
50   cL    = d.cLAlpha*alpha;
51   cD    = d.cD0 + d.k*cL^2;
52
53   drag  = pD*d.s*cD;
54   lift  = pD*d.s*cL;
55
56   z     = -lift*cA - drag*sA;
57   m     = d.c*z;
58   qDot  = m/d.inertia;
```

The closed-loop simulation incorporating the controls is AircraftSim. There is a flag to turn on the control and another to turn on the learning control. For this recipe, set addControl to true and addLearning to false. The simulation setup is shown as follows.

AircraftSim.m

```
 9   %% Options for control
10   addControl      = true;
11   addLearning     = true;
12
13   %% Initialize the simulation
14   nSim            = 1000;     % Number of time steps
15   dT              = 0.1;      % Time step (sec)
16   dRHS            = RHSAircraft;    % Get the default data structure has F
          -16 data
17   h               = 10000;
18   gamma           = 0.0;
19   v               = 250;
20   nPulse          = 10;
21   pitchDesired    = 0.2;
22   dL              = load('PitchNNWeights');
23   [x,  dRHS.thrust, deltaEq, cost] = EquilibriumState( gamma, v, h, dRHS
          );
24   fprintf(1,'Finding Equilibrium: Starting Cost %12.4e Final Cost %12.4e\
          n',cost);
25
26   if( addLearning )
27     temp   = load('DRHSL');
28     dRHSL = temp.dRHSL;
29     temp   = load('DNN');
30     dNN   = temp.d;
31   else
32     temp   = load('DRHSL');
33     dRHSL = temp.dRHSL;
```

219

```
34   end
35
36   accel = [0.0;0.0;0.0];
37
38   % Design the PID Controller
39   [aC, bC, cC, dC]   = PID(  1, 0.1, 100, 0.5, dT );
40   dRHS.delta         = deltaEq;
41   xDotEq             = RHSAircraft( 0, x, dRHS );
42   aEq                = xDotEq(3);
43   xC                 = [0;0];
```

The simulation loop is shown in the next listing. We don't show the plotting code.

AircraftSim.m

```
45   %% Simulation
46   xPlot = zeros(length(x)+8,nSim);
47   for k = 1:nSim
48
49     % Control
50         [~,L,D,pD]        = RHSAircraft( 0, x, dRHS );
51
52     % Measurement
53     pitch       = x(4);
54
55     % PID control
56     if( addControl )
57       pitchError  = pitch - pitchDesired;
58       xC          = aC*xC + bC*pitchError;
59       aDI         = PitchDynamicInversion( x, dRHSL );
60       aPID        = -(cC*xC + dC*pitchError);
61     else
62       pitchError  = 0;
63       aPID        = 0;
64     end
65
66     % Learning
67     if( addLearning )
68       xNN       = [x(4);x(1)^2 + x(2)^2];
69       aLearning = SigmaPiNeuralNet( 'output', xNN, dNN );
70     else
71       aLearning = 0;
72     end
73
74     if( addControl )
75       aTotal     = aPID - (aDI + aLearning);
76
77       % Convert acceleration to elevator angle
78       gain       = dRHS.inertia/(dRHS.rE*dRHS.sE*pD);
79       dRHS.delta = asin(gain*aTotal);
80     else
81       dRHS.delta = deltaEq;
```

```
82     end
83
84          % Plot storage
85          xPlot(:,k)   = [x;L;D;aPID;pitchError;dRHS.delta;aPID;aDI;
                 aLearning];
86
87     % Propagate (numerically integrate) the state equations
88     if( k > nPulse )
89       dRHS.externalAccel = [0;0;0];
90     else
91       dRHS.externalAccel = accel;
92     end
93     x        = RungeKutta( @RHSAircraft, 0, x, dT, dRHS );
94
95     % A crash
96     if( x(5) <= 0 )
97       break;
98     end
99   end
```

We command a 0.2-radian pitch angle using the PID control. The results are shown in Figure 7.10, Figure 7.11, Figure 7.12, and Figure 7.13. Note that the PID acceleration, a_{PID}, is much lower than the dynamic inversion acceleration, a_{DI}. The last plot in Figure 7.13, a_L, is for the learning acceleration, which is zero in this case.

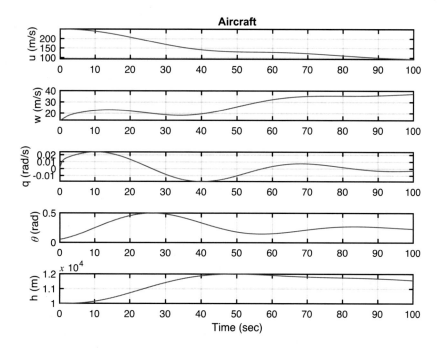

Figure 7.10: *Aircraft states during pitch angle change. The aircraft oscillates due to the pitch dynamics*

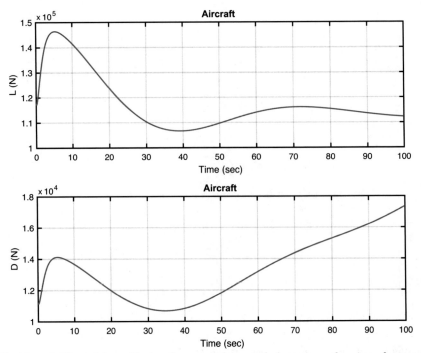

Figure 7.11: *Aircraft lift and drag. Notice the very substantial changes as the aircraft rotates*

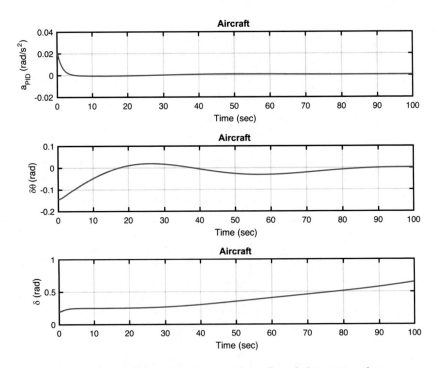

Figure 7.12: *Acceleration magnitude, change in aircraft pitch and elevator angle*

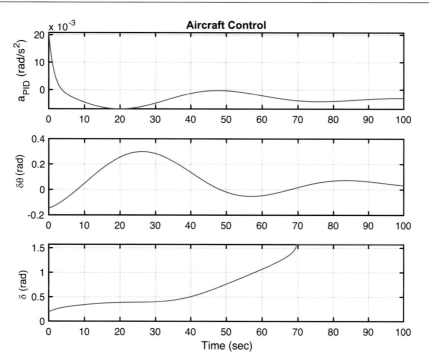

Figure 7.13: *Acceleration magnitude, PID vs. dynamic inversion (DI)*

The maneuver increases the drag, and we don't adjust the throttle to compensate. This will cause the airspeed to drop. In implementing the controller, we neglected to consider the coupling between states, but this can be added easily.

7.10 Neural Net for Pitch Dynamics

7.10.1 Problem

We want a nonlinear inversion controller with a PID controller and the sigma-pi neural net.

7.10.2 Solution

Train the neural net with a script that takes the angle and velocity squared input and computes the pitch acceleration error.

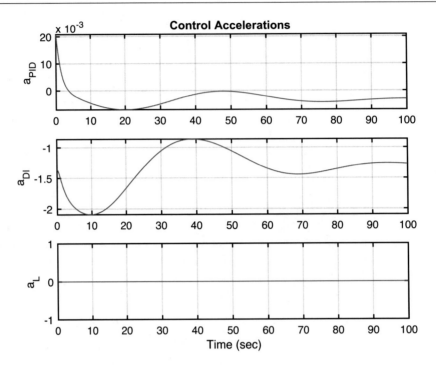

Figure 7.14: *Neural net fit to the delta acceleration*

7.10.3 How It Works

The `PitchNeuralNetTraining` script computes the pitch acceleration for a slightly different set of parameters. It then processes the delta acceleration. The script passes a range of pitch angles to the `PitchDynamicInversion` function and learns the acceleration. We use the velocity squared as an input because the dynamic pressure is proportional to the velocity squared. The base acceleration (in `dRHSL`) is for our "a priori" model. `dRHS` is the measured value. We assume that these are obtained during flight testing (Figure 7.14).

PitchNeuralNetTraining.m

```
11   % This is from flight testing
12   dRHS            = RHSAircraft;      % Get the default data (F-16 model)
13   h               = 10000;
14   gamma           = 0.0;
15   v               = 250;
16
17   % Get the equilibrium state
18   [x,  dRHS.thrust, deltaEq, cost] = EquilibriumState( gamma, v, h, dRHS
        );
19
20   % Angle of attack
21   alpha           = atan(x(2)/x(1));
22   cA              = cos(alpha);
```

```
23   sA             = sin(alpha);
24
25   % Create the assumed properties (truth)
26   dRHSL          = dRHS;
27   dRHSL.cD0      = 2.2*dRHS.cD0; % zero lift drag coefficient
28   dRHSL.k        = 1.0*dRHS.k;   % lift coupling with drag
29
30   % 2 inputs
31   xNN     = zeros(2,1);
32   d       = SigmaPiNeuralNet;
33   [~, d]  = SigmaPiNeuralNet( 'initialize', xNN, d );
35
36   theta   = linspace(0,pi/8);
37   v       = linspace(300,200);
38   n       = length(theta);
39   aT      = zeros(1,n);
40   aM      = zeros(1,n);
41
42   for k = 1:n
43     x(4)  = theta(k);
44     x(1)  = cA*v(k);
45     x(2)  = sA*v(k);
46     aT(k) = PitchDynamicInversion( x, dRHSL ); % truth
47     aM(k) = PitchDynamicInversion( x, dRHS  ); % model
48   end
49
50   % The delta pitch acceleration
51   dA  = aM - aT;
52
53   % Inputs to the neural net
54   v2   = v.^2;
55   xNN = [theta;v2];
56
57   % Outputs for training
58   d.y      = dA';
59   [aNN, d] = SigmaPiNeuralNet( 'batch learning', xNN, d );
60
61   % Save the data for the aircraft simulation
62   thisPath = fileparts(mfilename('fullpath'));
63   save( fullfile(thisPath,'DRHSL'),'dRHSL' );
64   save( fullfile(thisPath,'DNN'), 'd'   );
65
66   for j = 1:size(xNN,2)
67     aNN(j,:) = SigmaPiNeuralNet( 'output', xNN(:,j), d );
68   end
69
70   % Plot the results
71   yL        = {'\Delta a', '\Delta a_{NN}', '\theta', 'v^2'};
72   PlotSet(1:n,[dA;aNN';theta;v2],'x label','Input','y label',yL,'figure
         title','Neural Net Delta Pitch Acceleration');
```

The script first finds the equilibrium state using `EquilibriumState`. It then sets up the sigma-pi neural net using `SigmaPiNeuralNet`. `PitchDynamicInversion` is called twice during each step through the loop, once to get the model aircraft acceleration `aM` (the way we want the aircraft to behave, from `dRHS`) and once to get the true acceleration `aT` (from `dRHSL`). The delta acceleration, `dA`, is used to train the neural net. The neural net produces `aNN`. The resulting weights are saved in a MAT-file for use in `AircraftSim`. The simulation uses `dRHS`, but our pitch acceleration model uses `dRHSL`. The latter is saved in another MAT-file.

```
>> PitchNeuralNetTraining
Velocity               250.00 m/s
Altitude             10000.00 m
Flight path angle        0.00 deg
Z speed                 13.84 m/s
Thrust               11148.95 N
The angle of attack      3.17 deg
Elevator                11.22 deg
Initial cost          9.62e+01
Final cost            1.17e-17
```

As can be seen, the neural net reproduces the model very well. The script also outputs `DNN.mat` which contains the trained neural net data.

7.11   Nonlinear Simulation

7.11.1 Problem

We want to demonstrate our learning control system for controlling the longitudinal dynamics of an aircraft.

7.11.2 Solution

Enable learning with the sigma-pi neural net in the simulation script described in `AircraftSim`.

7.11.3 How It Works

After training the neural net in the previous recipe, we set `addLearning` to true. The weights are read in by loading the stored MAT-files. We command a 0.2-radian pitch angle using the PID learning control. The results are shown in Figure 7.15, Figure 7.16, and Figure 7.17. The figures show without learning control on the left, the same as Recipe 7.9, and with learning control on the right. In the lower-left plot of Figure 7.16, we see that without learning, the elevator saturates, becoming constant at the maximum value.

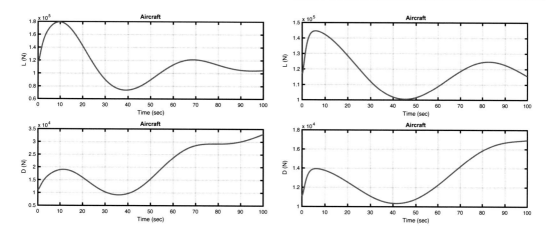

Figure 7.15: *Aircraft pitch angle change. Lift and drag variations are shown*

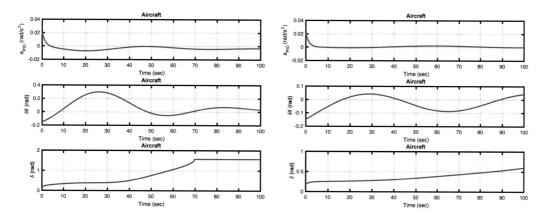

Figure 7.16: *Acceleration magnitude and angles. Without learning control, left, the elevator saturates*

Learning control helps the performance of the controller. However, the neural net weights are fixed throughout the simulation, as generated and saved by the prior recipe. Learning occurs prior to the controller becoming active. The control system is still sensitive to parameter changes since the learning part of the control was computed for a predetermined trajectory. Our weights were determined only as a function of the pitch angle and velocity squared. Additional inputs would improve the performance. There are many opportunities for you to try to expand and improve the learning system.

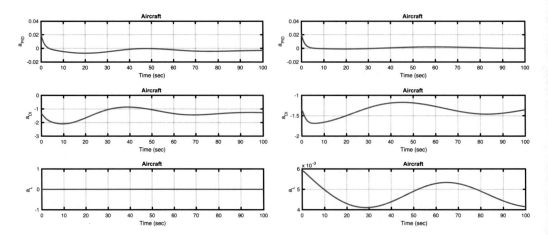

Figure 7.17: *Aircraft accelerations. On the bottom right is the learning acceleration*

7.12 Summary

This chapter has demonstrated adaptive or learning control for an aircraft. You learned about model tuning, model reference adaptive control, adaptive control, and gain scheduling. Table 7.3 lists the functions and scripts included in the companion code.

Table 7.3: *Chapter Code Listing*

File	Description
Combinations	Enumerates n integers for 1:n taken k at a time
AircraftSim	Simulation of the longitudinal dynamics of an aircraft
AtmDensity	Atmospheric density using a modified exponential model
EquilibriumState	Finds the equilibrium state for an aircraft
PID	Implements a PID controller
PitchDynamicInversion	Pitch angular acceleration
PitchNeuralNetTraining	Trains the pitch acceleration neural net
RecursiveLearning	Demonstrates recursive neural net training or learning
RHSAircraft	Right-hand side for aircraft longitudinal dynamics
SigmaPiNeuralNet	Implements a sigma-pi neural net
Sigmoid	Plots a sigmoid function

CHAPTER 8

■ ■ ■

Introduction to Neural Nets

Neural networks, or neural nets, are a popular way of implementing machine "intelligence." The idea is that they behave like the neuron in a brain. In our taxonomy, neural nets fall in the category of true machine learning, as shown on the right.

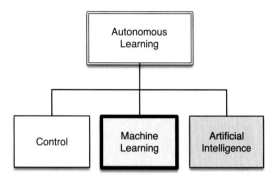

In this chapter, we will explore how neural nets work, starting with the most fundamental idea with a single neuron and working our way up to a multilayer neural net. Our example for this will be a pendulum. We will show how a neural net can be used to solve the prediction problem. This is one of the two uses of a neural net, prediction, and categorization. We'll start with a simple categorization example. We'll do a more sophisticated categorization of neural nets in Chapters 9 and 11.

8.1 Daylight Detector

8.1.1 Problem

We want to use a simple neural net to detect daylight.

8.1.2 Solution

Historically, the first neuron was the perceptron. This is a neural net with an activation function that is a threshold with an output of either 0 or 1. It is well suited for categorization problems. We will use a single perceptron in this example.

8.1.3 How It Works

Suppose our input is a light level measured by a photocell. If you weigh the input so that 1 is the value defining the brightness level at twilight, you get a sunny day detector. This is shown in the following script, SunnyDay. The script is named after a story (which may be apocryphal) of a neural net that was supposed to detect tanks but instead detected sunny days; this was due to all the training photos of tanks being taken, unknowingly, on a sunny day, while all the photos without tanks were taken on a cloudy day.

In this problem, solar flux is modeled using cosine and scaled so that it is 1 at noon. Any value greater than 0 is daylight.

SunnyDay.m

```
8   %% The data
9   t = linspace(0,24);         % time, in hours
10  d = zeros(1,length(t));
11  s = cos((2*pi/24)*(t-12)); % solar flux model
12
13  %% The activation function
14  % The nonlinear activation function which is a threshold detector
15  j     = s < 0;
16  s(j) = 0;
17  j     = s > 0;
18  d(j) = 1;
19
20  %% Plot the results
21  PlotSet(t,[s;d],'x label','Hour', 'y label',...
22      {'Solar Flux', 'Day/Night'}, 'figure title','Daylight Detector',...
23      'plot title', 'Daylight Detector');
24  set([subplot(2,1,1) subplot(2,1,2)],'xlim',[0 24],'xtick',[0 6 12 18 24]);
```

Figure 8.1 shows the detector results. The set(gca, ...) code sets the x-axis ticks to end at exactly 24 hours. This shows how categorization works.

If we had multiple neurons with thresholds set to detect sunlight levels within bands of solar flux, we would have a neural net sun clock.

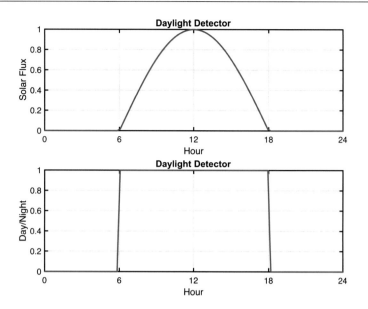

Figure 8.1: *The daylight detector*

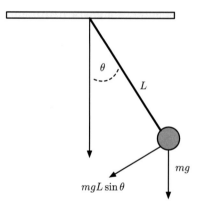

Figure 8.2: *A pendulum. The motion is driven by the acceleration of gravity*

8.2 Modeling a Pendulum

8.2.1 Problem

We want to implement the dynamics of a pendulum as shown in Figure 8.2. The pendulum will be modeled as a point mass with a rigid connection to its pivot. The rigid connection is a rod that cannot contract or expand.

231

8.2.2 Solution

The solution is to write a pendulum dynamics function in MATLAB. The dynamics will be written in torque form, that is, we will model it as rigid body rotation. Rigid body rotation is what happens when you spin a wheel. The entire object rotates together, hence the term rigid body. It will use the `RungeKutta` numerical integration function in the General folder of the included toolbox to integrate the equations of motion. Numerical integration of a differential integration allows one to predict the next value of the state, which in this case is angle and angular velocity, from previous values of the state and any inputs.

8.2.3 How It Works

Figure 8.2 shows the pendulum. The easiest way to get the equations is to write it as a torque problem, that is, as rigid body rotation. When you look at a two-dimensional pendulum, it moves in a plane, and its location has x and y coordinates. However, these two coordinates are constrained by the fixed pendulum of length L. We can write

$$L^2 = x^2 + y^2 \tag{8.1}$$

where L is the length of the rod and a constant and x and y are the coordinates in the plane. They are also the degrees of freedom in the problem. This shows that x is uniquely determined by y. If we write

$$x = L \sin \theta \tag{8.2}$$
$$y = L \cos \theta \tag{8.3}$$

where θ is the angle from the vertical, that is, it is zero when the pendulum is hanging straight down, we see that we need only one degree of freedom, θ, to model the motion. So our force problem becomes a rigid body rotational motion problem. The torque is related to the angular acceleration by the inertia as

$$T = I \frac{d^2\theta}{dt^2} \tag{8.4}$$

where I is the inertia and T is the torque. The inertia is constant and depends on the square of the pendulum length and the mass m:

$$I = mL^2 \tag{8.5}$$

The torque is produced by the component of the gravitational force, mg, that is perpendicular to the pendulum, where g is the acceleration of gravity. Recall that torque is the applied force, $mg \sin \theta$, times the moment arm, in this case, L. The torque is therefore

$$T = -mgL \sin \theta \tag{8.6}$$

The equations of motion are then

$$-gL \sin \theta = L^2 \frac{d^2\theta}{dt^2} \tag{8.7}$$

232

or simplifying

$$\frac{d^2\theta}{dt^2} + \left(\frac{g}{L}\right)\sin\theta = 0 \tag{8.8}$$

We set

$$\frac{g}{mL} = \Omega^2 \tag{8.9}$$

where Ω is the frequency of the pendulum's oscillation. This equation is nonlinear due to the $\sin\theta$. We can linearize it about small angles, θ, about vertical. For small angles,

$$\sin\theta \approx \theta \tag{8.10}$$

$$\cos\theta \approx 1 \tag{8.11}$$

to get the linear constant coefficient equation. The linear version of sine comes from Taylor's series expansion:

$$\sin\theta = \theta - \frac{\theta^3}{6} + \frac{\theta^5}{120} - \frac{\theta^7}{5040} + \cdots \tag{8.12}$$

You can see that the first term is a pretty good approximation around $\theta = 0$ which is when the pendulum is hanging vertically. We can apply this to any angle. Let $\theta = \theta + \theta_k$ where θ_k is our current angle and θ is now small. We can expand the sine term:

$$\sin(\theta + \theta_k) = \sin\theta\cos\theta_k + \sin\theta_k\cos\theta \approx \theta\cos\theta_k + \sin\theta_k \tag{8.13}$$

We get a linear equation with a new torque term and a different coefficient for θ:

$$\frac{d^2\theta}{dt^2} + \cos\theta_k\Omega^2\theta = -\Omega^2\sin\theta_k \tag{8.14}$$

This tells us that a linear approximation is useful regardless of the current angle.

Our final equations are (nonlinear and linear)

$$\frac{d^2\theta}{dt^2} + \Omega^2\sin\theta \;\; = \;\; 0 \tag{8.15}$$

$$\frac{d^2\theta}{dt^2} + \Omega^2\theta \;\; \approx \;\; 0 \tag{8.16}$$

The dynamic model is in the following code, with an excerpt from the header. This can be called by the `RungeKutta` function or any MATLAB integrator. There is an option to use either the full nonlinear dynamics or the linearized form of the dynamics, using a boolean field called `linear`. The state vector has the angle as the first element and the angle derivative, or angular velocity ω, as the second element. Time, the first input, is not used because it only appears in the equations as dt, so it is replaced with a tilde. The output is the derivative, `xDot`, of the state `x`. If no inputs are specified, the function will return the default data structure d.

RHSPendulum.m

```
14  %   x           (2,1)  State vector [theta;theta dot]
15  %   d           (.)    Data structure
16  %                      .linear  (1,1) If true use a linear model
17  %                      .omega   (1,1) Input gain
24  %
25  function xDot = RHSPendulum( ~, x, d )
26
27  if( nargin < 1 )
28    xDot = struct('linear',false,'omega',0.5);
29    return
30  end
31
32  if( d.linear )
33    f = x(1);
34  else
35    f = sin(x(1));
36  end
37
38  xDot = [x(2);-d.omega^2*f];
```

The code for xDot has two elements. The first element is just the second element of the state because the derivative of the angle is the angular velocity. The second term is the angular acceleration computed using our equations. The set of differential equations that is implemented is the set of first-order differential equations:

$$\frac{d\theta}{dt} = \omega \tag{8.17}$$

$$\frac{d\omega}{dt} = -\Omega^2 \sin\theta \tag{8.18}$$

First order means there are only first derivatives of time on the left-hand side.

The script PendulumSim, shown as follows, simulates the pendulum by integrating the dynamical model. Setting the data structure field linear to true gives the linear model. Note that the state is initialized with a large initial angle of three radians to highlight the differences between the models.

PendulumSim.m

```
1   %% Pendulum simulation
7   %% Initialize the simulation
8   n            = 1000;        % Number of time steps
9   dT           = 0.1;         % Time step (sec)
10  dRHS         = RHSPendulum;  % Get the default data structure
11  dRHS.linear  = false;       % true for linear model
12
13  %% Simulation
14  xPlot        = zeros(2,n);
15  theta0       = 3;           % radians
```

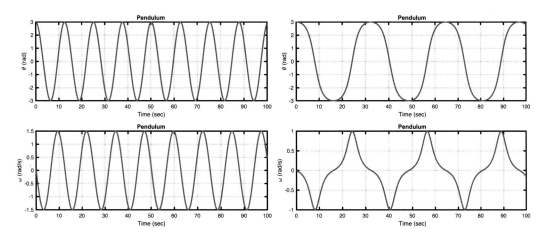

Figure 8.3: *A pendulum modeled by the linear and nonlinear equations. The period for the nonlinear model is not the same as for the linear model. The left-hand plot is linear and the right nonlinear*

```
16  x                = [theta0;0];  % [angle;velocity]
17
18  for k = 1:n
19    xPlot(:,k)     = x;
20    x              = RungeKutta( @RHSPendulum, 0, x, dT, dRHS );
21  end
22
23  %% Plot the results
24  yL       = {'\theta (rad)' '\omega (rad/s)'};
25  [t,tL]   = TimeLabel(dT*(0:n-1));
26
27  PlotSet( t, xPlot, 'x label', tL, 'y label', yL, ...
28           'plot title', 'Pendulum', 'figure title', 'Pendulum State' );
```

Figure 8.3 shows the results of the two models. The period of the nonlinear model is not the same as that of the linear model.

8.3 Single Neuron Angle Estimator

8.3.1 Problem

We want to use a neural net to estimate the angle between the rigid pendulum and the vertical.

235

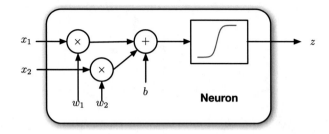

Figure 8.4: *A two-input neuron*

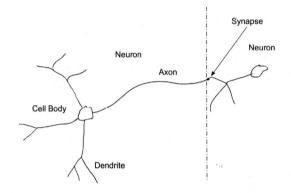

Figure 8.5: *A real neuron can have 10,000 inputs!*

8.3.2 Solution

We will derive the equations for a linear estimator and then replicate it with a neural net consisting of a single neuron.

8.3.3 How It Works

Let's first look at a single neuron with two inputs. This is shown in Figure 8.4. This neuron has inputs x_1 and x_2, a bias b, weights w_1 and w_2, and a single output z. The activation function σ takes the weighted input and produces the output:

$$z = \sigma(w_1 x_1 + w_2 x_2 + b) \tag{8.19}$$

Compare this with a real neuron as shown in Figure 8.5. A real neuron has multiple inputs via the dendrites. These branches mean that multiple inputs can connect to the cell body through the same dendrite. The output is via the axon. Each neuron has one output. The axon connects to a dendrite through the synapse. Signals pass from the axon to the dendrite via a synapse.

There are numerous commonly used activation functions. We show three:

$$\sigma(y) = \tanh(y) \tag{8.20}$$

$$\sigma(y) = \frac{2}{1 - e^{-y}} - 1 \tag{8.21}$$

$$\sigma(y) = y \tag{8.22}$$

236

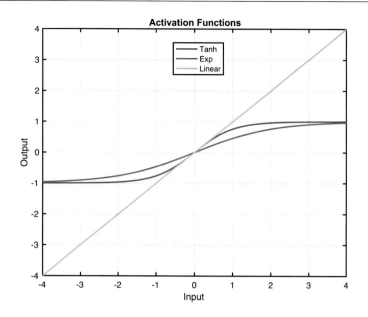

Figure 8.6: *The three activation functions*

The exponential one is normalized and offset from zero, so it ranges from -1 to 1. The following code in the script OneNeuron computes and plots these three activation functions for an input q.

OneNeuron.m

```
7   %% Look at the activation functions
8   q        = linspace(-4,4);
9   v1       = tanh(q);
10  v2       = 2./(1+exp(-q)) - 1;
11
12  PlotSet(q,[v1;v2;q],'x label','Input', 'y label',...
13    'Output', 'figure title','Activation Functions','plot title', '
         Activation Functions',...
14    'plot set',{[1 2 3]},'legend',{{'Tanh','Exp','Linear'}});
```

Figure 8.6 shows the three activation functions on one plot.

Activation functions that saturate model a biological neuron that has a maximum firing rate. These particular functions also have good numerical properties that are helpful in learning. An important property is that they have analytical derivatives.

Now that we have defined our neuron model, let's return to the pendulum dynamics. The solution to the linear pendulum equation is

$$\theta = a \sin \Omega t + b \cos \Omega t \qquad (8.23)$$

237

Given initial angle θ_0 and angular rate $\dot{\theta}_0$, we get the angle as a function of time:

$$\theta(t) = \frac{\dot{\theta}_0}{\Omega} \sin \Omega t + \theta_0 \cos \Omega t \tag{8.24}$$

For small Ωt

$$\theta(t) = \dot{\theta}_0 t + \theta_0 \tag{8.25}$$

which is a linear equation. Change this to a discrete-time problem:

$$\theta_{k+1} = \dot{\theta}_k \Delta t + \theta_k \tag{8.26}$$

where Δt is the time step between measurements, θ_k is the current angle, and θ_{k+1} is the angle at the next step. The linear approximation to the angular rate is

$$\dot{\theta}_k = \frac{\theta_k - \theta_{k-1}}{\Delta t} \tag{8.27}$$

so combining Equations 8.26 and 8.27, our "estimator" is

$$\theta_{k+1} = 2\theta_k - \theta_{k-1} \tag{8.28}$$

It does not need to know the time step.

Let's do the same thing with a neural net. Our neuron inputs are x_1 and x_2. If we set

$$
\begin{aligned}
x_1 &= \theta_k & (8.29) \\
x_2 &= \theta_{k-1} & (8.30) \\
w_1 &= 2 & (8.31) \\
w_2 &= -1 & (8.32) \\
b &= 0 & (8.33)
\end{aligned}
$$

we get

$$z = \sigma(2\theta_k - \theta_{k-1}) \tag{8.34}$$

which is, aside from the activation function σ, our estimator.

Continuing through OneNeuron, the following code implements the estimators. We input a pure sine wave that is only valid for small pendulum angles. We then compute the neuron with the linear activation function and then the tanh activation function. Note that the variable thetaN is equivalent to using the linear activation function.

OneNeuron.m

```
16   %% Look at the estimator for a pendulum
17   omega    = 1;              % pendulum frequency in rad/s
18   t        = linspace(0,20);
```

```
19  theta    = sin(omega*t);
20  thetaN   = 2*theta(2:end) - theta(1:end-1); % linear estimator for "next
       " theta
21  truth    = theta(3:end);
22  tOut     = t(3:end);
23  thetaN   = thetaN(1:end-1);
24
25  % Apply the activation function
26  z = tanh(thetaN);
27
28  PlotSet(tOut, [truth;thetaN;z],'x label','Time (s)', 'y label',...
29    'Next angle', 'figure title','One neuron','plot title', 'One neuron'
       ,...
30    'plot set',{[1 2 3]},'legend',{{'True','Estimate','Neuron'}});
```

Figure 8.7 shows the two neuron outputs, linear and `tanh`, compared with the truth. The one with the linear activation function matches the truth very well. The `tanh` does not, but that is to be expected because it saturates.

The one-neuron function with the linear activation function is the same as the estimator by itself. Usually, output nodes and this neural net have only an output node and have linear activation functions. This makes sense; otherwise, the output would be limited to the saturation value of the activation functions, as we have seen with `tanh`. With any other activation function, the output does not produce the desired result. This particular example is one in which a neural net doesn't give us any advantage and was chosen because it reduces to a simple linear estimator.

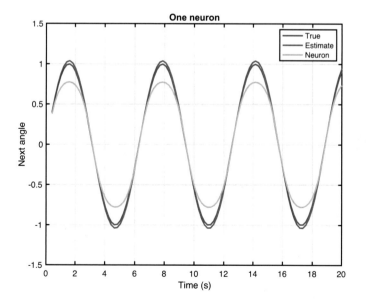

Figure 8.7: *The true pendulum dynamics compared to the linear and* tanh *neuron output*

For more general problems, with more inputs and nonlinear dependencies among the inputs, activation functions that have saturation may be valuable.

For this, we will need a multi-neuron net that will be discussed in the last section of the chapter. Note that even the neuron with the linear activation function does not quite match the truth value. If we were to use the linear activation function with the nonlinear pendulum, it would not work very well. A nonlinear estimator would be complicated, but a neural net with multiple layers (deep learning) could be trained to cover a wider range of conditions.

8.4 Designing a Neural Net for the Pendulum

8.4.1 Problem

We want to estimate angles for a nonlinear pendulum.

8.4.2 Solution

We will use `NeuralNetMLFF` to build a neural net from training sets. (MLFF stands for multi-layer, feedforward.) We will run the net using `NeuralNetMLFF`. The code for `NeuralNetMLFF` is included with the neural net developer GUI in the next chapter.

8.4.3 How It Works

The script for this recipe is `NNPendulumDemo`. The first part generates the test data running the same simulation as `PendulumSim.m` in Recipe 8.2. We calculate the period of the pendulum to set the simulation time step at a small fraction of the period. Note that we will use `tanh` as the activation function for the net.

NNPendulumDemo.m

```
9   % Demo parameters
10  nSamples    = 800;       % Samples in the simulation
11  nRuns       = 2000;      % Number of training runs
12  activation  = 'tanh';    % activation function
13
14  omega       = 0.5;       % frequency in rad/s
15  tau         = 2*pi/omega; % period in secs
16  dT          = tau/100;   % sample at a rate of 20*omega
17
18  rng(100);                % consistent random number generator
19
20  %% Initialize the simulation RHS
21  dRHS        = RHSPendulum; % Get the default data structure
22  dRHS.linear = false;
23  dRHS.omega  = omega;
24
25  %% Simulation
26  nSim    = nSamples + 2;
27  x       = zeros(2,nSim);
28  theta0 = 0.1;                % starting position (angle)
```

```
29  x(:,1) = [theta0;0];
30  for k = 1:nSim-1
31    x(:,k+1) = RungeKutta( @RHSPendulum, 0, x(:,k), dT, dRHS );
32  end
```

The next block defines the network and trains it using `NeuralNetTraining`. `NeuralNetTraining` and `NeuralNetMLFF` are described in the next chapter. Briefly, we define a first layer with three neurons and a second output layer with a single neuron; the network has two inputs, which are the previous two angles.

NNPendulumDemo.m

```
41  %% Define a network with two inputs, three inner nodes, and one output
42  layer           = struct;
43  layer(1,1).type = activation;
44  layer(1,1).alpha = 1;
45  layer(2,1).type = 'sum'; %'sum';
46  layer(2,1).alpha = 1;
47
48  % Thresholds
49  layer(1,1).w0 = rand(3,1) - 0.5;
50  layer(2,1).w0 = rand(1,1) - 0.5;
51
52  % Weights w(i,j) from jth input to ith node
53  layer(1,1).w  = rand(3,2) - 0.5;
54  layer(2,1).w  = rand(1,3) - 0.5;
55
56  %% Train the network
57  % Order the samples using a random list
58  kR              = ceil(rand(1,nRuns)*nSamples);
59  thetaE          = x(1,kR+2); % Angle to estimate
60  theta           = [x(1,kR);x(1,kR+1)]; % Previous two angles
61  e               = thetaE - (2*theta(1,:) - theta(2,:));
62  [w,e,layer] = NeuralNetTraining( theta, thetaE, layer );
63
64  PlotSet(1:length(e), e.^2, 'x label','Sample', 'y label','Error^2',...
65    'figure title','Training Error','plot title','Training Error','plot
         type','ylog');
66
67  % Assemble a new network with the computed weights
68  layerNew              = struct;
69  layerNew(1,1).type    = layer(1,1).type;
70  layerNew(1,1).w       = w(1).w;
71  layerNew(1,1).w0      = w(1).w0;
72  layerNew(2,1).type    = layer(2,1).type; %'sum';
73  layerNew(2,1).w       = w(2).w;
74  layerNew(2,1).w0      = w(2).w0;
75  network.layer         = layerNew;
```

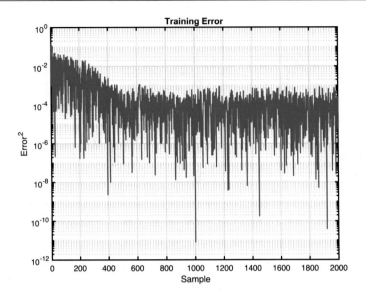

Figure 8.8: *Training error*

The training data structure includes the weights to be computed. It defines the number of layers and the type of activation function. The initial weights are random. Training returns the new weights and the training error. We pass the training data in random order to the function using the index array k. This gives better results than if we passed it in order. We also send the same training data multiple times using the parameter nRuns. Figure 8.8 shows the training error. It looks good. To see the weights that were calculated, just display w at the command line. For example, the weights of the output node are now

```
>> w(2)
ans =
  a struct with fields:

       w: [-0.67518 -0.21789 -0.065903]
      w0: -0.014379
    type: 'tanh'
```

We test the neural net in the last block of code. We rerun the simulation and then run the neural net using NeuralNetMLFF. Note that you may choose to initialize the simulation with a different starting point than in the training data by changing the value of thetaD.

NNPendulumDemo.m

```
71  layerNew(1,1).w0     = w(1).w0;
72  layerNew(2,1).type   = layer(2,1).type; %'sum';
73  layerNew(2,1).w      = w(2).w;
74  layerNew(2,1).w0     = w(2).w0;
75  network.layer        = layerNew;
76
```

```
77  %% Simulate the pendulum with a different starting point
78  x(:,1)          = [0.1;0];
79
80  %% Simulate the pendulum and test the trained network
81  % Choose the same or a different starting point and simulate
```

The results in Figure 8.9 look good. The neural net estimated angle is quite close to the true angle. Note, however, that we ran the same magnitude pendulum oscillation (thetaD = theta0), which is exactly what we trained it to recognize. If we run the test with a different starting point, such as 0.5 radians compared to 0.1 of the training data, there is more error in the estimated angles as shown in Figure 8.10.

If we want the neural net to predict angles for other magnitudes, it needs to be trained with a diverse set of data that models all conditions. When we trained the network, we let it see the same oscillation magnitude several times. This is not productive. It might also be necessary to add more nodes to the net or more layers to make a more general-purpose estimator.

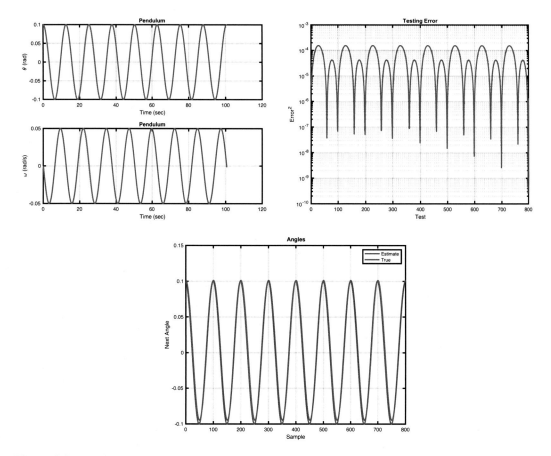

Figure 8.9: *Neural net results: the simulated state, the testing error, and the truth angles compared to the neural net's estimate*

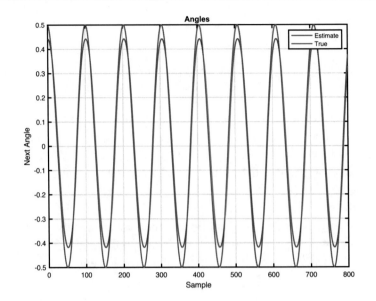

Figure 8.10: *Neural estimated angles for a different magnitude oscillation*

8.5   XOR Example

We'll give many examples of the Deep Learning Toolbox in subsequent chapters. We'll do one example just to get you going. This example doesn't even unlock a fraction of the power in the Deep Learning Toolbox. We will implement the XOR example. The DLXOR.m script is shown in the following, using the MATLAB functions feedforwardnet, configure, train, and sim.

DLXOR.m

```
1  %% Use the Deep Learning Toolbox to create the XOR neural net
2  % See also feedforwardnet, randi, configure, train, sim
3
4  %% Create the network
5  % 2 layers
6  % 2 inputs
7  % 1 output
8
9  net = feedforwardnet(2);
10
11  % XOR Truth table
12  a   = [1 0 1 0];
13  b   = [1 0 0 1];
14  c   = [0 0 1 1];
15
16  % How many sets of inputs
17  n   = 600;
18
```

```
19  % This determines the number of inputs and outputs
20  x       = zeros(2,n);
21  y       = zeros(1,n);
22
23  % Create training pairs
24  for k = 1:n
25    j         = randi([1,4]);
26    x(:,k)    = [a(j); b(j)];
27    y(k)      = c(j);
28  end
29
30  net         = configure(net, x, y);
31  net.name    = 'XOR';
32  net         = train(net,x,y);
33  c           = sim(net,[a;b]);
34
35  fprintf('\n    a      b    c\n');
36  for k = 1:4
37    fprintf('%5.0f %5.0f %5.2f\n',a(k),b(k),c(k));
38  end
39
40  % This only works for feedforwardnet(2);
41  fprintf('\nHidden layer biases %6.3f %6.3f\n',net.b{1});
42  fprintf('Output layer bias   %6.3f\n',net.b{2});
43  fprintf('Input layer weights  %6.2f %6.2f\n',net.IW{1}(1,:));
44  fprintf('                     %6.2f %6.2f\n',net.IW{1}(2,:));
45  fprintf('Output layer weights %6.2f %6.2f\n',net.LW{2,1}(1,:));
46
47  fprintf('Hidden layer activation function %s\n',net.layers{1}.
          transferFcn);
```

Running the script produces the MATLAB GUI shown in Figure 8.11.

As you can see, we have two inputs, one hidden layer and one output layer. The diagram indicates that our hidden layer activation function is nonlinear, while the output layer is linear. The GUI is interactive, and you can study the learning process by clicking the buttons. For example, if you click the performance button, you get Figure 8.12. Just about everything in the network development is customizable. The GUI is a real-time display. You can watch the training in progress. If you just want to look at the layout, type `view(net)`.

The three major boxes in the GUI are Algorithms, Progress, and Plots. Under **Algorithms**, we have

- Data division: Data division divides the data into training, validation, and test sets. "Random" means that the division between the three categories is done randomly.

- Training: This shows the training method to be used.

- Performance: This says that the mean squared error (MSE) is used to determine how well the network works. Other methods, such as maximum absolute error, could be used.

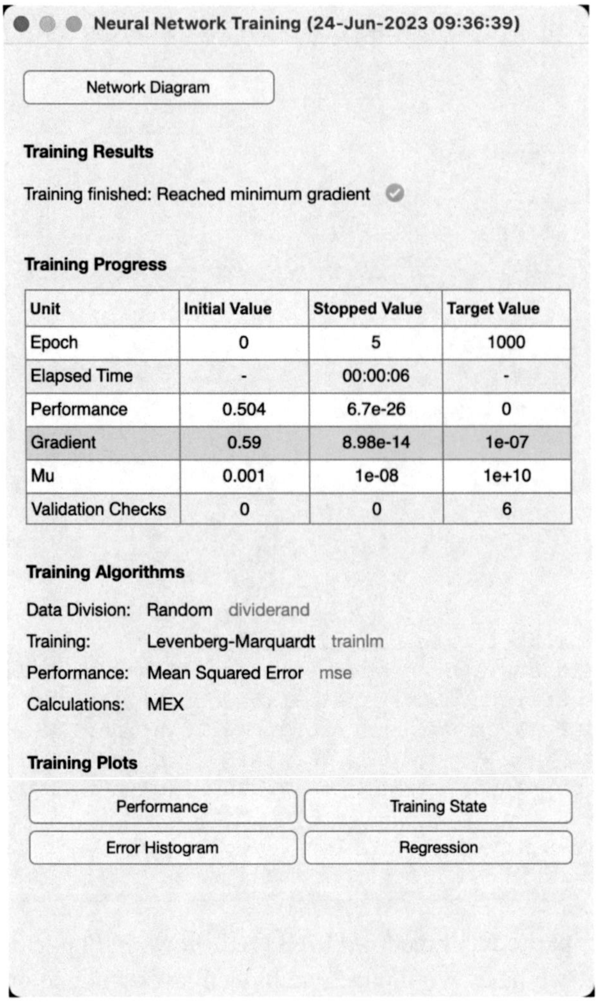

Figure 8.11: *Deep learning network GUI*

Mean squared is useful because the error grows as the square of the deviation, meaning that large errors are more heavily weighted.

- Calculations: This shows that the calculations are done via a mex file, that is, in a C or C++ program.

The **Progress** of the GUI is useful to watch during long training sessions. We are seeing it at the end:

- Epoch: This says five epochs were used. The range is 0 to 1000 epochs.

- Time: This gives you the clock time during training.

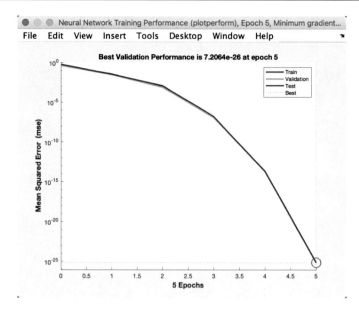

Figure 8.12: *Network training performance*

- Performance: This shows you the MSE performance during training.

- Gradient: This shows the gradient that shows the speed of training as discussed earlier.

- Mu: This is the control parameter for training the neural network.

- Validation checks: This shows that no validation checks failed.

The last section is **Plots**. There are four figures we can study to understand the process.

Figure 8.12 shows the training performance as a function of epoch. The mean squared error is the criteria. The test, validation, and training sets have their own lines. In this training, all have the same values.

Figure 8.13 shows the training state as a function of epoch. Five epochs are used. The titles show the final values in each plot. The top plot shows the progression of the gradient. It decreases with each epoch. The next shows mu decreasing linearly with epoch. The bottom plot shows that there were no validation failures during the training.

Figure 8.14 gives a training histogram. This shows the number of instances when one of the sets shows the error value on the x-axis. The bars are divided into training, validation, and test sets. Each number on the x-axis is a bin. Only three bins are occupied, in this case. The histogram shows that the training sets are more numerous than the validation or test sets.

Figure 8.15 gives a training regression. There are four subplots: one for training sets, one for validation sets, one for test sets, and one for all sets. There are only two targets, 0 and 1. The linear fit doesn't give much information in this case since we can only have a linear fit with two points. The plot title says we reached the minimum gradient after five epochs, that is, after

247

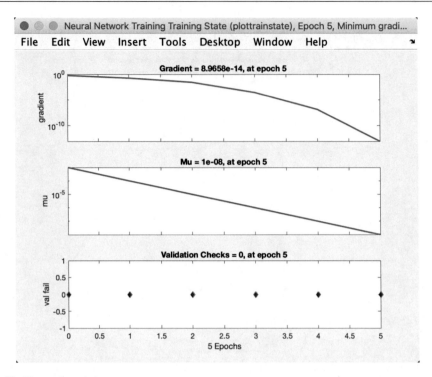

Figure 8.13: *Network training state*

passing all the cases through the training five times. The legend shows the data, the fit, and the Y=T plot which is the same as the linear fit in this system.

Typing

```
>> net = feedforwardnet(2);
```

creates the neural network data structure which is quite flexible and complex. The "2" means two neurons in one layer. If we wanted two layers with two neurons each, we would type

```
>> net = feedforwardnet([2 2]);
```

We create 600 training sets. `net = configure(net, x, y);` configures the network. The configure function determines the number of inputs and outputs from the `x` and `y` arrays. The network is trained with `net = train(net,x,y);` and simulated with `c = sim(net, [a;b]);`. We extract the weights and biases from the cell arrays `net.IW`, `net.LW`, and `net.b`. "I" stands for input and "IW" for layer. The input is from the single input node to the two hidden nodes, and the layer is from the two hidden nodes to the one output node.

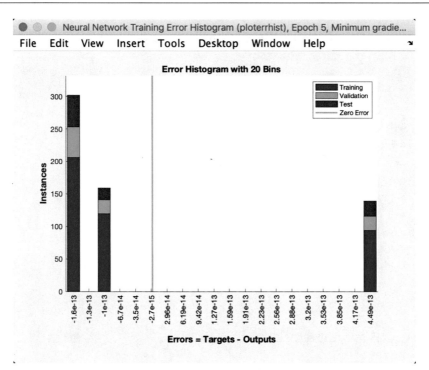

Figure 8.14: *Network training histogram*

Now the training sets are created randomly from the truth table. You can run this script many times, and usually you will get the right result, but not always. This is an example where it worked well:

```
>> DLXOR

    a     b    c
    1     1    0.00
    0     0    0.00
    1     0    1.00
    0     1    1.00

Hidden layer biases    1.735 -1.906
Output layer bias      1.193
Input layer weights    -2.15    1.45
                       -1.83    1.04
Output layer weights   -1.16    1.30
Hidden layer activation function tansig
Hidden layer activation function purelin
```

`tansig` is a hyperbolic tangent sigmoid function.

249

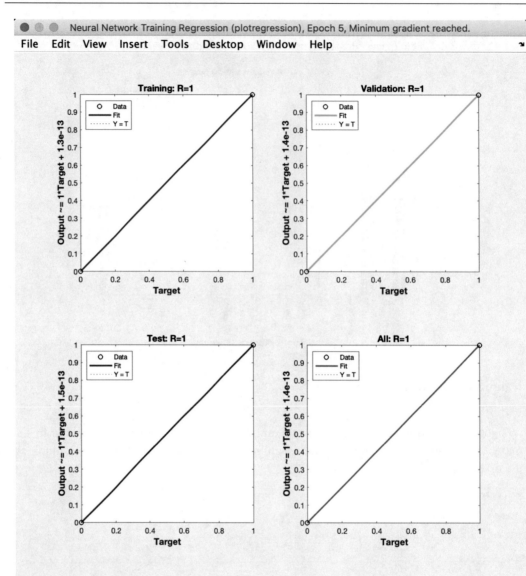

Figure 8.15: *Regression*

Each run will result in different weights, even when the network gives the correct results. For example:

```
>> DLXOR

    a     b     c
    1     1   0.00
    0     0  -0.00
    1     0   1.00
```

```
     0      1   1.00

 Hidden layer biases  4.178  0.075
 Output layer bias    -1.087
 Input layer weights   -4.49  -1.36
                       -3.79  -3.67
 Output layer weights   2.55  -2.46
```

There are many possible sets of weights and biases because the weight/bias sets are not unique. Note that the 0.00 are really not 0. This means that used operationally, we would need to set a threshold such as

```
if( abs(c) < tol )
  c = 0;
end
```

You might be interested in what happens if we add another layer to the network, by creating it with net = feedforwardnet([2 2]). Figure 8.16 shows the network in the GUI.

The additional hidden layer makes it easier for the neural net to fit the data from which it is learning. On the left are the two inputs, a and b. In each hidden layer, there is a weight w and bias b. Weights are always needed, but biases are not always used. Both hidden layers have nonlinear activation functions. The output layer produces the one output using a linear activation function.

```
>> DLXOR
     a    b   c
     1    1   0.00
     0    0   0.00
     1    0   1.00
     0    1   1.00
```

This produces good results too. We haven't explored all the diagnostic tools available when using feedforwardnet. There is a lot of flexibility in the software. You can change activation functions, change the number of hidden layers, and customize it in many different ways. This particular example is very simple as the input sets are limited to four possibilities.

We can explore what happens when the inputs are noisy, not necessarily all ones or zeros. We do this in DLXORNoisy.m, and the only difference from the original script is in lines 33–35 where we add Gaussian noise to the inputs:

DLXORNoisy.m

```
33   net       = train(net,x,y);
34   a         = a + 0.01*randn(1,4);
35   b         = b + 0.01*randn(1,4);
```

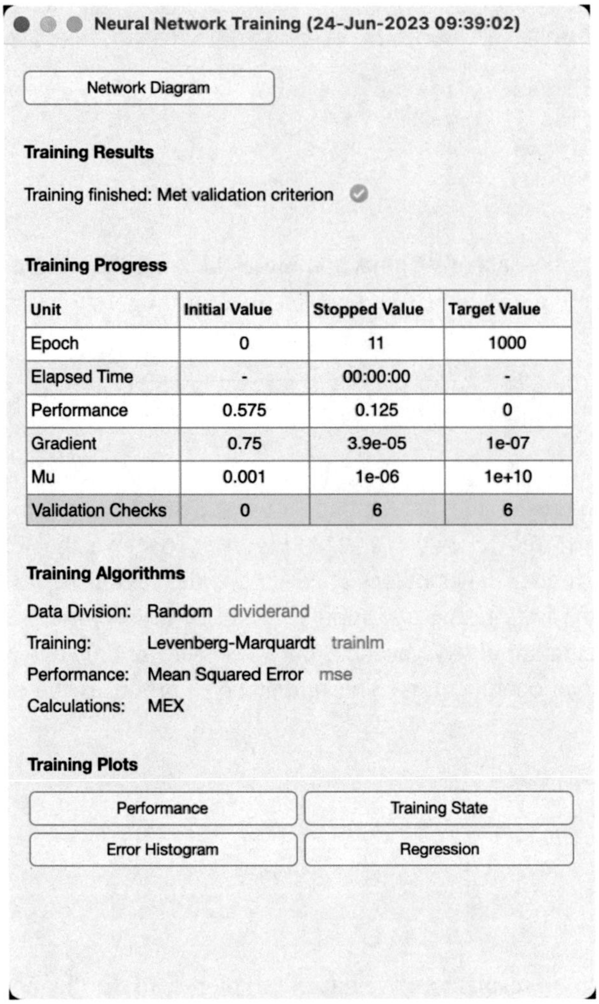

Figure 8.16: *Deep learning network GUI with two hidden layers*

The output from running this script is shown as follows:

```
>> DLXORNoisy

      a       b       c
  0.991   1.019  -0.003
  0.001  -0.005  -0.002
  0.996   0.009   0.999
 -0.001   1.000   1.000

Hidden layer biases  -1.793    2.135
Output layer bias     -1.158
Input layer weights      1.70     1.54
                         1.80     1.52
```

```
Output layer weights   -1.11    1.15
Hidden layer activation function tansig
Output layer activation function purelin
```

As one might expect, the outputs are not exactly one or zero.

8.6 Training

The neural net is a nonlinear system due to the nonlinear activation functions. The Levenberg-Marquardt training algorithm is one way of solving a nonlinear least squares problem. This algorithm only finds a local minimum which may or may not be a global minimum. Other algorithms, such as genetic algorithms, downhill simplex, simulated annealing, and so on, could also be used for finding weights and biases. To achieve second-order training speeds, one has to compute the Hessian matrix. The Hessian matrix is a square matrix of the second-order partial derivative of a scalar-valued function. Suppose we have a nonlinear function:

$$f(x_1, x_2) \tag{8.35}$$

then the Hessian is

$$H = \begin{bmatrix} \frac{\partial^2 f}{\partial x_1^2} & \frac{\partial^2 f}{\partial x_1 \partial x_2} \\ \frac{\partial^2 f}{\partial x_2 \partial x_1} & \frac{\partial^2 f}{\partial x_2^2} \end{bmatrix} \tag{8.36}$$

x_k are weights and biases. This can be very expensive to compute. In the Levenberg-Marquardt algorithm, we make an approximation:

$$H = J^T J \tag{8.37}$$

where

$$J = \begin{bmatrix} \frac{\partial f}{\partial x_1} & \frac{\partial f}{\partial x_2} \end{bmatrix} \tag{8.38}$$

The approximate Hessian is

$$H = \begin{bmatrix} (\frac{\partial f}{\partial x_1})^2 & \frac{\partial f}{\partial x_1}\frac{\partial f}{\partial x_2} \\ \frac{\partial f}{\partial x_1}\frac{\partial f}{\partial x_2} & (\frac{\partial f}{\partial x_2})^2 \end{bmatrix} \tag{8.39}$$

This is an approximation of the second derivative that only requires direct computation of first derivatives. This approach reduces the overall computational burden by eliminating the need to compute the second derivatives. The gradient is

$$g = J^T e \tag{8.40}$$

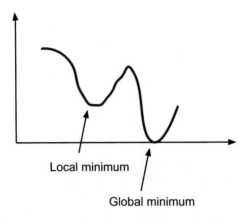

Figure 8.17: *Local and global minimums*

where e is a vector of errors. The Levenberg-Marquardt uses the following algorithm to update the weights and biases:

$$x_{k+1} = x_k - \left[J^T J + \mu I \right]^{-1} J^T e \qquad (8.41)$$

I is the identity matrix (a matrix with all diagonal elements equal to 1). If the parameter μ is zero, this is Newton's method. With a large μ, this becomes gradient descent which is faster. Thus, μ is a control parameter. After a successful step, we decrease μ since we are in less need of the advantages of the faster gradient descent.

Why are gradients so important and why can they get us into trouble? Figure 8.17 shows a curve with a local and a global minimum. If our search first enters the local minimum, the gradient is steep and will drive us to the bottom from which we might not get out. Thus, we would not have found the best solution.

The cost can be very complex even for simple problems.

8.7   Summary

This chapter has demonstrated neural learning to predict pendulum angles. It introduces the concept of a neuron. It demonstrates a one-neuron network for a pendulum and shows how it compares with a linear estimator. A perceptron example and a multilayer pendulum angle estimator are also given. Table 8.1 lists the functions and scripts included in the companion code. The last two functions are borrowed from the next chapter, which will cover multilayer neural nets in more depth.

Table 8.1: *Chapter Code Listing*

File	Description
DLXOR	XOR neural net
DLXORNoisy	XOR neural net with noise
NNPendulumDemo	Trains a neural net to track a pendulum
OneNeuron	Explores a single neuron
PendulumSim	Simulates a pendulum
RHSPendulum	Right-hand side of a nonlinear pendulum
SunnyDay	Recognizes daylight
Chapter 9 Functions	
NeuralNetMLFF	Computes the output of a multilayer feedforward neural net
NeuralNetTraining	Training with backpropagation

CHAPTER 9

■ ■ ■

Classification of Numbers Using Neural Networks

Pattern recognition in images is a classic application of neural nets. This chapter builds upon the previous one by exploring multilayer networks, which fall into the Machine Learning branch of our Autonomous Learning taxonomy. In this case, we will look at images of computer-generated digits and the problem of identifying the digits correctly. These images will represent numbers from scanned documents. Attempting to capture the variation in

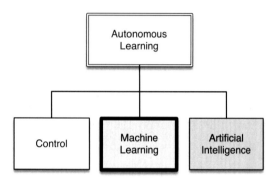

digits with algorithmic rules, considering fonts and other factors, quickly becomes impossibly complex, but with a large number of examples, a neural net can readily perform the task. We allow the weights in the net to perform the job of inferring rules about how each digit may be shaped, rather than codifying them explicitly.

For this chapter, we will limit ourselves to images of a single digit. The process of segmenting a series of digits into individual images may be solved by many techniques, not just neural nets.

9.1 Generate Test Images with Defects

9.1.1 Problem

The first step in creating our classification system is to generate sample data. In this case, we want to load images of numbers from zero to nine and generate test images with defects. For our purposes, defects will be introduced with simple Poisson or shot noise (a random number with a standard deviation of the square root of the pixel values).

© The Author(s), under exclusive license to APress Media, LLC, part of Springer Nature 2024
M. Paluszek, S. Thomas, *MATLAB Machine Learning Recipes*,
https://doi.org/10.1007/978-1-4842-9846-6_9

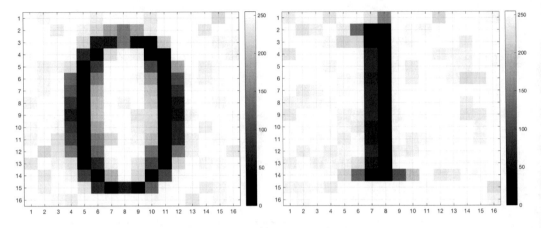

Figure 9.1: *A sample image of the digits 0 and 1 with noise added*

9.1.2 Solution

We will generate the images in MATLAB by writing a digit to axes using `text`, then creating an image using `print`. There is an option to capture the pixel data directly from print without creating an interim file, which we will utilize. We will extract the 16 by 16 pixel area with our digit, then apply the noise. We will also allow the font to be an input. See Figure 9.1 for examples.

9.1.3 How It Works

The code listing for the `CreateDigitImage` function is shown as follows. The inputs are the digit and the desired font. It creates a 16×16 pixel image of a single digit. The intermediate figure used to display the digit text is invisible. We will use the `'RGBImage'` option for `print` to get the pixel values without creating an image file. The function has options for a built-in demo that will create pixels for the digit 0 and display the image in a figure if no inputs or outputs are given. The default font if none is given is Courier.

CreateDigitImage.m

```
15  function pixels = CreateDigitImage( num, fontname )
16
17  if nargin < 1
18    num = 0;
19    CreateDigitImage( num );
20    return;
21  end
22  if nargin < 2
23    fontname = 'courier';
24  end
25
26  fonts = listfonts;
27  avail = strcmpi(fontname,fonts);
28  if ~any(avail)
```

```
29    error('MachineLearning:CreateDigitImage',...
30      'Sorry, the font ''%s'' is not available.',fontname);
31  end
32
33  f = figure('Name','Digit','visible','off');
34  a1 = axes( 'Parent', f, 'box', 'off', 'units', 'pixels', 'position', [0
        0 16 16] );
35
36  % 20 point font digits are 15 pixels tall (on Mac OS)
37  % text(axes,x,y,string)
38  text(a1,4,10,num2str(num),'fontsize',19,'fontunits','pixels','unit','
        pixels',...
39    'fontname',fontname)
40
41  % Obtain image data using print and convert to grayscale
42  cData = print('-RGBImage','-r0');
43  iGray = rgb2gray(cData);
44
45  % Print image coordinate system starts from upper left of the figure,
        NOT the
46  % bottom, so our digit is in the LAST 16 rows and the FIRST 16 columns
47  pixels = iGray(end-15:end,1:16);
48
49  % Apply Poisson (shot) noise; must convert the pixel values to double
        for the
50  % operation and then convert them back to uint8 for the sum. the uint8
        type will
51  % automatically handle overflow above 255 so there is no need to apply
        a limit.
52  noise = uint8(sqrt(double(pixels)).*randn(16,16));
53  pixels = pixels - noise;
54
55  close(f);
56
57  if nargout == 0
58    h = figure('name','Digit Image');
59    imagesc(pixels);
60    colormap(h,'gray');
61    grid on
62    set(gca,'xtick',1:16)
63    set(gca,'ytick',1:16)
64    colorbar
65  end
```

■ **TIP**   Note that we check that the font exists using `listfonts` before trying to use it, and throw an error if it's not found.

Now, we can create the training data using images generated with our new function. In the following recipes, we will use data for both a single-digit identification and a multiple-digit

259

identification net. We use a `for` loop to create a set of images and save them to a MAT-file using the helper function `SaveTS`. This saves the training sets with their input and output, and indices for training and testing, in a special structure format. Note that we scale the pixel values, which are nominally integers with a value from 0 to 255, to have values between 0 and 1.

Our data-generating script `DigitTrainingData` uses a `for` loop to create a set of noisy images for each desired digit (between 0 and 9). It saves the data along with indices for data to use for training. The pixel output of the images is scaled from 0 (black) to 1 (white), so it is suitable for neuron activation in the neural net. It has two flags at the top, one for a one-digit mode and a second to automatically change fonts.

DigitTrainingData.m

```matlab
1    %% Generate the training data
10
11   % Control switches
12   oneDigitMode = true;   % the first digit is the desired output
13   changeFonts = true;    % randomly select a font
14
15   % Number of training data sets
16   digits      = 0:5;
17   nImagesPer = 20;
18
19   % Prepare data
20   nDigits     = length(digits);
21   nImages     = nDigits*nImagesPer;
22   input       = zeros(256,nImages);
23   output      = zeros(1,nImages);
24   trainSets = [];
25   testSets  = [];
26   if (changeFonts)
27     fonts = {'times','helvetica','courier'};
28   else
29     fonts = 'times';
30     kFont = 1;
31   end
32
33   % Loop through digits
34   kImage = 1;
35   for j = 1:nDigits
36     fprintf('Digit %d\n', digits(j));
37     for k = 1:nImagesPer
38       if (changeFonts)
39         % choose a font randomly
40         kFont = ceil(rand*3);
41       end
42       pixels = CreateDigitImage( digits(j), fonts{kFont} );
43       % scale the pixels to a range 0 to 1
44       pixels = double(pixels);
45       pixels = pixels/255;
46       input(:,kImage) = pixels(:);
```

```
47      if (oneDigitMode)
48        if (j == 1)
49          output(j,kImage) = 1;
50        end
51      else
52        output(j,kImage) = 1;
53      end
54      kImage = kImage + 1;
55    end
56    sets = randperm(10);
57    trainSets = [trainSets (j-1)*nImages+sets(1:5)]; %#ok<AGROW>
58    testSets = [testSets (j-1)*nImages+sets(6:10)]; %#ok<AGROW>
59  end
60
61  % Use 75% of the images for training and save the rest for testing
62  trainSets = sort(randperm(nImages,floor(0.75*nImages)));
63  testSets = setdiff(1:nImages,trainSets);
64
65  % Save the training set to a MAT-file (dialog window will open)
66  SaveTS( input, output, trainSets, testSets );
```

The helper function will ask for a filename and save the training set. You can load it at the command line to verify the fields. Here's an example with the training and testing sets truncated:

```
>> trainingData = load('Digit0TrainingTS')
trainingData =
  a struct with fields:
    Digit0TrainingTS: [1x1 struct]

>> trainingData.Digit0TrainingTS
ans =
  a struct with fields:
         inputs: [256x120 double]
     desOutputs: [1x120 double]
      trainSets: [1 3 4 5 6 8 9 ...  115 117 118 120]
       testSets: [2 7 16 20 28 33 37 ... 112 114 116 119]
```

Note that the output field is a boolean with a value of 1 when the image is of the desired digit and 0 when it is not. In the single-digit data sets, selected by using the boolean flag oneDigitMode, the output is a single row. In a multi-digit set, it has as many rows as there are digits in the set. The images use a randomly selected font from among Times, Helvetica, and Courier if the changeFonts boolean is true. Table 9.1 shows the three training sets created using this script.

Table 9.1: *Digit Training Sets*

'Digit0TrainingTS'	Single-digit set with 120 images of the digits 0 through 5, all in the same font
'Digit0FontsTS'	Single-digit set of 0 through 5 with random fonts
'DigitTrainingTS'	Multi-digit set with 200 images of the digits 0 through 9, same font

Table 9.2: *Digit Neural Network Setup Files*

'Digit0Net'	Single-digit network
'DigitNet'	Multi-digit network

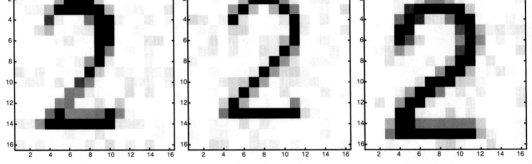

Figure 9.2: *Images of the digit 2 in different fonts*

We have created the following sets for use in these recipes:

There are also two files that define the neural net structures. These are given in Table 9.2.

Figure 9.2 shows example images of the digit 2 in the three different fonts, from Digit0TrainingTS.

9.2 Create the Neural Net Functions

9.2.1 Problem

We want to create a neural net tool that can be trained to identify digits. In this recipe, we will discuss the functions underlying the NeuralNetDeveloper tool, shown in the next recipe. This interface does not use the latest GUI-building features of MATLAB, so we will not get into detail about the GUI code itself although the full GUI is available in the companion code.

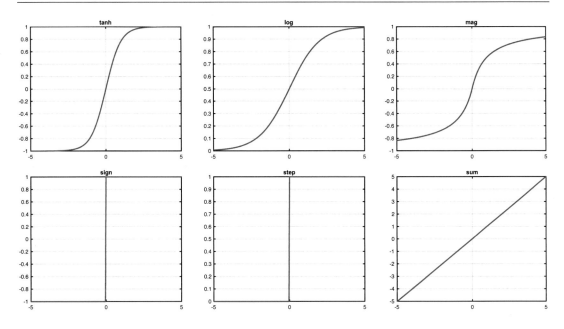

Figure 9.3: *Available neuron activation functions: sign, sigmoid mag, step, logistic (log), tanh, and sum*

9.2.2 Solution

The GUI uses a multilayer feedforward (MLFF) neural network function to classify digits. In this type of network, each neuron depends only on the inputs it receives from the previous layer. We will discuss the function which implements the neuron.

9.2.3 How It Works

The basis of the neural net is the Neuron function. Our neuron function provides six different activation types: sign, sigmoid mag, step, logistic, tanh, and sum [26]. This can be seen in Figure 9.3.

The default type of activation function is tanh. Two other functions useful in multilayer networks are exponential (sigmoid logistic function):

$$\frac{1}{1 + e^{-x}} \tag{9.1}$$

or sigmoid magnitude

$$\frac{x}{1 + |x|} \tag{9.2}$$

where "sigmoid" refers to a function with an S-shape.

It is a good idea to try different activation functions for any new problem. The activation function is what distinguishes a neural network, and machine learning, from curve fitting. The input x would be the sum of all inputs plus a bias.

■ **TIP**   The sum activation function is linear, and the output is just the sum of the inputs.

The following code shows `Neuron` which implements a single neuron in the neural net. It has as an input the type, or activation function, and the outputs include the derivative of this function. A default type of `log` is enabled (for the sigmoid logistic function).

Neuron.m

```
29  function [y, dYDX] = Neuron( x, type, t )
30
31  % Input processing
32  if( nargin < 1 )
33    x = [];
34  end
35  if( nargin < 2 )
36    type = [];
37  end
38  if( nargin < 3 )
39    t = 0;
40  end
41  if( isempty(type) )
42    type = 'log';
43  end
44  if( isempty(x) )
45    x = sort( [linspace(-5,5) 0 ]);
46  end
47
48  % Compute the function value and the derivative
49  switch lower( deblank(type) )
50    case 'tanh'
51      yX   = tanh(x);
52      dYDX = sech(x).^2;
53
54    case 'log'
55      % sigmoid logistic function
56      yX   = 1./(1 + exp(-x));
57      dYDX = yX.*(1 - yX);
58
59    case 'mag'
60      % sigmoid magnitude function
61      d    = 1 + abs(x);
62      yX   = x./d;
63      dYDX = 1./d.^2;
64
65    case 'sign'
66      yX           = ones(size(x));
67      yX(x < 0)    = -1;
68      dYDX         = zeros(size(yX));
69      dYDX(x == 0) = inf;
70
```

```
71   case 'step'
72     yX              = ones(size(x));
73     yX(x < t)     = 0;
74     dYDX           = zeros(size(yX));
75     dYDX(x == t) = inf;
76
77   case 'sum'
78     yX   = x;
79     dYDX = ones(size(yX));
80
81   otherwise
82     error([type ' is not recognized'])
83   end
84
85   % Output processing
86   if( nargout == 0 )
87     PlotSet( x, yX, 'x label', 'Input', 'y label', 'Output',...
88       'plot title', [type ' Neuron'] );
89     PlotSet( x, dYDX, 'x label','Input', 'y label','dOutput/dX',...
90       'plot title',['Derivative of ' type ' Function'] );
91   else
92     y = yX;
93   end
```

Neurons are combined into the feedforward neural network using a simple data structure of layers and weights. The input to each neuron is a combination of the signal y, the weight w, and the bias w_0, as in this line:

```
1   y   = Neuron( w*y - w0, type );
```

The output of the network is calculated by the function `NeuralNetMLFF`. This computes the output of a multilayer feedforward neural net. Note that this also outputs the derivatives as obtained from the neuron activation functions, for use in training. The function is described as follows:

NeuralNetMLFF.m

```
1   %% NEURALNETMLFF Computes the output of a multilayer feed-forward
        neural net.
2   % The input layer is a data structure that contains the network data.
3   % This data structure must contain the weights and activation functions
4   % for each layer. Calls the Neuron function.
5   %
6   % The output layer is the input data structure augmented to include
7   % the inputs, outputs, and derivatives of each layer for each run.
8   %% Form
9   %   [y, dY, layer] = NeuralNetMLFF( x, network )
```

The input and output layers are data structures containing the weights and activation functions for each layer. Our network will use backpropagation as a training method [22]. This is a gradient descent method, and it uses the output of the derivative by the network directly. Due to this use of derivatives, any threshold functions such as a step function are substituted with a sigmoid function for the training to make it continuous and differentiable. The main parameter is the learning rate α, which multiplies the gradient changes applied to the weights in each iteration. This is implemented in `NeuralNetTraining`.

The `NeuralNetTraining` function performs training, that is, computes the weights in the neurons, using backpropagation. If no inputs are given, it will do a demo for the network where node 1 and node 2 use `exp` functions for the activation functions. The function form is given as follows:

NeuralNetTraining.m

```
1   %% NEURALNETTRAINING Training using back propagation.
2   % Computes the weights for a neural net using back propagation. If no
        inputs are
3   % given it will do a demo for the network where node 1 and node 2 use
        exp
4   % functions. Calls NeuralNetMLFF which implements the network.
5   %
6   %   sin(    x) -- node 1
7   %              \ /      \
8   %               \      ---> Output
9   %              / \      /
10  %   sin(0.2*x) -- node 2
11  %
12  %% Form
13  %   [w, e, layer] = NeuralNetTraining( x, y, layer )
```

The backpropagation is performed by calling `NeuralNetMLFF` in a loop for the number of runs requested. A wait bar is displayed since training can take some time. Note that this can handle any number of intermediate layers. The field `alpha` contains the learning rate for the method.

NeuralNetTraining.m

```
137  % Perform back propagation
138  h = waitbar(0, 'Neural Net Training in Progress' );
139  for j = 1:nRuns
140     % Work backward from the output layer
141     [yN, dYN,layerT] = NeuralNetMLFF( x(:,j), temp );
142     e(:,j)           = y(:,j) - yN(:,1); % error
143
144     for k = 1:nLayers
145        layer(k,j).w  = temp.layer(k,1).w;
146        layer(k,j).w0 = temp.layer(k,1).w0;
147        layer(k,j).x  = layerT(k,1).x;
148        layer(k,j).y  = layerT(k,1).y;
```

```
149      layer(k,j).dY = layerT(k,1).dY;
150    end
151
152    % Last layer delta is calculated first
153    layer(nLayers,j).delta = e(:,j).*dYN(:,1);
154    % Intermediate layers use the subsequent layer's delta
155    for k   = (nLayers-1):-1:1
156      layer(k,j).delta = layer(k,j).dY.*(temp.layer(k+1,1).w'*layer(k+1,j
              ).delta);
157    end
158    % Now that we have all the deltas, update the weights (w) and biases
            (w0)
159    for k = 1:nLayers
160      temp.layer(k,1).w   = temp.layer(k,1).w   + layer(k,1).alpha*layer(k,
              j).delta*layer(k,j).x';
161      temp.layer(k,1).w0 = temp.layer(k,1).w0 - layer(k,1).alpha*layer(k,
              j).delta;
162    end
163
164    waitbar(j/nRuns);
165  end
166  w = temp.layer;
167  close(h);
```

9.3   Train a Network with One Output Node

9.3.1 Problem

We want to train the neural network to classify numbers. The first step is to identify a single number. In this case, we will have a single output node, and our training data will include our desired digit, starting with 0, plus a few other digits (1–5).

9.3.2 Solution

We can create this neural network with our GUI, shown in Figure 9.4. The network flows from left to right in the graphic. We can try training the net with the output node having different types, such as `sign` and `logistic`. In our case, we start with a `sigmoid` function for the hidden layer and a `step` function for the output node. Type `NeuralNetDeveloper` in the command line and hit enter or return to open the GUI. Hit the "Open" button and load the file `Digit0Net`.

If you enter the numbers in the figure, you sill see the neural net display in the figure. The box on the upper left of the GUI lets you set up the network with the number of inputs, in this case, one per pixel; the number of outputs, one because we want to identify one digit; and the number of hidden layers. The box to the right lets us design each layer. All neurons in a layer are identical. The box on the far right lets us set the weight for each input to the node and the bias

267

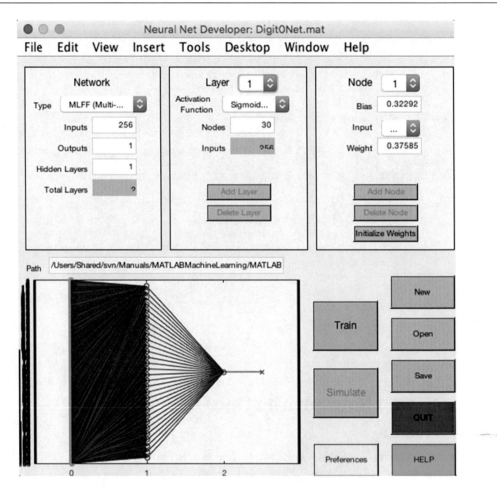

Figure 9.4: *A neural net with 256 inputs, one per pixel, an intermediate layer with 30 nodes, and one output. The image shows how the inputs are connected to the output node*

for the node. The path is the path to the training data. The display shows the resulting network. The graphic is useful, but the number of nodes in the hidden layer makes it hard to read.

Our GUI has a separate training window, Figure 9.5. It has buttons for loading and saving training sets, training, and testing the trained neural net. It will plot results automatically based on the preferences selected. In this case, we have loaded the training set from Recipe 9.1 that uses multiple fonts, `Digit0FontsTS`, which is displayed at the top of the figure window.

■ **TIP**   `close all force` closes GUIs even when they are hung in a callback.

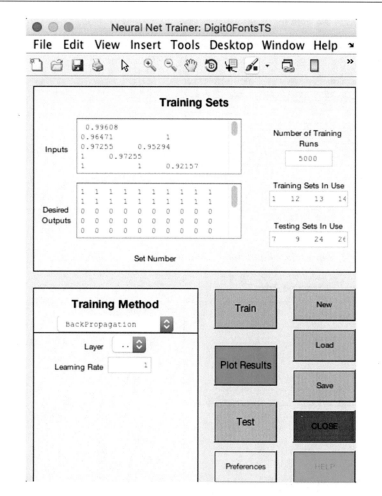

Figure 9.5: *The Neural Net Trainer GUI opens when the train button is clicked in the developer*

9.3.3 How It Works

We build the network using the GUI with 256 inputs, one for each pixel; 30 nodes in one hidden layer; and 1 output node. We load the training data from the first recipe into the Trainer GUI and must select the number of training runs. 2000 runs should be sufficient if our neuron functions are selected properly. We have an additional parameter to select, the learning rate for the backpropagation; it is reasonable to start with a value of 1.0. Note that our training data script assigned 75% of the images for training and reserved the remainder for testing, using `randperm` to extract a random set of images. The training records the weights and biases for each run and generates plots on completion. We can easily plot these for the output node, which has just 30 nodes and one bias. See Figure 9.6. To train the network, click "Train" on the GUI. This will open a new window. Then load in the file `DigitTrainingTS`. You can then train the network. Since the training starts with random weights, your results may not be exactly the same as in these plots.

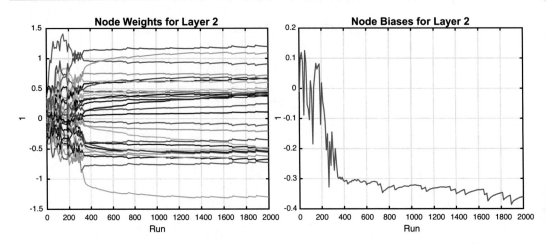

Figure 9.6: *Layer 2 node weights and the evolution of biases during training*

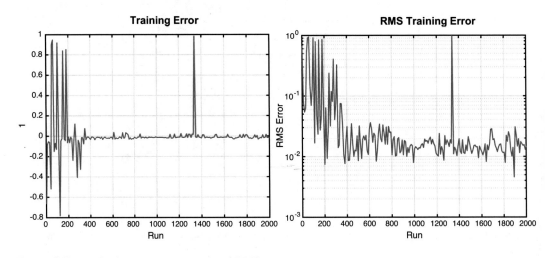

Figure 9.7: *Single-digit training error and RMS error*

The training function also outputs the training error as the net evolves and the RMS of the error, which has dropped off to near 0.01 by about run 1000 shown in Figure 9.7.

Since we have a large number of input neurons, a line plot is not very useful for visualizing the evolution of the weights for the hidden layer. However, we can view the weights at any given iteration as an image. Figure 9.8 shows the weights for the network with 30 nodes after training visualized using `imagesc`. We may wonder if we need all 30 nodes in the hidden layer or if we could extract the necessary number of features identifying our chosen digit with less. In the image on the right, the weights are shown sorted along the dimension of the input pixel for each node; we can see that only a few nodes seem to have much variation from the random values they are initialized with, especially nodes 14, 18, and 21. That is, many of our node seem to be having no impact.

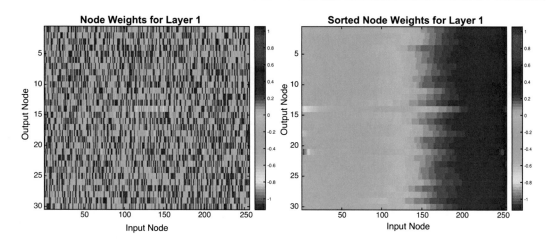

Figure 9.8: *Single-digit network, 30 node hidden layer weights. The plot on the left shows the weight value. The plot on the right shows the weights sorted by pixel for each node*

Since this visualization seems helpful, we add the code to the training GUI after the generation of the weight line plots. We create two images in one figure, the initial value of the weights on the left and the training values on the right. The HSV colormap looks more striking here than the default Parula map. The code that generates the images in `NeuralNetTrainer` looks like this:

```
% New figure: weights as image
newH = figure('name',['Node Weights for Layer ' num2str(j)]);
endWeights = [h.train.network(j,1).w(:);h.train.network(j,end).w(:)];
minW = min(endWeights);
maxW = max(endWeights);
subplot(1,2,1)
imagesc(h.train.network(j,1).w,[minW maxW])
colorbar
ylabel('Output Node')
xlabel('Input Node')
title('Weights Before Training')
subplot(1,2,2)
imagesc(h.train.network(j,end).w,[minW maxW])
colorbar
xlabel('Input Node')
title('Weights After Training')
        colormap hsv
h.resultsFig = [newH; h.resultsFig];
```

Note that we compute the minimum and maximum weight values among both the initial and final iterations, for scaling the two colormaps the same. Now, since many of our 30 initial nodes seemed unneeded, we reduce the number of nodes in that layer to 10, reinitialize the weights (randomly), and train again. Now we get our new figure with the weights displayed as an image before and after the training (Figure 9.9).

271

Node weights for Layer 1:

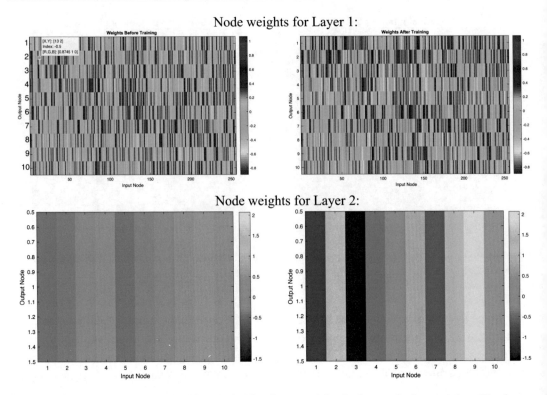

Figure 9.9: *Single-digit network, 10 node hidden layer weights before and after training. The first row shows the data for the first layer, and the second for the second layer, which has just one output*

Now we can see more patches of colors that have diverged from the initial random weights in the images for the 256 pixel weights, and we see the clear variation in the weights for the second layer as well. The GUI allows you to save the trained net for future use.

9.4 Testing the Neural Network

9.4.1 Problem

We want to test the single-digit neural net that we trained in the previous recipe.

9.4.2 Solution

We can test the network with inputs that were not used in training. This is explicitly allowed in the GUI as it has separate indices for the training data and testing data. We selected 75% of our sample images for training and saved the remaining images for testing in our `DigitTrainingData` script from Recipe 9.1.

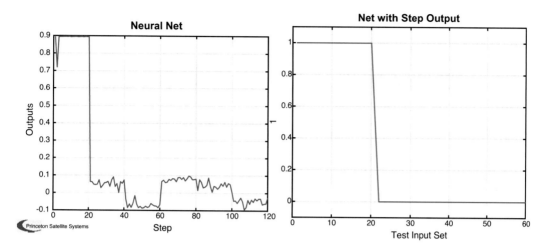

Figure 9.10: *Neural net results with sigmoid (left) and step (right) activation functions*

9.4.3 How It Works

In the case of our GUI, simply click the test button to run the neural network with each of the cases selected for testing.

Figure 9.10 shows the results for a network with the output node using the sigmoid magnitude function and another case with the output node using a step function – that is, the output is limited to 0 or 1. Note that the first 20 images in the data set are the digit 0, with an output value of 1, and the rest are the digits 1 to 5, with an output value of 0. For the step function, the output is 1 for the first 20 sets and 0 for all other sets, as desired. The sigmoid is similar except that instead of being 0 after 20 sets, the output varies between $+0.1$ and -0.1. Between 20 and 120, it almost averages to 0, the same as the result from the step. This shows that the activation functions are similarly interpreting the data.

9.5   Train a Network with Many Outputs

9.5.1 Problem

We want to build a neural net that can detect all ten digits separately.

9.5.2 Solution

Add nodes so that the output layer has ten nodes, each of which will be 0 or 1 when the representative digit (0–9) is input. Try the output nodes with different functions, like logistic and step. Now that we have more digits, we will go back to having 30 nodes in the hidden layer.

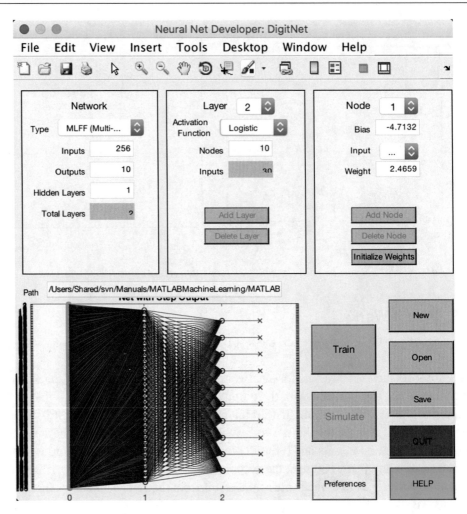

Figure 9.11: *Net with multiple outputs*

9.5.3 How It Works

Our training data now consists of all ten digits, with a binary output of zeros with a one in the correct slot. The network setup is shown in Figure 9.11. For example, the digit 1 will be represented as

$$[0\ 1\ 0\ 0\ 0\ 0\ 0\ 0\ 0]$$

The digit 3 would have a 1 in the fourth element. We follow the same procedure for training. We initialize the net, load the training set into the GUI, and specify the number of training runs for the backpropagation.

The training data, in Figure 9.12, shows that much of the learning is achieved in the first 3000 runs.

The test data, in Figure 9.13, shows that each set of digits (in sets of 20 in this case, for 200 total tests) is correctly identified.

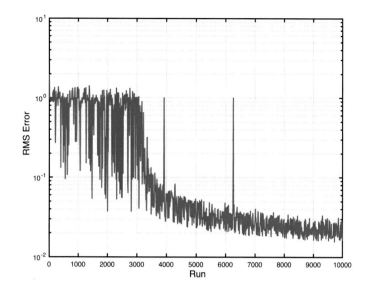

Figure 9.12: *Training RMS for multiple digit neural net*

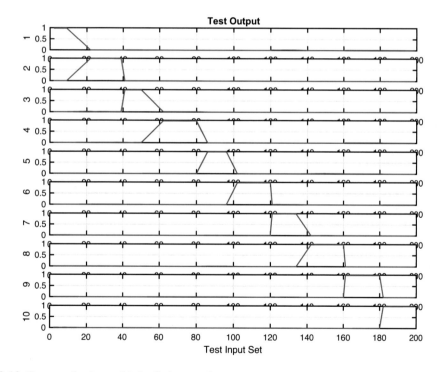

Figure 9.13: *Test results for multiple-digit neural net*

Once you have saved a net that is working well to a MAT-file, you can call it with new data using the function `NeuralNetMLFF`.

```
>> data = load('NeuralNetMat');
>> network = data.DigitsStepNet;
>> y = NeuralNetMLFF( DigitTrainingTS.inputs(:,1), data.DigitsStepNet )
y =
     1
     0
     0
     0
     0
     0
     0
     0
     0
     0
```

It is valuable to use visualization of the neural net weights, to gain insight into the problem, and our problem is small enough that we can do so with images. We can view a single set of 256 weights for one hidden neuron as a 16×16 pixel image, and view the whole set with each neuron in its row as before (Figure 9.14), to see the patterns emerging.

You can see parts of digits as mini-patterns in the individual node weights. Simply use `imagesc` with `reshape` like this:

```
>> figure;
>> imagesc(reshape(net.DigitsStepNet.layer(1).w(23,:),16,16));
>> title('Weights to Hidden Node 23')
```

and see images as in Figure 9.15. These three nodes (chosen at random) show a 1, 2, and 3. We would expect the 30 nodes to each have "noisy" replicas of the digits.

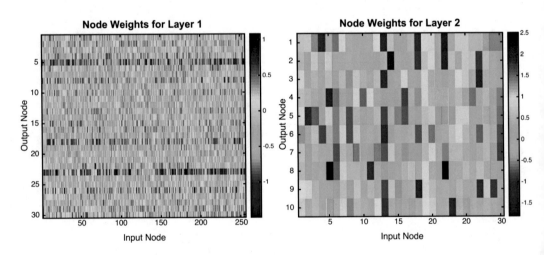

Figure 9.14: *Multiple-digit neural net weights*

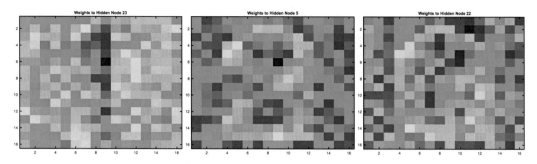

Figure 9.15: *Multiple-digit neural net weights*

9.6 Summary

This chapter has demonstrated neural learning to classify digits. An interesting extension to our tool would be the use of image datastores, rather than a matrix representation of the input data. Table 9.3 lists the functions and scripts included in the companion code. The digits were trained with data in which the digits were in the same orientation. If the training data had data in different orientations, a neural net could still identify the digits, but a bigger net and much more training data would be required. The final chapter on star-based attitude determination has a neural network that does not care about rotation angle.

Table 9.3: *Chapter Code Listing*

File	Description
DigitTrainingData	Creates a training set of digit images
CreateDigitImage	Creates a noisy image of a single digit
Neuron	Model an individual neuron with multiple activation functions
NeuralNetMLFF	Computes the output of a multilayer feedforward neural net
NeuralNetTraining	Training with backpropagation
DrawNeuralNet	Displays a neural net with multiple layers
SaveTS	Saves a training set MAT-file with index data

CHAPTER 10

■ ■ ■

Data Classification with Decision Trees

In this chapter, we will develop the theory for binary decision trees. Decision trees can be used to classify data and fall into the Learning category in our Autonomous Learning taxonomy. Binary trees are easiest to implement because each node branches to two other nodes, or none. We will create functions for the decision trees and to generate sets of data to classify. Figure 10.1 shows a simple binary tree. Point "a" is in the upper-left quadrant. The first

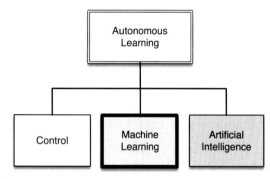

binary test finds that its x value is greater than one. The next test finds that its y value is greater than one and puts it in set 2. Although the boundaries show square regions, the binary tree tests for regions that go to infinity in both x and y.

A binary decision tree is a decision tree in which at each decision node there are only two decisions to make. Once you make a decision, the next decision node provides you with two additional options. Each node accepts a binary value of zero or one. A zero sends you down one path, and one sends you down the other path. At each decision node, you are testing a new variable. When you get to the bottom, you will have found a path where all of the values are true. The problem with a binary tree of n variables is that it will have $2^n - 1$ nodes. Four variables would require 15 decision nodes. Eight variables would require 65 decision nodes and so forth. If the order of testing variables is fixed, we call it an ordered tree.

For classification, we are assuming that we can make a series of binary decisions to classify something. If we can, we can implement the reasoning in a binary tree.

© The Author(s), under exclusive license to APress Media, LLC, part of Springer Nature 2024
M. Paluszek, S. Thomas, *MATLAB Machine Learning Recipes*,
https://doi.org/10.1007/978-1-4842-9846-6_10

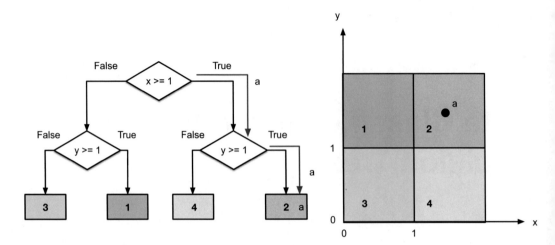

Figure 10.1: *A simple binary tree with one point to classify*

10.1   Generate Test Data

10.1.1 Problem

We want to generate a set of training and testing data for classification.

10.1.2 Solution

Write a function using `rand` to generate data over a selected range in two dimensions, x and y.

10.1.3 How It Works

The function `ClassifierSets` generates random data and assigns them to classes. The function call is

ClassifierSets.m

```
21   function p = ClassifierSets( n, xRange, yRange, name, v, f, setName )
```

The first argument to `ClassifierSets` is the square root of several points. The second, `xRange`, gives the x range for the data, and the third, `yRange`, gives the y range. The n^2 points will be placed randomly in this region. The next argument is a cell array with names of the sets, `name`. These are used for plot labels. The remaining input is a list of vertices, `v`, and the faces, `f`. The faces select the vertices to use in each polygon. The faces connect the vertices into specific polygons. `f` is a cell array since each face array can be any length. A triangle has a length of 3; a hexagon has a length of 6. Triangles, rectangles, and hexagons can be easily meshed so that there are no gaps.

Classes are defined by adding polygons that divide the data into regions. Any polygon can be used. You should pick polygons so that there are no gaps. Rectangles are easy, but you could also use uniformly sized hexagons. The following code is the built-in demo. The demo is the last subfunction in the function. This specifies the vertices and faces.

ClassifierSets.m

```
107   function Demo
108
109   v = [0 0;0 4; 4 4; 4 0; 0 2; 2 2; 2 0;2 1;4 1;2 1];
110   f = {[5 6 7 1] [5 2 3 9 10 6] [7 8 9 4]};
111   ClassifierSets( 5, [0 4], [0 4], {'width', 'length'}, v, f );
```

In this demo, there are three polygons. All points are defined in a square ranging from 0 to 4 in both the x and y directions.

The other subfunctions are `PointInPolygon` and `Membership`. `Membership` determines if a point is in a polygon. `Membership` calls `PointInPolygon` to assign points to sets. `ClassifierSets` randomly puts points in the regions. It figures out which region each point is in using this code in the function, `PointInPolygon`.

ClassifierSets.m

```
86    function r = PointInPolygon( p, v )
87
88    m = size(v,2);
89
90    % All outside
91    r = 0;
92
93    % Put the first point at the end to simplify the looping
94    v = [v v(:,1)];
95
96    for i = 1:m
97       j   = i + 1;
98       v2J = v(2,j);
99       v2I = v(2,i);
100      if (((v2I > p(2)) ~= (v2J > p(2))) && ...
101          (p(1) < (v(1,j) - v(1,i)) * (p(2) - v2I) / (v2J - v2I) + v(1,i)))
102         r = ~r;
103      end
104   end
```

This code can determine if a point is inside a polygon defined by a set of vertices. It is used frequently in computer graphics and games when you need to know if one object's vertex is in another polygon. You could correctly argue that this function could replace our decision tree logic for this type of problem. However, a decision tree can compute membership for more complex sets of data. Our classifier set is simple and makes it easy to validate the results.

Run `ClassifierSets` to see the demo shown in Figure 10.2. Given the input ranges, it determines the membership of randomly selected points. `p` is a data structure that holds the vertices and the membership. It plots the points after creating a new figure using `NewFigure`. It then uses `patch` to create the rectangular regions.

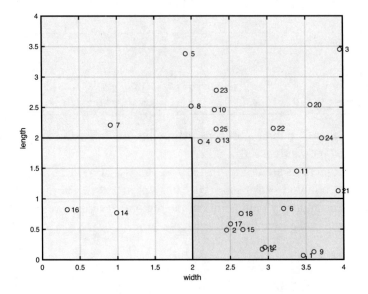

Figure 10.2: *Classifier set with three regions from the demo. Two are rectangles, and one is L shaped*

ClassifierSets.m

```
34   p.x      = (xRange(2) - xRange(1))*(rand(n,n)-0.5) + mean(xRange);
35   p.y      = (yRange(2) - yRange(1))*(rand(n,n)-0.5) + mean(yRange);
36   p.m      = Membership( p, v, f );
38
39   NewFigure(setName);
40   i = 0;
41   drawNum = n^2 < 50;
42   for j = 1:n
43     for k = 1:n
44       i = i + 1;
45       plot(p.x(k,j),p.y(k,j),'marker','o','MarkerEdgeColor','k')
46       if( drawNum )
47         text(p.x(k,j),p.y(k,j),sprintf(' %3d',i));
48       end
49       hold on
50     end
51   end
52
53   m = length(f);
54   a = linspace(0,2*pi-2*pi/m,m)';
55   c = abs(cos([a a+pi/6 a+3*pi/5]));
56
57   for k = 1:m
58     patch('vertices',v,'faces',f{k},'facecolor',c(k,:),'facealpha',0.1)
59   end
60
```

```
61  xlabel(name{1});
62  ylabel(name{2});
63  grid on
```

The function shows the data numbers if there are fewer than 50 points. The MATLAB function `patch` is used to generate the polygons. The code shows a range of graphics coding including the use of graphics parameters. Notice the way we create m colors.

■ **TIP**   You can create an unlimited number of colors for plots using `linspace` and `cos`.

`ClassifierTestSet` can generate test sets or demonstrate a trained decision tree. Figure 10.2 shows that the classification regions are regions with sides parallel to the x- or y-axis. The regions should not overlap.

10.2 Drawing Trees

10.2.1 Problem

We want to draw a binary decision tree to show decision tree thinking.

10.2.2 Solution

The solution is to use MATLAB graphics functions – `patch`, `text`, and `line` – to draw a tree.

10.2.3 How It Works

The function `DrawBinaryTree` draws any binary tree. The function call is

DrawBinaryTree.m

```
25  function d = DrawBinaryTree( d, name )
```

You pass it a data structure, d, with the decision criteria in a cell array. The `name` input is optional. It has a default option for the name. The boxes start from the left and go row by row. In a binary tree, the number of rows is related to the number of nodes through the formula for a geometric series:

$$m = \log_2(n) \tag{10.1}$$

where m is the number of rows and n is the number of boxes. Therefore, the function can compute the number of rows.

The function starts by checking the number of inputs and either running the demo or returning the default data structure. When you write a function, you should always have defaults for anything where one is possible.

■ **TIP**   Whenever possible, have default inputs for function arguments.

It immediately creates a new figure with that name. It then steps through the boxes, assigning them to rows based on it being a binary tree. The first row has one box, the next two boxes, the following four boxes, etc. As this is a geometric series, it will soon get unmanageable! This points to a problem with decision trees. If they have a depth of more than four, even drawing them is impossible. As it draws the boxes, it computes the bottom and top points that will be the anchors for the lines between the boxes. After drawing all the boxes, it draws all the lines. All of the drawing functionality is in the subfunction `DrawBox`.

DrawBinaryTree>DrawBox

```
93   function DrawBox( t, x, y, w, h, d )
94   %% DrawBinaryTree>DrawBox
95   % Draw boxes and text
96   % DrawBox( t, x, y, w, h, d )
97
98   v = [x y 0;x y+h 0; x+w y+h 0;x+w y 0];
99
100  patch('vertices',v,'faces',[1 2 3 4],'facecolor',[1;1;1]);
101
102  text(x+w/2,y + h/2,t,'fontname',d.font,'fontsize',...
103      d.fontSize,'HorizontalAlignment','center');
```

This draws a box using the `patch` function and the text using the `text` function. `'facecolor'` is white. RGB numbers go from zero to one. Setting `'facecolor'` to [1 1 1] makes the face white and leaves the edges black. As with all MATLAB graphics, there are dozens of properties that you can edit to produce beautiful graphics. Notice the extra arguments in `text`. The most interesting is `'HorizontalAlignment'` in the last line. It allows you to center the text in the box. MATLAB does all the figuring of font sizes for you.

The following listing shows the code in `DrawBinaryTree`, for drawing the tree, starting after checking for demos. The function returns the default data structure if one output is specified and no inputs are specified. The first part of the code creates a new figure and draws the boxes at each node. It also creates arrays for the box locations for use in drawing the lines that connect the boxes. It starts with the default argument for the name. The code in the first set of loops draws the boxes for the trees. `rowID` is a cell array. Each row in the cell is an array. A cell array allows each cell to be different. This makes it easy to have different length arrays in the cell. If you used a standard matrix, you would need to resize rows as new rows were added.

DrawBinaryTree.m

```
37   if( nargin < 2 )
38      name = 'Binary Tree';
39   end
40
41   NewFigure(name);
42   m        = length(d.box);
43   nRows    = ceil(log2(m+1));
44   w        = d.w;
```

```
45  h         = d.h;
46  i         = 1;
47  x         = -w/2;
48  y         = 1.5*nRows*h;
49  nBoxes    = 1;
50  bottom    = zeros(m,2);
51  top       = zeros(m,2);
52  rowID     = cell(nRows,1);
53  % Draw a box at each node
54  for k = 1:nRows
55    for j = 1:nBoxes
56      bottom(i,:)     = [x+w/2 y ];
57      top(i,:)        = [x+w/2 y+h];
58      DrawBox(d.box{i},x,y,w,h,d);
59      rowID{k}        = [rowID{k} i];
60      i               = i + 1;
61      x               = x + 1.5*w;
62      if( i > length(d.box) )
63        break;
64      end
65    end
66    nBoxes  = 2*nBoxes;
67    x       = -(0.25+0.5*(nBoxes/2-1))*w - nBoxes*w/2;
68    y       = y - 1.5*h;
69  end
```

The remaining code draws the lines between the boxes.

DrawBinaryTree.m

```
71  % Draw the lines between boxes
72  for k = 1:length(rowID)-1
73    iD = rowID{k};
74    i0 = 0;
75    % Work from left to right of the current row
76    for j = 1:length(iD)
77      x(1) = bottom(iD(j),1);
78      y(1) = bottom(iD(j),2);
79      iDT  = rowID{k+1};
80      if( i0+1 > length(iDT) )
81        break;
82      end
83      for i = 1:2
84        x(2) = top(iDT(i0+i),1);
85        y(2) = top(iDT(i0+i),2);
86        line(x,y);
87      end
88      i0 = i0 + 2;
89    end
90  end
91  axis off
```

The following built-in demo draws a binary tree. The demo creates three rows. It starts with the default data structure. You only have to add strings for the decision points. The boxes are in a flat list.

DrawBinaryTree.m

```
116  function Demo
117  %% DrawBinaryTree>Demo
118  % Draw a simple binary data tree
119
120  d                 = DefaultDataStructure;
121  d.box{1}          = 'a > 0.1';
122  d.box{2}          = 'b > 0.2';
123  d.box{3}          = 'b > 0.3';
124  d.box{4}          = 'a > 0.8';
125  d.box{5}          = 'b > 0.4';
126  d.box{6}          = 'a > 0.2';
127  d.box{7}          = 'b > 0.3';
128
129  DrawBinaryTree( d );
```

Notice that it calls the subfunction `DefaultDataStructure` to initialize the demo.

DrawBinaryTree>DefaultDataStructure

```
105  function d = DefaultDataStructure
106  %% DrawBinaryTree>DefaultDataStructure
107  % Default data structure
108
109  d                 = struct();
110  d.fontSize        = 12;
111  d.font            = 'courier';
112  d.w               = 1;
113  d.h               = 0.5;
114  d.box             = {};
```

■ **TIP**   Always have the function return its default data structure. The default should have values that work.

It starts with the default argument for the name. The loops draw the boxes for the trees `rowID` is a cell array. Each row in the cell is an array. A cell array allows each cell to be different. This makes it easy to have different length arrays in the cell. If you used a standard matrix, you would need to resize rows as new rows were added. The binary tree resulting from the demo is shown in Figure 10.3. The text in the boxes could be anything you want.

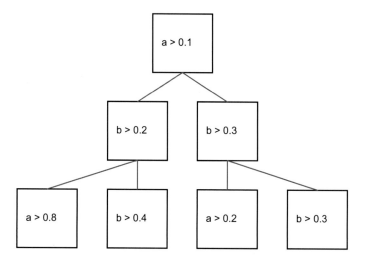

Figure 10.3: *Binary tree from the demo in* `DrawBinaryTree`

The inputs for `box` could have been done in a loop. You could create them using `sprintf`. For example, for the first box, you could write

```
d.box{1}= sprintf('%s %s %3.1f','a','>',0.1);
```

and put similar code in a loop.

10.3   Implementation

Decision trees are the main focus of this chapter. We'll start by looking at how we determine if our decision tree is working correctly. We'll then hand-build a decision tree and finally write learning code to generate the decisions for each block of the tree.

10.3.1 Problem

We need to measure the homogeneity of a set of data at different nodes on the decision tree. A data set is homogeneous if the points are similar to each other. For example, if you were trying to study grade points in a school with an economically diverse population, you would want to know if your sample was all children from wealthy families. Our goal in the decision tree is to end up with homogeneous sets.

10.3.2 Solution

The solution is to implement the Gini impurity measure for a set of data. The function will return a single number as the homogeneity measure. The Gini impurity tells you how often you would pick the wrong element of a set if you randomly selected that element.

10.3.3 How It Works

The homogeneity measure is called the information gain (IG). The information gain is defined as the increase in information by splitting at the node. This is

$$\Delta I = I(p) - \frac{N_{c_1}}{N_p} I(c_1) - \frac{N_{c_2}}{N_p} I(c_2) \tag{10.2}$$

where I is the impurity measure and N is the number of samples at that node. If our tree is working, it should go down, eventually to zero or a very small number. In our training set, we know the class of each data point. Therefore, we can determine the information gained. Essentially, we have gained information if the mixing decreases in the child nodes. For example, in the first node in a decision tree, all the data is mixed. There are two child nodes for the first node. After the decision in the first node, we expect that each child node will have more of one class than the other child node. We look at the percentages of classes in each node and look for the maximum increase in non-homogeneity.

There are three impurity measures:

- Gini impurity

- Entropy

- Classification error

Gini impurity, I_G, is the criterion to minimize the probability of misclassification. We don't want to push a sample into the wrong category.

$$I_G = 1 - \sum_{1}^{c} p(i|t)^2 \tag{10.3}$$

$p(i|t)$ is the proportion of the samples in class c_i at node t. For a binary class entropy, I_E, is either zero or one.

$$I_E = 1 - \sum_{1}^{c} p(i|t) \log_2 p(i|t) \tag{10.4}$$

The classification error, I_C, is

$$I_C = 1 - \max p(i|t) \tag{10.5}$$

We will use Gini impurity in the decision tree. The following code implements the Gini measure. The first part just decides whether it is initializing the function or updating it. All data is saved in the data structure d. This is often easier than using global data. One advantage is that you can use the function multiple times in the same script or function without mixing up the persistent data in the function.

HomogeneityMeasure.m

```
23  function [i, d] = HomogeneityMeasure( action, d, data )
24
25  if( nargin == 0 )
26    if( nargout == 1 )
27      i = DefaultDataStructure;
28    else
29      Demo;
30    end
31    return
32  end
33
34  switch lower(action)
35    case 'initialize'
36      d = Initialize( d, data );
37      i = d.i;
38    case 'update'
39      d = Update( d, data );
40      i = d.i;
41    otherwise
42      error('%s is not an available action',action);
43  end
```

Initialize initializes the data structure and computes the impurity measures for the data. There is one class for every different value of the data. For example, [1 2 3 3] would have three classes.

HomogeneityMeasure.m

```
64  function d = Initialize( d, data )
65  %% HomogeneityMeasure>Initialize
66
67  m       = reshape(data, [],1);
68  c       = 1:max(m);
69  n       = length(m);
70  d.dist  = zeros(1,c(end));
71  d.class = c;
72  if( n > 0 )
73    for k = 1:length(c)
74      j         = find(m==c(k));
75      d.dist(k) = length(j)/n;
76    end
77  end
78  d.i = 1 - sum(d.dist.^2);
```

The demo is shown as follows. We try different four sets of data and get the measures. Zero is homogeneous. One means there is no data.

HomogeneityMeasure.m

```
88   function d = Demo
89   % Demonstrate the homogeniety measure for a data set.
90
91   data    = [1 2 3 4 3 1 2 4 4 1 1 1 2 2 3 4]; fprintf('%2.0f',data);
92   d       = HomogeneityMeasure;
93   [i, d]  = HomogeneityMeasure( 'initialize', d, data );
94   fprintf('\nHomogeneity Measure %6.3f\n',i);
95   fprintf('Classes            [%1d %1d %1d %1d]\n',d.class);
96   fprintf('Distribution       [%5.3f %5.3f %5.3f %5.3f]\n',d.dist);
97
98   data = [1 1 1 2 2]; fprintf('%2.0f',data);
99   [i, d] = HomogeneityMeasure( 'update', d, data );
100  fprintf('\nHomogeneity Measure %6.3f\n',i);
101  fprintf('Classes            [%1d %1d %1d %1d]\n',d.class);
102  fprintf('Distribution       [%5.3f %5.3f %5.3f %5.3f]\n',d.dist);
103
104  data = [1 1 1 1]; fprintf('%2.0f',data);
105  [i, d] = HomogeneityMeasure( 'update', d, data );
106  fprintf('\nHomogeneity Measure %6.3f\n',i);
107  fprintf('Classes            [%1d %1d %1d %1d]\n',d.class);
108  fprintf('Distribution       [%5.3f %5.3f %5.3f %5.3f]\n',d.dist);
109
110  data = []; fprintf('%2.0f',data);
111  [i, d] = HomogeneityMeasure( 'update', d, data );
112  fprintf('\nHomogeneity Measure %6.3f\n',i);
113  fprintf('Classes            [%1d %1d %1d %1d]\n',d.class);
114  fprintf('Distribution       [%5.3f %5.3f %5.3f %5.3f]\n',d.dist);
```

`i` is the homogeneity measure. `d.dist` is the fraction of the data points that have the value of that class. The class is the distinct values. The outputs of the demo are shown as follows:

```
>> HomogeneityMeasure
 1 2 3 4 3 1 2 4 4 1 1 1 2 2 3 4
Homogeneity Measure  0.742
Classes            [1 2 3 4]
Distribution       [0.312 0.250 0.188 0.250]
 1 1 1 2 2
Homogeneity Measure  0.480
Classes            [1 2 3 4]
Distribution       [0.600 0.400 0.000 0.000]
 1 1 1 1
Homogeneity Measure  0.000
Classes            [1 2 3 4]
Distribution       [1.000 0.000 0.000 0.000]

Homogeneity Measure  1.000
```

```
Classes              [1 2 3 4]
Distribution         [0.000 0.000 0.000 0.000]
```

The second to last set has a zero which is the desired value. If there are no inputs, it returns a one since by definition for a class to exist it must have members.

10.4   Creating a Tree

10.4.1 Problem

We want to implement a decision tree for classifying data with two parameters.

10.4.2 Solution

The solution is to write a binary decision tree function in MATLAB called `DecisionTree`.

10.4.3 How It Works

A decision tree [24] breaks down data by asking a series of questions about the data. Our decision trees will be binary in that there will be a yes or no answer to each question. For each feature in the data, we ask one question per decision node. This always splits the data into two child nodes. We will be looking at two parameters that determine class membership. The parameters will be numerical measurements.

At the following nodes, we ask additional questions further splitting the data. Figure 10.4 shows the parent/child structure. We continue this process until the samples at each node are in one of the classes. At each node, we want to ask the question that provides us with the most information about the class in which our samples reside. In constructing our decision tree for a two-parameter classification, we have two decisions at each node:

• Which parameter (x or y) to check

• What value of the parameter to use in the decision

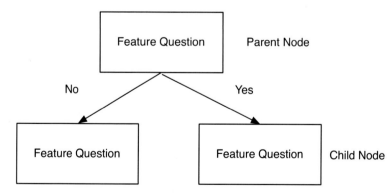

Figure 10.4: *Parent/child nodes*

Training is done using the Gini values given in the previous recipe. We use the MAT-LAB function `fminbnd` at each node, once for each of the two parameters. `fminbnd` is a one-dimensional local minimizer that finds the minimum of a function between two specified endpoints. If you know the range of interest, then this is a very effective way to find the minimum.

$$\min_x f(x) \text{ such that } x_1 < x < x_2 \tag{10.6}$$

There are two actions, "train" and "test." The "train" action creates the decision tree, and "test" runs the generated decision tree. You can also input your decision tree. `FindOptimalAction` finds the parameter that minimizes the inhomogeneity on both sides of the division. The function called by `fminbnd` is `RHSGT`. We only implement the greater than action. The function call is

```
21   function [d, r] = DecisionTree( action, d, t )
```

`action` is a string that is either "train" or "test." `d` is the data structure that defines the tree. `t` is the inputs for either training or testing. The outputs are the updated data structure and `r` with the results.

The function is first called with training data, and the action is "train." The main function is short.

DecisionTree.m

```
32   switch lower(action)
33     case 'train'
34       d = Training( d, t );
35       d.box(1)
36
37     case 'test'
38       for k = 1:length(d.box)
39         d.box(k).id = [];
40       end
41       [r, d] = Testing( d, t );
42       for k = 1:length(d.box)
43         d.box(k)
44       end
45     otherwise
46       error('%s is not an available action',action);
47   end
```

We added the error case `otherwise` for completeness. We use `lower` to eliminate case sensitivity. `Training` creates the decision tree. A decision tree is a set of boxes connected by lines. A parent box has two child boxes if it is a decision box. A class box has no children. The subfunction `Training` trains the tree. It adds boxes at each node.

DecisionTree.m

```
49  %% DecisionTree>Training
50  function d = Training( d, t )
51  [n,m]    = size(t.x);
52  nClass   = max(t.m);
53  box(1)   = AddBox( 1, 1:n*m, [] );
54  box(1).child = [2 3];
55  [~, dH] = HomogeneityMeasure( 'initialize', d, t.m );
56
57  class    = 0;
58  nRow     = 1;
59  kR0      = 0;
60  kNR0     = 1; % Next row;
61  kInRow   = 1;
62  kInNRow  = 1;
63  while( class < nClass )
64    k     = kR0 + kInRow;
65    idK   = box(k).id; % Data that is in the box and to use to compute
          the next action
66    % Enter this loop if it not a non-decision box
67    if( isempty(box(k).class) )
68      [action, param, val, cMin] = FindOptimalAction( t, idK, d.xLim, d.
          yLim, dH );
69      box(k).value             = val;
70      box(k).param             = param;
71      box(k).action            = action;
72      x                        = t.x(idK);
73      y                        = t.y(idK);
74      if( box(k).param == 1 ) % x
75        id  = find(x >    d.box(k).value );
76        idX = find(x <=   d.box(k).value );
77      else % y
78        id  = find(y >  d.box(k).value );
79        idX = find(y <=  d.box(k).value );
80      end
81      % Child boxes
82      if( cMin < d.cMin) % Means we are in a class box
83        class          = class + 1;
84        kN             = kNR0 + kInNRow;
85        box(k).child   = [kN kN+1];
86        box(kN)        = AddBox( kN, idK(id), class  );
87        class          = class + 1;
88        kInNRow        = kInNRow + 1;
89        kN             = kNR0 + kInNRow;
90        box(kN)        = AddBox( kN, idK(idX), class );
91        kInNRow        = kInNRow + 1;
92      else
93        kN             = kNR0 + kInNRow;
94        box(k).child   = [kN kN+1];
95        box(kN)        = AddBox( kN, idK(id) );
96        kInNRow        = kInNRow + 1;
```

```
97          kN                = kNR0 + kInNRow;
98          box(kN)           = AddBox( kN, idK(idX) );
99          kInNRow           = kInNRow + 1;
100       end
101     end
102           % Update current row
103     kInRow    = kInRow + 1;
104     if( kInRow > nRow )
105       kR0         = kR0 + nRow;
106       nRow        = 2*nRow; % Add two rows
107       kNR0        = kNR0 + nRow;
108       kInRow      = 1;
109       kInNRow     = 1;
110     end
111   end
112
113   for k = 1:length(box)
114     if( ~isempty(box(k).class) )
115       box(k).child = [];
116     end
117     box(k).id = [];
118     fprintf(1,'Box %3d action %2s Value %4.1f\n',k,box(k).action,box(k).
            value);
119   end
120
121   d.box = box;
```

We use `fminbnd` to find the optimal switch point. We need to compute the homogeneity on both sides of the switch and sum the values. The sum is minimized by `fminbnd` in the subfunction `FindOptimalAction`. This code is designed for rectangular region classes. Other boundaries won't necessarily work correctly. The code is fairly involved. It needs to keep track of the box numbering to make the parent-child connections. When the homogeneity measure is low enough, it marks the boxes as containing the classes.

The data structure `box` has multiple fields. One is the action to be taken in a decision box. The `param` is 1 for x and anything else for y. That determines if it is making the decision based on x or y. The `value` is the value used in the decision. `child` are indexed to the box children. The remaining code determines which row the box is in. `class` boxes have no children. The fields are shown in Table 10.1.

Table 10.1: *Box Data Structure Fields*

Field	Decision Box	Class Box
action	String	Not used
value	Value to be used in the decision	Not used
param	x or y	Not used
child	Array with two children	Empty
id	Empty	ID of data in the class
class	Class ID	Not used

10.5   Handmade Tree

10.5.1 Problem

We want to test a handmade decision tree.

10.5.2 Solution

The solution is to write a script to test a handmade decision tree.

10.5.3 How It Works

We write the test script `SimpleClassifierDemo` shown as follows. It uses the `'test'` action for `DecisionTree`. It generates 5^2 points. We create rectangular regions so that the face arrays have four elements for each polygon. `DrawBinaryTree` draws the tree.

SimpleClassifierDemo.m

```
11   d = DecisionTree;
12
13   % Vertices for the sets
14   v = [ 0 0; 0 4; 4 4; 4 0; 2 4; 2 2; 2 0; 0 2; 4 2];
15
16   % Faces for the sets
17   f = { [6 5 2 8] [6 7 4 9] [6 9 3 5] [1 7 6 8] };
18
19   % Generate the testing set
20   pTest  = ClassifierSets( 5, [0 4], [0 4], {'width', 'length'}, v, f, '
         Testing Set' );
21
22   % Test the tree
23   [d, r] = DecisionTree( 'test',  d, pTest  );
24
25   q = DrawBinaryTree;
26   c = 'xy';
27   for k = 1:length(d.box)
28     if( ~isempty(d.box(k).action) )
29       q.box{k} = sprintf('%c %s %4.1f',c(d.box(k).param),d.box(k).action,
             d.box(k).value);
```

```
30    else
31      q.box{k} = sprintf('Class %d',d.box(k).class);
32    end
33  end
34  DrawBinaryTree(q);
35
36  m = reshape(pTest.m,[],1);
37
38  for k = 1:length(r)
39    fprintf(1,'Class %d\n',m(r{k}(1)));
40    for j = 1:length(r{k})
41      fprintf(1,'%d ',r{k}(j));
42    end
43    fprintf(1,'\n')
44  end
```

`SimpleClassifierDemo` uses the handmade example in the default data of `DecisionTree`.

DecisionTree.m

```
203  function d = DefaultDataStructure
204  %% DecisionTree>DefaultDataStructure
205  % Generate a default data structure
206  d.tree         = DrawBinaryTree;
207  d.threshold    = 0.01;
208  d.xLim         = [0 4];
209  d.yLim         = [0 4];
210  d.data    = [];
211  d.cMin         = 0.01;
212  d.box(1)    = struct('action','>','value',2,'param',1,'child',[2 3],'id
               ',[],'class',[]);
213  d.box(2)    = struct('action','>','value',2,'param',2,'child',[4 5],'id
               ',[],'class',[]);
214  d.box(3)    = struct('action','>','value',2,'param',2,'child',[6 7],'id
               ',[],'class',[]);
215
216  for k = 4:7
217    d.box(k) = struct('action','','value',0,'param',0,'child',[],'id',[],
               'class',[]);
218  end
```

Figure 10.5 shows the results from `SimpleClassifierDemo`. There are four rectangular areas which are our sets.

We can create a decision tree by hand as shown Figure 10.6.

The decision tree sorts the samples into four sets. In this case, we know the boundaries and can use them to write the inequalities. In software, we will have to determine what values

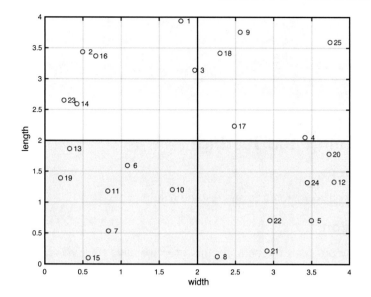

Figure 10.5: *Data and classes in the test set*

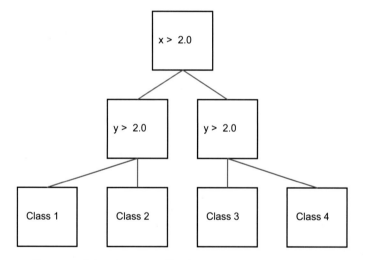

Figure 10.6: *A manually created decision tree. The drawing is generated by* `DecisionTree`. *The last row of boxes is the data sorted into the four classes. The last nodes are the classes. Each box is a decision tree node*

provide the shortest branches. The following is the output of `SimpleClassifierDemo`. The decision tree properly classifies all of the data:

```
>> SimpleClassifierDemo

Class 3
4 6 9 13 18
```

```
Class 2
7 14 17 21
Class 1
1 2 5 8 10 11 12 23 25
Class 4
3 15 16 19 20 22 24
```

10.6 Training and Testing

10.6.1 Problem

We want to train our decision tree and test the results.

10.6.2 Solution

We replicated the previous recipe, only this time we have DecisionTree create the decision tree instead of creating it by hand.

10.6.3 How It Works

TestDecisionTree trains and tests the decision tree. It is very similar to the code for the handmade decision tree demo, SimpleClassifierDemo. Once again, we use rectangles for the regions.

TestDecisionTree.m

```
8   % Vertices for the sets
9   v = [ 0 0; 0 4; 4 4; 4 0; 2 4; 2 2; 2 0; 0 2; 4 2];
10
11  % Faces for the sets
12  f = { [6 5 2 8] [6 7 4 9] [6 9 3 5] [1 7 6 8] };
13
14  % Generate the training set
15  pTrain = ClassifierSets( 40, [0 4], [0 4], {'width', 'length'},...
16    v, f, 'Training Set' );
17
18  % Create the decision tree
19  d     = DecisionTree;
20  d     = DecisionTree( 'train', d, pTrain );
21
22  % Generate the testing set
23  pTest = ClassifierSets( 5, [0 4], [0 4], {'width', 'length'},...
24    v, f, 'Testing Set' );
25
26  % Test the tree
27  [d, r] = DecisionTree( 'test',  d, pTest  );
28
29  q = DrawBinaryTree;
30  c = 'xy';
31  for k = 1:length(d.box)
32    if( ~isempty(d.box(k).action) )
```

```
33      q.box{k} = sprintf('%c %s %4.1f',c(d.box(k).param),...
34        d.box(k).action,d.box(k).value);
35    else
36      q.box{k} = sprintf('Class %d',d.box(k).class);
37    end
38  end
39  DrawBinaryTree(q);
40
41  m = reshape(pTest.m,[],1);
42
43  for k = 1:length(r)
44    fprintf(1,'Class %d\n',m(r{k}(1)));
45    for j = 1:length(r{k})
46      fprintf(1,'%d ',r{k}(j));
47    end
48    fprintf(1,'\n')
49  end
```

It uses `ClassifierSets` to generate the training data. The output includes the coordinates and the sets in which they fall. We then create the default data structure and call `DecisionTree` in training mode.

The tree is shown in Figure 10.7. The training data is shown in Figure 10.8 and the testing data in Figure 10.9. We need enough testing data to fill the classes. Otherwise, the decision tree generator may draw the lines to encompass just the data in the training set.

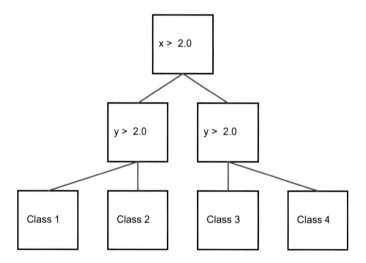

Figure 10.7: *The tree derived from the training data. It is essentially the same as the handmade tree. The values in the generated tree are not exactly 2.0*

299

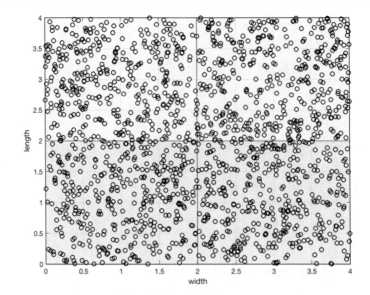

Figure 10.8: *The training data. A large amount of data is needed to fill the classes*

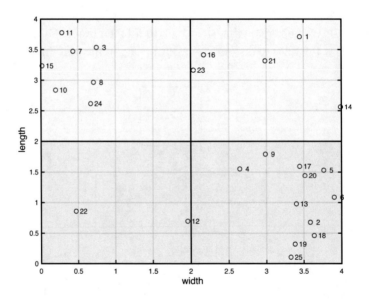

Figure 10.9: *The testing data*

The results are similar to the simple test:

```
Class 3
1 14 16 21 23
Class 2
2 4 5 6 9 13 17 18 19 20 25
Class 1
3 7 8 10 11 15 24
Class 4
12 22
```

The generated tree separates the data effectively.

10.7   Summary

This chapter has demonstrated data classification using decision trees in MATLAB. We also wrote a new graphics function to draw decision trees. The decision tree software is not general purpose but can serve as a guide to more general-purpose code. Table 10.2 lists the functions and scripts included in the companion code.

Table 10.2: *Chapter Code Listing*

File	Description
ClassifierSets	Generates data for classification or training
DecisionTree	Implements a decision tree to classify data
DrawBinaryTree	Generates data for classification or training
HomogeneityMeasure	Computes Gini impurity
SimpleClassifierDemo	Demonstrates decision tree testing
SimpleClassifierExample	Generates data for a simple problem
TestDecisionTree	Tests a decision tree

CHAPTER 11

■ ■ ■

Pattern Recognition with Deep Learning

Neural nets fall into the Learning category of our taxonomy. In this chapter, we will expand our neural net toolbox with convolution and pooling layers. A general neural net is shown in Figure 11.1. This is a "deep learning" neural net because it has multiple internal layers. Each layer may have a distinct function and form. In the previous chapter, our network had multiple layers, but they were all functionally similar and fully connected. In this chapter, we will also introduce another convolutional neural network.

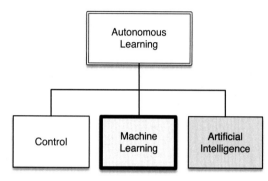

A convolutional neural network is a type of deep learning network that is a pipeline with multiple stages [21]. There are three types of layers:

- Convolutional layers (hence the name): Convolve a feature with the input matrix so that the output emphasizes that feature. Effectively find patterns.

- Pooling layers: These reduce the number of inputs to be processed in layers further down the chain.

- Fully connected layers in which each input node is connected to each output node.

A convolutional neural net is shown in Figure 11.2. This is also a "deep learning" neural net because it has multiple internal layers, but now the layers are of the three types described earlier.

We can have as many layers as we want. The following recipes will detail each step in the chain. We will start by showing how to gather image data online. We won't use online data, but the process may be useful for your work.

M. Paluszek, S. Thomas, *MATLAB Machine Learning Recipes*,
https://doi.org/10.1007/978-1-4842-9846-6_11

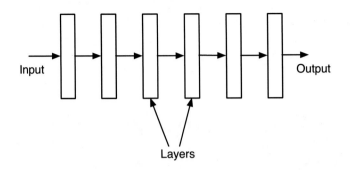

Figure 11.1: *Deep learning neural net*

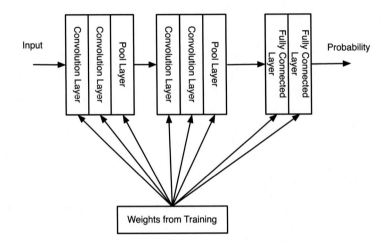

Figure 11.2: *Deep learning convolutional neural net [15]*

We will then describe the convolution process. The convolution process helps accent features in an image. For example, if a circle is a key feature, convolving a circle with an input image will emphasize circles.

The next recipe will implement pooling. This is a way of condensing the data. For example, if you have an image of a face, you might not need every pixel. You need to find the major features, mouth and eyes, for example, but might not need details of the person's iris. This is the reverse of what people do with sketching. A good artist can use a few strokes to represent a face. They then fill in detail in successive passes over the drawing. Pooling, at the risk of losing information, reduces the number of pixels to be processed.

We will then demonstrate the full network using random weights. Finally, we will train the network using a subset of our data and test it on the remaining data, as before.

For this chapter, we are going to use pictures of cats. Our network will produce a probability that a given image is a picture of a cat. We will train networks using cat images and also reuse some of our digit images from the previous chapter.

11.1   Obtain Data Online for Training a Neural Net

11.1.1   Problem

We want to find photographs online for training a cat recognition neural net.

11.1.2   Solution

Use the online database ImageNet to search for images of cats.

11.1.3   How It Works

ImageNet, www.image-net.org, is an image database organized according to the WordNet hierarchy. Each meaningful concept in WordNet is called a "synonym set." There are more than 100,000 sets and 14 million images in ImageNet. For example, type in "siamese cat." Click the link. You will see 445 images. You'll notice that there are a wide variety of shots from many angles and a wide range of distances.

```
Synset: Siamese cat, Siamese
Definition: a slender short-haired blue-eyed breed of cat having a pale
     coat with dark ears paws face and tail tip.
Popularity percentile: 57%
Depth in WordNet: 8
```

This is a great resource! However, we are going to instead use pictures of our cats for our test to avoid copyright issues. The database of photos on ImageNet may prove to be an excellent resource for you to use in training your neural nets. However, you should review the ImageNet license agreement to determine whether your application can use these images without restrictions.

11.2   Generating Training Images of Cats

11.2.1   Problem

We want grayscale photographs for training a cat recognition neural net.

11.2.2   Solution

Take photographs using a digital camera. Crop them to a standard size manually, then process them using native MATLAB functions to create grayscale images.

11.2.3 How It Works

We first take pictures of several cats. We'll use them to train the net. The photos are taken using an iPhone 6. We limit the photos to facial shots of the cats. We then frame the shots so that they are reasonably consistent in size and minimize the background. We then convert them to grayscale.

We use the function `ImageArray` to read the images. It takes a path to a folder containing the images to be processed. A lot of the code has nothing to do with image processing, just with dealing with unix files in the folder that are not images. `ScaleImage` is in the file reading loop to scale them. We flip them upside down so that they are right side up from our viewpoint. We then average the color values to make grayscale. This reduces an n by n by 3 array to n by n. The rest of the code displays the images packed into a frame. Finally, we scale all the pixel values down by 256 so that each value is from 0 to 1. The body of `ImageArray` is shown in the following listing.

ImageArray.m

```
1   %% IMAGEARRAY Read an array of images from a directory
17  function [s, sName] = ImageArray( folderPath, scale )
24
25  c = cd;
26  cd(folderPath)
27
28  d = dir;
29  n = length(d);
30  j = 0;
31
32  s     = cell(n-2,1);
33  sName = cell(1,length(n));
34  for k = 1:n
35    name = d(k).name;
36    if( ~strcmp(name,'.') && ~strcmp(name,'..') )
37      j         = j + 1;
38      sName{j}  = name;
39      t         = ScaleImage(flipud(imread(name)),scale);
40      s{j}      = (t(:,:,1)+ t(:,:,2) + t(:,:,3))/3;
41    end
42  end
43
44  del = size(s{1},1);
45  lX  = 3*del;
46
47  % Draw the images
48  NewFigure(folderPath);
49  colormap(gray);
50  n = length(s);
51  x = 0;
52  y = 0;
53  for k = 1:n
54    image('xdata',[x;x+del],'ydata',[y;y+del],'cdata', s{k} );
```

306

```
55    hold on
56    x = x + del;
57    if ( x == lX )
58      x = 0;
59      y = y + del;
60    end
61  end
62  axis off
63  axis image
64
65  for k = 1:length(s)
66    s{k} = double(s{k})/256;
67  end
68
69  cd(c)
```

The function has a built-in demo with our local folder of cat images. The images are scaled down by a factor of 2^4, or 16, so that they are displayed as 64 by 64 pixel images.

ImageArray.m

```
72  %%% ImageArray>Demo
73  % Generate an array of cat images
74
75  c0 = cd;
76  p = mfilename('fullpath');
77  cd(fileparts(p));
78  ImageArray( fullfile('..','Cats'), 4 );
79  cd(c0);
```

The full set of images in the `Cats` folder, as loaded and scaled in the demo, are shown in Figure 11.3.

`ImageArray` averages the three colors to convert the color images to grayscale. It flips them upside down since the image coordinates are opposite that of MATLAB. We used the

Figure 11.3: *64 by 64 pixel grayscale cat images*

GraphicConverter application to crop the images around the cat's face and make them all 1024 by 1024 pixels. One of the challenges of image matching is to do this process automatically. Also, typically training uses thousands of images. We will be using just a few to see if our neural net can determine if the test image is a cat or even one we have used in training! ImageArray scales the image using the function ScaleImage, shown as follows.

ScaleImage.m

```
1   %% SCALEIMAGE Scale an image by powers of 2.
20  function s2 = ScaleImage( s1, q )
21
22  % Demo
23  if( nargin < 1 )
24     Demo
25     return
26  end
27
28  n = 2^q;
29
30  [mR,~,mD]  = size(s1);
31
32  m = mR/n;
33
34  s2 = zeros(m,m,mD,'uint8');
35
36  for i = 1:mD
37    for j = 1:m
38      r = (j-1)*n+1:j*n;
39      for k = 1:m
40        c          = (k-1)*n+1:k*n;
41        s2(j,k,i) = mean(mean(s1(r,c,i)));
42      end
43    end
44  end
```

Notice that it creates the new image array as uint8. Figure 11.4 shows the results of scaling a full-color image.

11.3 Matrix Convolution

11.3.1 Problem

We want to implement convolution as a technique to emphasize key features in images, to make learning more effective. This will then be used in the next recipe to create a convolving layer for the neural net.

Figure 11.4: *Image scaled from 1024 by 1024 to 256 by 256*

11.3.2 Solution

Implement convolution using MATLAB matrix operations.

11.3.3 How It Works

We create an n-by-n mask that we apply to an m-by-m where m is greater than n. We start in the upper-left corner of the matrix, as shown in Figure 11.5. We multiply the mask times the corresponding elements in the input matrix and do a double sum. That is the first element of the convolved output. We then move it column by column until the highest column of the mask is aligned with the highest column of the input matrix. We then return it to the first column and increment the row. We continue until we have traversed the entire input matrix and our mask is aligned with the maximum row and maximum column. For example, the number 8 in the result is gotten by adding elements (1,1), (1,2), and (2,2), the element for which the mask has one as the element.

The mask represents a feature. In effect, we are seeing if the feature appears in different areas of the image. We can have multiple masks. There is one bias and one weight for each element of the mask for each feature. In this case, instead of 16 sets of weights and biases, we only have 4. For large images, the savings can be substantial. In this case, the convolution works on the image itself. Convolutions can also be applied to the output of other convolutional layers or pooling layers as shown in Figure 11.2.

Convolution is implemented in `Convolve.m`. The mask is input `a` and the matrix to be convolved is input `b`.

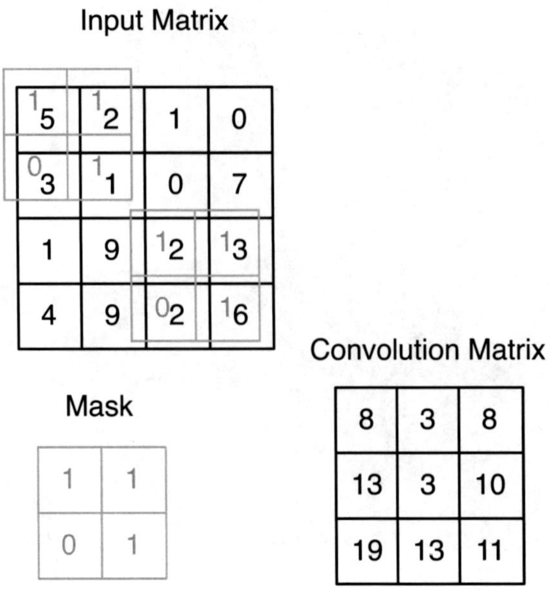

Input Matrix

Mask

Convolution Matrix

Figure 11.5: *Convolution process showing the mask at the beginning and end of the process*

Convolve.m

```
23   function c = Convolve( a, b )
24
25   % Demo
26   if( nargin < 1 )
27     Demo
28     return
29   end
30
31   [nA,mA]  = size(a);
32   [nB,mB]  = size(b);
33   nC       = nB - nA + 1;
34   mC       = mB - mA + 1;
35   c        = zeros(nC,mC);
36   for j = 1:mC
37     jR = j:j+nA-1;
38     for k = 1:nC
39       kR = k:k+mA-1;
40       c(j,k) = sum(sum(a.*b(jR,kR)));
41     end
42   end
```

The demo, which convolves a 3×3 mask with a 6×6 matrix, produces the following 4×4 matrix output:

```
>> Convolve
a =

     1      0      1
     0      1      0
     1      0      1
b =
     1      1      1      0      0      0
     0      1      1      1      0      1
     0      0      1      1      1      0
     0      0      1      1      0      1
     0      1      1      0      0      1
     0      1      1      0      0      1
ans =
     4      3      4      1
     2      4      3      5
     2      3      4      2
     3      3      2      3
```

11.4　Convolution Layer

11.4.1　Problem

We want to implement a convolution-connected layer. This will apply a mask to an input image.

11.4.2　Solution

Use code from `Convolve` to implement the layer. It slides the mask across the image, and the number of outputs is reduced.

11.4.3　How It Works

The "convolution" neural net scans the input with the mask. Each input to the mask passes through an activation function that is identical to a given mask. `ConvolutionLayer` has its built-in neuron function shown in the listing.

ConvolutionLayer.m

```
1   %% CONVOLUTIONLAYER Convolution layer for a neural net
23  function y = ConvolutionLayer( x, d )
24
25  % Demo
26  if( nargin < 1 )
27    if( nargout > 0 )
28      y = DefaultDataStructure;
29    else
30      Demo;
```

```
31      end
32      return
33   end
34
35   a         = d.mask;
36   aFun      = str2func(d.aFun);
37   [nA,mA]   = size(a);
38   [nB,mB]   = size(x);
39   nC        = nB - nA + 1;
40   mC        = mB - mA + 1;
41   y         = zeros(nC,mC);
42   scale     = nA*mA;
43   for j = 1:mC
44      jR = j:j+nA-1;
45      for k = 1:nC
46         kR = k:k+mA-1;
47         y(j,k) = sum(sum(a.*Neuron(x(jR,kR),d, aFun)));
48      end
49   end
50
51   y = y/scale;
52
53   %% ConvolutionLayer>Neuron
54   function y = Neuron( x, d, afun )
55   % Neuron function
56   y = afun(x.*d.w + d.b);
```

Figure 11.6 shows the inputs and outputs from the demo (not shown in the listing). The tanh activation function is used in this demo. The weights and biases are random. The convolution of the mask, which is all ones, is just the sum of all the points for which the mask has a nonzero value. The output is scaled by the number of elements in the mask.

11.5 Pooling to Outputs of a Layer

11.5.1 Problem

We want to pool the outputs of the convolution layer to reduce the number of points we need to process in further layers. This uses the Convolve function created in the previous recipe.

11.5.2 Solution

Implement a new function to take the output of the convolution function.

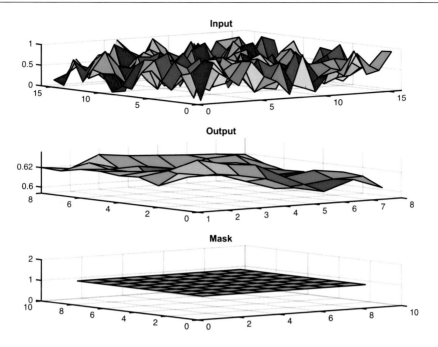

Figure 11.6: *Inputs and outputs for the convolution layer*

11.5.3 How It Works

Pooling layers take a subset of the outputs of the convolutional layers and pass that on. They do not have any weights. Pooling layers can use the maximum value of the pool or take the median or mean value. Our pooling function has all three options. The pooling function divides the input into n-by-n subregions and returns an n-by-n matrix.

Pooling is implemented in `Pool.m`. Notice we use `str2func` instead of a switch statement. `a` is the matrix to be pooled, `n` is the number of pools, and `type` is the name of the pooling function.

Pool.m

```
24  function b = Pool( a, n, type )
25
26  % Demo
27  if( nargin < 1 )
28    Demo
29    return
30  end
31
32  if( nargin <3 )
33    type = 'mean';
34  end
35
36  n = n/2;
```

313

```
37  p = str2func(type);
38
39  nA = size(a,1);
40
41  nPP = nA/n;
42
43  b = size(n,n);
44  for j = 1:n
45    r = (j-1)*nPP +1:j*nPP;
46    for k = 1:n
47      c = (k-1)*nPP +1:k*nPP;
48      b(j,k) = p(p(a(r,c)));
49    end
50  end
```

These two demos create four pools from a 4×4 matrix. Each number in the output matrix is a pool of one-quarter of the input matrix. It uses the default `'mean'` pool method.

```
>> Pool([1:4;3:6;6:9;7:10],4)
ans =
    2.5000    4.5000
    7.0000    9.0000
>> Pool([1:4;3:6;6:9;7:10],4,'max')
ans =
    4    6
    8   10
```

The pool is a neural layer whose activation function is effectively the argument passed to `Pool`.

11.6   Fully Connected Layer

11.6.1 Problem

We want to implement a fully connected layer.

11.6.2 Solution

Use `FullyConnectedNN` to implement the network.

11.6.3 How It Works

The "fully connected" neural net layer is the traditional neural net where every input is connected to every output as shown in Figure 11.7. We implement the fully connected network with n inputs and m outputs. Each output path can have a different weight and bias. `FullyConnectedNN` can handle any number of inputs or outputs. The following listing shows the data structure function as well as the function body.

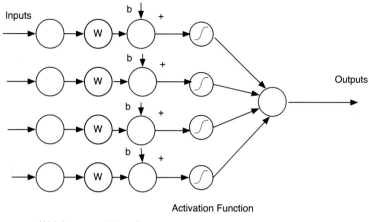

Figure 11.7: *Fully connected neural net. This shows only one output*

FullyConnectedNN.m

```
23   function y = FullyConnectedNN( x, d )
24
25   % Demo
26   if( nargin < 1 )
27     if( nargout > 0 )
28       y = DefaultDataStructure;
29     else
30       Demo;
31     end
32     return
33   end
34
35   y = zeros(d.m,size(x,2));
36
37   aFun = str2func(d.aFun);
38
39   n = size(x,1);
40   for k = 1:d.m
41     for j = 1:n
42       y(k,:) = y(k,:) + aFun(d.w(j,k)*x(j,:) + d.b(j,k));
43     end
44   end
45
46   function d = DefaultDataStructure
47   %%% FullyConnectedNN>DefaultDataStructure
48   % Default Data Structure
49
50   d = struct('w',[],'b',[],'aFun','tanh','m',1);
```

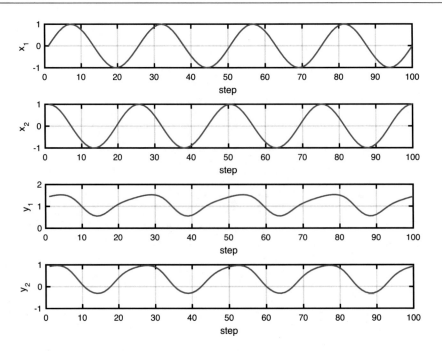

Figure 11.8: *The two outputs from the* `FullyConnectedNN` *demo function are shown vs. the two inputs*

Figure 11.8 shows the outputs from the built-in function demo. The `tanh` activation function is used in this demo. The weights and biases are random, so the plot will be different every time you run the demo. The change in shape from the input to the output is the result of the activation function.

11.7 Determining the Probability

11.7.1 Problem

We want to calculate the probability that the output is what we expect from neural net outputs.

11.7.2 Solution

Implement the Softmax function. Given a set of inputs, it calculates a set of positive values that add to 1. This will be used for the output nodes of our network.

11.7.3 How It Works

The *Softmax* function is a generalization of the logistic function. The equation is

$$p_j = \frac{e^{q_j}}{\sum_{k=1}^{N} e^{q_k}} \tag{11.1}$$

where q is a vector of inputs, N is the number of inputs, and p is the output values that sum to 1. The function is implemented in Softmax.m.

Softmax.m

```
18  function [p, pMax, kMax] = Softmax( q )
25
26  q = reshape(q,[],1);
27  n = length(q);
28  p = zeros(1,n);
29
30  den = sum(exp(q));
31
32  for k = 1:n
33    p(k) = exp(q(k))/den;
34  end
35
36  [pMax,kMax] = max(p);
```

The built-in demo passes in a short list of outputs.

Softmax.m

```
38  function Demo
39  %% Softmax>Demo
40  q = [1,2,3,4,1,2,3];
41  [p, pMax, kMax] = Softmax( q )
42  sum(p)
```

The results of the demo are

```
>> Softmax
p =
    0.0236    0.0643    0.1747    0.4748    0.0236    0.0643    0.1747
pMax =
    0.4748
kMax =
    4
ans =
    1.0000
```

The last number is the sum of p which should be (and is) 1.

11.8   Test the Neural Network

11.8.1 Problem

We want to integrate convolution, pooling, a fully connected layer, and Softmax so that our network outputs a probability.

11.8.2 Solution

The solution is to write a convolutional neural net. We integrate the convolution, pooling, fully connected net, and Softmax functions. We then test it with randomly generated weights.

11.8.3 How It Works

Figure 11.9 shows the image processing neural network. It has one convolutional layer, one pooling layer, and a fully connected layer, and the final layer is the Softmax.

ConvolutionNN implements the network. It uses the functions ConvolutionLayer, Pool, FullyConnectedNN, and Softmax that we have implemented in the prior recipes. The code in ConvolutionNN that implements the network is shown as follows, in the subfunction NeuralNet. It can generate plots if requested using mesh.

ConvolutionalNN.m

```
108  function r = NeuralNet( d, t, ~ )
109  %%% ConvolutionalNN>NeuralNet
110  % Execute the neural net. Plot if there are three inputs.
111
112  % Convolve the image
113  yCL   = ConvolutionLayer( t, d.cL );
114
115  % Pool outputs
116  yPool = Pool( yCL, d.pool.n, d.pool.type );
117
```

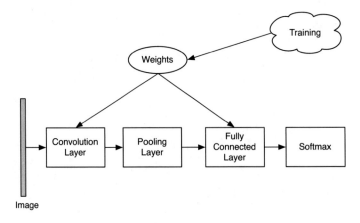

Figure 11.9: *Neural net for image processing*

```
118  % Apply a fully connected layer
119  yFC    = FullyConnectedNN( yPool, d.fCNN );
120  [~,r] = Softmax( yFC );
121
122  % Plot if requested
123  if( nargin > 2 )
124    NewFigure('ConvolutionNN');
125    subplot(3,1,1);
126    mesh(yCL);
127    title('Convolution Layer')
128    subplot(3,1,2);
129        mesh(yPool);
130    title('Pool Layer')
131    subplot(3,1,3);
132        mesh(yFC);
133    title('Fully Connected Layer')
134  end
```

ConvolutionNN has additional subfunctions for defining the data structure and training and testing the network.

We begin by testing the neural net initialized with random weights, using TestNN. This is a script that loads the cat images using ImageArray, initializes a convolutional network with random weights, and then runs it with a selected test image:

```
1  >> TestNN
2  Image IMG_3886.png has a 13.1% chance of being a cat
```

As expected, an untrained neural net does not identify a cat! Figure 11.10 shows the output of the various stages of network processing.

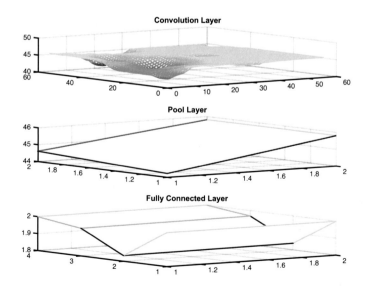

Figure 11.10: *Stages in the convolutional neural net processing*

11.9 Recognizing an Image

11.9.1 Problem

We want to determine if an image is that of a cat.

11.9.2 Solution

We train the neural network with a series of cat images. We then use one picture from the training set and a separate picture reserved for testing and compute the probabilities that they are cats.

11.9.3 How It Works

We run the script TrainNN to see if the input image is a cat. It trains the net from the images in the Cats folder. Many thousands of function evaluations are required for meaningful training, but allowing just a few function evaluations shows that the function is working.

TrainNN.m

```
1   %% Train a neural net on the Cats images
8   p   = mfilename('fullpath');
9   c0 = cd;
10  cd(fileparts(p));
11
12  folderPath = fullfile('..','Cats');
13  [s, name]  = ImageArray( folderPath, 4 );
14  d          = ConvolutionalNN;
15
16  % Use all but the last for training
17  s = s(1:end-1);
18
19  % This may take awhile
20  % Use at least 10000 iterations to see a higher change of being a cat!
21  disp('Start training...')
22  d.opt.Display = 'iter';
23  d.opt.MaxFunEvals = 500;
24  d =      ConvolutionalNN( 'train', d, s );
25
26  % Test the net using the last image that was not used in training
27  [d, r] = ConvolutionalNN( 'test', d, s{end} );
28
29  fprintf(1,'Image %s has a %4.1f%% chance of being a cat\n',name{end
        },100*r);
30
31  % Test the net using the first image
32  [d, r] = ConvolutionalNN( 'test', d, s{1} );
33
34  fprintf(1,'Image %s has a %4.1f%% chance of being a cat\n',name{1},100*
        r);
```

The script returns that the probability of either being a cat is now 38.8%. This is an improvement considering we only trained it with one image. It took a couple of hours to process.

```
>> TrainNN

Exiting: Maximum number of function evaluations has been exceeded
        - increase MaxFunEvals option.
        Current function value: 0.612029

Image IMG_3886.png has a 38.8% chance of being a cat
Image IMG_0191.png has a 38.8% chance of being a cat
```

fminsearch uses a direct search method (Nelder-Mead simplex), and it is very sensitive to initial conditions.

Using this search method poses a fundamental performance barrier for this neural net training, especially for deep learning where the combinatorics of different weight combos are so big. Better (and faster) results with a global optimization method are likely.

The training code from ConvolutionNN is shown as follows. It uses MATLAB fmin search. fminsearch tweaks the gains and biases until it gets a good fit between all the input images and the training images.

ConvolutionalNN.m

```
46  function d = Training( d, t )
47  %%% ConvolutionalNN>Training
48
49  d              = Indices( d );
50  x0             = DToX( d );
51  [x,d.fVal]      = fminsearch( @RHS, x0, d.opt, d, t );
52  d              = XToD( x, d );
```

We can improve the results with

- Adjusting fminsearch parameters

- More images

- More features (masks)

- Changing the connections in the fully connected layer

- Adding the ability of ConvolutionalNN to handle RGB images directly, rather than converting them to grayscale.

- Using a different search method such as a genetic algorithm.

11.10   Using AlexNet

11.10.1 Problem

We want to use the pretrained network AlexNet for image classification.

11.10.2 Solution

Depending on your version of MATLAB, install AlexNet from the Add-On Explorer or download the support package for GoogLeNet. Load some images and test. These are classification networks, so we will use `classify` to run them.

11.10.3 How It Works

First, we need to download the support packages with the Add-On Explorer. If you attempt to run `alexnet` or `googlenet` without having them installed, you will get a link directly to the package in the Add-On Explorer. You will need your MathWorks password.

AlexNet is a pretrained convolutional neural network (CNN) that has been trained on approximately 1.2 million images from the ImageNet data set (http://image-net.org/index). The model has 25 layers and can classify images into 1000 object categories. It can be used for all sorts of object classification. However, if an object was not in the training set, it won't be able to identify the object. If a banana was in the training set, you could expect the CNN to correctly identify a new picture of a banana. But if you gave it a picture of a plantain, and plantain was NOT in the CNN, then it might not find a match, or, more likely, it might incorrectly classify it like a banana.

AlexNetTest.m

```
8    %% Load the network
9    % Access the trained model. This is a SeriesNetwork.
10   net = alexnet;
11   net
12
13   % See details of the architecture
14   net.Layers
```

The network layers' printout is shown as follows:

```
>> AlexNetTest

ans =

  25x1 Layer array with layers:

     1   'data'    Image Input                    227x227x3 images with 'zerocenter' normalization
     2   'conv1'   Convolution                    96 11x11x3 convolutions with stride [4  4] and padding [0
             0  0  0]
     3   'relu1'   ReLU                           ReLU
     4   'norm1'   Cross Channel Normalization    cross channel normalization with 5 channels per element
     5   'pool1'   Max Pooling                    3x3 max pooling with stride [2  2] and padding [0  0  0  0]
     6   'conv2'   Grouped Convolution            2 groups of 128 5x5x48 convolutions with stride [1  1] and
             padding [2  2  2  2]
     7   'relu2'   ReLU                           ReLU
     8   'norm2'   Cross Channel Normalization    cross channel normalization with 5 channels per element
     9   'pool2'   Max Pooling                    3x3 max pooling with stride [2  2] and padding [0  0  0  0]
```

```
10   'conv3'    Convolution          384 3x3x256 convolutions with stride [1  1] and padding [1
              1  1  1]
11   'relu3'    ReLU                 ReLU
12   'conv4'    Grouped Convolution  2 groups of 192 3x3x192 convolutions with stride [1  1] and
              padding [1  1  1  1]
13   'relu4'    ReLU                 ReLU
14   'conv5'    Grouped Convolution  2 groups of 128 3x3x192 convolutions with stride [1  1] and
              padding [1  1  1  1]
15   'relu5'    ReLU                 ReLU
16   'pool5'    Max Pooling          3x3 max pooling with stride [2  2] and padding [0  0  0  0]
17   'fc6'      Fully Connected      4096 fully connected layer
18   'relu6'    ReLU                 ReLU
19   'drop6'    Dropout              50% dropout
20   'fc7'      Fully Connected      4096 fully connected layer
21   'relu7'    ReLU                 ReLU
22   'drop7'    Dropout              50% dropout
23   'fc8'      Fully Connected      1000 fully connected layer
24   'prob'     Softmax              softmax
25   'output'   Classification Output crossentropyex with 'tench' and 999 other classes
```

There are many layers in this convolutional network. ReLU and Softmax are the activation functions. In the first layer, "zero center" normalization is used. This means the images are normalized to have a mean of zero and a standard deviation of one. Two layers are new: cross-channel normalization and grouped convolution. Filter groups, also known as grouped convolution, were introduced with AlexNet in 2012. You can think of the output of each filter as a channel and filter groups as groups of the channels. Filter groups allowed more efficient parallelization across GPUs. They also improved performance. Cross-channel normalization normalizes across channels, instead of one channel at a time. We've discussed convolution in Chapter 3. The weights in each filter are determined during training. Dropout is a layer that ignores nodes, randomly, when training the weights. This prevents interdependencies between nodes.

For our first example, we load an image that comes with MATLAB, of a set of peppers. This image is larger than the input size of the network, so we use the top-left corner of the image. Note that each pretrained network has a fixed input image size that we can determine from the first layer.

AlexNetTest.m

```
16  %% Load a test image and classify it
17  % Read the image to classify
18  I = imread('peppers.png');  % ships with MATLAB
19
20  % Adjust size of the image to the net's input layer
21  sz = net.Layers(1).InputSize;
22  I = I(1:sz(1),1:sz(2),1:sz(3));
23
24  % Classify the image using AlexNet
25  [label, scorePeppers] = classify(net, I);
26
27  %  Show the image and the classification results
28  NewFigure('Pepper'); ax = gca;
29  imshow(I);
30  title(ax,label);
31
32  PlotSet(1:length(scorePeppers),scorePeppers,'x label','Category',...
33          'y label','Score','plot title','Peppers');
```

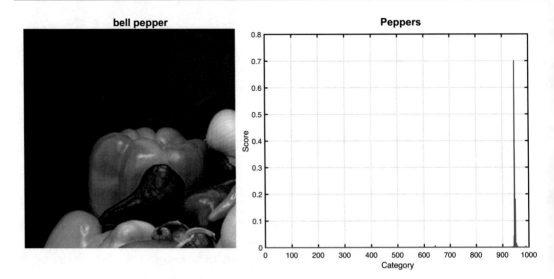

Figure 11.11: *Test image labeled with the classification and the scores. The image is classified as a "bell pepper"*

The images and results for the AlexNet example are shown in Figure 11.11. The pepper scores are tightly clustered.

For fun, and to learn more about this network, we print out the categories that had the next highest scores, sorted from high to low. The categories are stored in the last layer of the net in its Classes.

AlexNetTest.m

```
35  % What other categories are similar?
36  disp('Categories with highest scores for Peppers:')
37  kPos = find(scorePeppers>0.01);
38  [vals,kSort] = sort(scorePeppers(kPos),'descend');
39  for k = 1:length(kSort)
40    fprintf('%13s:\t%g\n',net.Layers(end).Classes(kPos(kSort(k))),vals(k)
          );
41  end
```

The results show that the net was considering all fruits and vegetables. The Granny Smith had the next highest score, followed by cucumber, while the fig and lemon had much smaller scores. This makes sense since Granny Smiths and cucumbers are also usually green.

```
Categories with highest scores for Peppers:

  bell pepper:  0.700013
Granny Smith:  0.180637
    cucumber:  0.0435253
         fig:  0.0144056
       lemon:  0.0100655
```

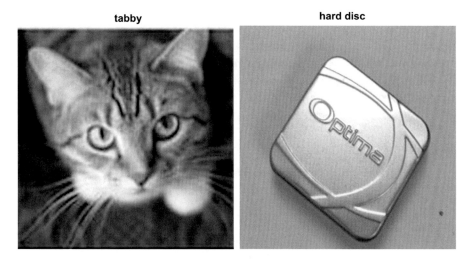

tabby   **hard disc**

Figure 11.12: *Test images and the classification by AlexNet. They are classified as "tabby" and "hard disc"*

We also have two of our test images. One is of a cat and one of a metal box. The scores for the cat classification are shown as follows:

```
Categories with highest scores for Cat:
         tabby:    0.805644
   Egyptian cat:   0.15372
      tiger cat:   0.0338047
```

The selected label is *tabby*. The net can recognize that the photo is of a cat, as the other highest scored categories are also kinds of cats. Although what a tiger cat might be, as distinguished from a tabby, we can't say...

The metal box proves the biggest challenge to the net. The category scores above 0.05 are shown as follows, and the images with their label are shown in Figure 11.12:

```
Categories with highest scores for Box:
    hard disc:   0.291533
        loupe:   0.0731844
        modem:   0.0702888
         pick:   0.0610284
         iPod:   0.0595867
    CD player:   0.0508571
```

In this case, the hard disc is by far the highest score, but the score is much lower than that of the tabby cat – roughly 0.3 vs. 0.8. The summary of scores is

```
AlexNet results summary:

Pepper        0.7000
Cat           0.8056
Box           0.2915
```

Summary

This chapter has demonstrated the steps for implementing a convolutional neural network using MATLAB. Convolutional neural nets were used to process pictures of and numbers of cats for learning. When trained, the neural net was asked to identify other pictures to determine if they were pictures of a cat or a number. Table 11.1 lists the functions and scripts included in the companion code.

Table 11.1: *Chapter Code Listing*

File	Description
AlexNetTest	Uses AlexNet
Activation	Generates activation functions
ConvolutionalNN	Implements a convolutional neural net
ConvolutionLayer	Implements a convolutional layer
Convolve	Convolves a 2D array using a mask
FullyConnectedNN	Implements a fully connected neural network
ImageArray	Reads in images in a folder and converts to grayscale
Pool	Pools a 2D array
ScaleImage	Scale and image
Softmax	Implements the Softmax function
TrainNN	Trains the convolutional neural net with cat images
TestNN	Tests the convolutional neural net on a cat image
TrainingData.mat	Data from TestNN.

CHAPTER 12

■ ■ ■

Multiple Hypothesis Testing

12.1 Overview

Tracking is the process of determining the position of other objects as their position changes with time. Air traffic control radar systems are used to track aircraft. Aircraft in flight must track all nearby objects to avoid collisions and to determine if they are threats. Automobiles with radar cruise control use their radar to track cars in front of them so that the car can maintain safe spacing and avoid a collision.

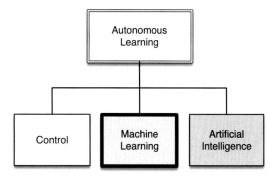

When you are driving, you maintain situation awareness by identifying nearby cars and figuring out what they are going to do next. Your brain processes data from your eyes to characterize its car. You track objects by their appearance since, in general, the cars around you all look different. Of course, at night you only have tail lights, so the process is harder. You can often guess what each car is going to do, but sometimes you guess wrong and that can lead to collisions.

Radar systems just see blobs. Lidar (laser radar) has a narrower beam, so it can scan cars around you and potentially track a particular car by its shape. Cameras should be able to do what your eyes and brain do, but that requires a lot of processing. License plate readers have gotten very good and are used routinely for tolls, but humans can look at a car and identify its type and color as a way of tracking a particular car. As noted, at night it is hard to reliably identify a car with a camera. As the blobs are measured by a radar, we want to collect all blobs, as they vary in position and speed, and attach them to a particular car's track. This way, we can reliably predict where it will go next. This leads to the topic of this chapter, Track-Oriented Multiple Hypothesis Testing.

M. Paluszek, S. Thomas, *MATLAB Machine Learning Recipes*,
https://doi.org/10.1007/978-1-4842-9846-6_12

Table 12.1: MHT terms

Term	Definition
Clutter	Transient objects of no interest to the tracking system
Cluster	A collection of tracks that are linked by common observations
Error ellipsoid	An ellipsoidal volume around an estimated position
Family	A set of tracks with a common root node. At most, one track per family can be included in a hypothesis. A family can at most represent one target
Gate	A region around an existing track position. Measurements within the gate are associated with the track
Hypothesis	A set of tracks that do not share any common observations
N-Scan pruning	Using the track scores from the last N scans of data to prune tracks. The count starts from a root node. When the tracks are pruned, a new root node is established
Observation	A measurement that indicates the presence of an object. The observation may be of a target or be spurious
Pruning	Removal of low-score tracks
Root node	An established track to which observations can be attached and which may spawn additional tracks
Scan	A set of data taken simultaneously
Target	An object being tracked
Trajectory	The path of a target
Track	A trajectory that is propagated
Track branch	A track in a family that represents a different data association hypothesis. Only one branch can be correct
Track score	The log-likelihood ratio for a track

Track-Oriented Multiple Hypothesis Testing (MHT) is a powerful technique for assigning measurements to tracks of objects when the number of objects is unknown or changing. It is useful for accurate tracking of multiple objects. MHT uses statistics to determine the probability that the object your radar just measured is one for which you already have a track and the probability that the object is a new one, for example, a car that just cut in front of you. MHT terms are defined in Table 12.1.

Hypotheses are sets of tracks with consistent data, that is, where no measurements are assigned to more than one track. The track-oriented approach recomputes the hypotheses using the newly updated tracks after each scan of data is received. Rather than maintaining, and expanding, hypotheses from scan to scan, the track-oriented approach discards the hypotheses formed on scan $k - 1$. The tracks that survive pruning are propagated to the next scan k where new tracks are formed, using the new observations, and reformed into hypotheses. Except for the necessity to delete some tracks based on low probability, no information is lost because the track scores that are maintained contain all the relevant statistical data.

The software in this chapter uses a powerful track pruning algorithm that does the pruning in one step. Because of its speed, ad hoc pruning methods are not required, leading to more robust and reliable results. The track management software is, as a consequence, quite simple.

The MHT Module requires GLPK, the GNU Linear Programming Kit (www.gnu.org/ software/glpk/), and, specifically, the MATLAB mex wrapper GLPKMEX (http://glpkmex. sourceforge.net). Both are distributed under the GNU license. Both the GLPK library and the GLPKMEX program are operating system dependent and must be compiled from the source code on your computer. Once GLPK is installed, the mex must be generated from MATLAB from the GLPKMEX source code.

■ **TIP**   GNU is a recursive name for "GNU's Not Unix."

The command that is executed from MATLAB to create the mex should look like this:

```
mex -v -I/usr/local/include glpkcc. cpp /usr/local/lib/libglpk.a
```

where the "v" specifies verbose printout, and you should replace `/usr/local` with your operating system–dependent path to your installation of GLPK. The resulting mex file (Mac) is

```
glpkcc.mexmaci64
```

The MHT software was tested with GLPK version 4.47 and GLPKMEX version 2.11.

12.2   Theory

12.2.1   Introduction

Figure 12.1 shows the general tracking problem in the context of automobile tracking. Two scans of data are shown. When the first scan is done, there are two tracks. The uncertainty ellipsoids are shown, and they are based on all previous information. In the $k - 1$ scan (a scan is a set of measurements taken at the same time), three measurements are observed. Each scan has multiple measurements, the measurements in each new scan are numbered beginning with one, and the measurement numbers are not meant to imply any correlation across subsequent scans. One and three are within the ellipsoids of the two tracks, but two is in both. It may be a measurement of either of the tracks or a spurious measurement. In scan k, four measurements are taken. Only measurement 4 is in one of the uncertainty ellipsoids. Three might be interpreted as spurious, but it is actually due to a new track from a third vehicle that separates from the blue track. Measurement 1 is outside of the red ellipsoid but is a good measurement of the red track and (if correctly interpreted) indicates that the model is erroneous. Four is a good measurement of the blue track and indicates that the model is valid. Measurement 2 of scan k is outside

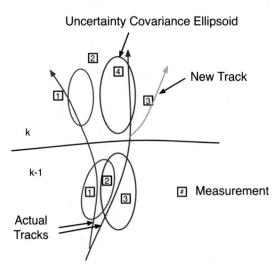

Figure 12.1: *Tracking problem*

both uncertainty ellipsoids. The illustration shows how the tracking system should behave, but without the tracks, it would be difficult to interpret the measurements. As shown, a measurement can be

1. Valid

2. Spurious

3. A new track

"Spurious" means that the measurement is not associated with any tracked object and isn't a new track. We can't determine the nature of any measurement without going through the MHT process.

We define a contact as an observation where the signal-to-noise ratio is above a certain threshold. The observation then constitutes a measurement. Low signal-to-noise ratio observations can happen in both optical and radar systems. Thresholding reduces the number of observations that need to be associated with tracks but may lose valid data. An alternative is to treat all observations as a contact but adjust the measurement error accordingly.

Valid measurements must then be assigned to tracks. An ideal tracking system would be able to categorize each measurement accurately and then assign them to the correct track. The system must also be able to identify new tracks and remove tracks that no longer exist. A tracking system may have to deal with hundreds of objects (perhaps after a collision or due to debris in the road).

A sophisticated system should be able to work with multiple objects as groups or clusters if the objects are more or less moving in the same direction. This reduces the number of states a system must handle. If a system handles groups, then it must be able to handle groups spawning from groups.

If we were confident that we were only tracking one vehicle, all of the data might be incorporated into the state estimate. An alternative is to incorporate only the data within the covariance ellipsoids and treat the remainder as outliers.

The covariance is a matrix representing the uncertainty of a vector. For a two-element vector, the covariance matrix is

$$\begin{bmatrix} \sigma_1^2 & \rho_{12} \\ \rho_{12} & \sigma_2^2 \end{bmatrix} \tag{12.1}$$

where σ_1 is the standard deviation of the first element and σ_2 is the standard deviation of the second. ρ_{12} is a measure of the cross-coupling between elements 1 and 2. If they are uncorrelated, it is zero.

If the latter strategy were taken, it would be sensible to remember that data in case future measurements also were "outliers" in which case the filter might go back and incorporate different sets of outliers into the solution. This could easily happen if the model were invalid. For example, if the vehicle, which had been cruising at a constant speed, suddenly began maneuvering and the filter model did not allow for maneuvers.

The multiple model filters help with the erroneous model problem and should be used anytime a vehicle might change mode. It does not tell us how many vehicles we are tracking, however. With multiple models, each model would have its own error ellipsoids, and the measurements would fit one better than the other, assuming that one of the models was a reasonable model for the tracked vehicle in its current mode.

12.2.2 Example

Referring to Figure 12.1, in the first scan we have three measurements. Measurements 1 and 3 are associated with existing tracks and are used to update those tracks. Measurement 2 could be associated with either. It might be a spurious measurement or could be a new track, so the algorithm forms a new hypothesis. In scan 2, measurement 4 is associated with the blue track. Measurements 1, 2, and 3 are not within the error ellipsoids of either track. Since the figure shows the true track, we can see that measurement 1 is associated with the red track. Both measurements 1 and 2 are just outside the error ellipsoid for the red track. Measurement 2 in scan 2 might be consistent with measurement 2 in scan 1 and could result in a new track. Measurement 3 in scan 2 is a new track.

12.2.3 Algorithm

In classical multiple target tracking [28], the problem is divided into two steps, association and estimation. Step 1 associates contacts with targets, and step 2 estimates each target's state. Complications arise when there is more than one reasonable way to associate contacts with targets. The Multiple Hypothesis Testing (MHT) approach is to form alternative hypotheses to explain the source of the observations. Each hypothesis assigns observations to targets or false alarms.

There are two basic approaches to MHT [5]. The first, following Reid [25], operates within a structure in which hypotheses are continually maintained and updated as observation data is received. In the second, the track-oriented approach to MHT, tracks are initiated, updated, and scored before being formed into hypotheses. The scoring process consists of comparing the likelihood that the track represents a true target vs. the likelihood that it is a collation of false alarms. Thus, unlikely tracks can be deleted before the next stage in which tracks are formed into hypotheses. It is a good thing to discard the old hypotheses and start from scratch each time because this approach maintains the important track data while preventing an explosion of an impractically large number of hypotheses.

The track-oriented approach recomputes the hypotheses using the newly updated tracks after each scan of data is received. Rather than maintaining, and expanding, hypotheses from scan to scan, the track-oriented approach discards the hypotheses formed on scan $k - 1$. The tracks that survive pruning are predicted to the next scan k where new tracks are formed, using the new observations, and reformed into hypotheses. Except for the necessity to delete some tracks based on low probability or N-scan pruning, no information is lost because the track scores that are maintained contain all the relevant statistical data.

Track scoring is done using log-likelihood ratios. LR is the likelihood ratio, LLR is the log-likelihood ratio, and L is the likelihood.

$$L(K) = \log[\mathrm{LR}(K)] = \sum_{k=1}^{K} [\mathrm{LLR}_K(k) + \mathrm{LLR}_S(k)] + \log[L_0] \qquad (12.2)$$

where the subscript K denotes kinematic (position) and the subscript S denotes signal (measurement). It is assumed that the two are statistically independent.

$$L_0 = \frac{P_0(H_1)}{P_0(H_0)} \qquad (12.3)$$

where H_1 and H_0 are the true targets and false alarm hypotheses. log is a natural logarithm. The likelihood ratio for the kinematic data is the probability that the data is a result of the true target divided by the probability that the data is due to a false alarm:

$$\mathrm{LR}_K = \frac{p(D_K|H_1)}{p(D_K|H_0)} = \frac{e^{-d^2/2}/((2\pi)^{M/2}\sqrt{|S|}}{1/V_C} \qquad (12.4)$$

where

1. M in the denominator of the third formula is the measurement dimension.

2. V_C is the measurement volume.

3. $S = HPT^T + R$ is the measurement residual covariance matrix.

4. $d^2 = y^T S^{-1} y$ is the normalized statistical distance for the measurement.

The statistical distance is defined by the residual y, the difference between the measurement and the estimated measurement, and the covariance matrix S. The numerator is the multivariate Gaussian.

12.2.4 Measurement Assignment and Tracks

The following are the rules for each measurement:

1. Each measurement creates a new track.

2. Each measurement in each gate updates the existing track. If there is more than one measurement in a gate, the existing track is duplicated with the new measurement.

3. All existing tracks are updated with a "missed" measurement, creating a new track.

Figure 12.2 gives an example. We are starting with two tracks. There are two tracks and three measurements. All three measurements are in the gate for track 1, but only one is in the gate for track 2. Each measurement produces a new track. The three measurements produce three tracks based on track 1, and the one measurement produces one track based on track 2.

There are three types of tracks created from each scan, in general:

1. An existing track is updated with a new measurement, assuming it corresponds to that track.

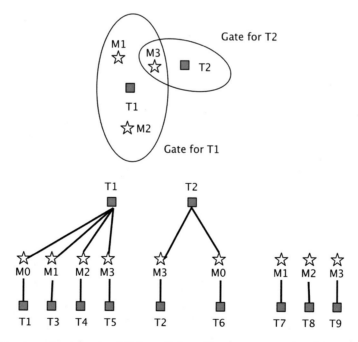

Figure 12.2: *Measurements and gates. M0 is an "absent" measurement. An absent measurement should exist but does not*

2. An existing track is carried along with no update, assuming that no measurement was made for it in that scan.

3. A completely new track is generated for each measurement, assuming that the measurement represents a new object.

Each track also spawns a new track assuming that there was no measurement for the track. Thus, in this case, three measurements and two tracks result in nine new tracks. Tracks 7–9 are initiated based only on the measurement, which may not be enough information to initiate the full-state vector. If this is the case, there would be an infinite number of tracks associated with each measurement, not just one new track. If we have a radar measurement, we have azimuth, elevation, range, and range rate. This gives all position states and one velocity state.

12.2.5 Hypothesis Formation

In MHT, a valid hypothesis is any compatible set of tracks. For two or more tracks to be compatible, they cannot describe the same object, and they cannot share the same measurement in any of the scans. The task in hypothesis formation is to find one or more combinations of tracks that (1) are compatible and (2) maximize some performance function.

Before discussing the method of hypothesis formation, it is useful to first consider track formation and how tracks are associated with unique objects. New tracks may be formed in one of two ways:

1. The new track is based on some existing track, with the addition of a new measurement.

2. The new track is NOT based on any existing tracks; it is based solely on a single new measurement.

Recall that each track is formed as a sequence of measurements across multiple scans. In addition to the raw measurement history, every track also contains a history of state and covariance data that is computed from a Kalman Filter. Kalman Filters were explored in Chapter 8. When a new measurement is appended to an existing track, we are spawning a new track that includes all of the original track's measurements, plus this new measurement. Therefore, the new track is describing the same object as the original track.

A new measurement can also be used to generate a completely new track that is independent of past measurements. When this is done, we are effectively saying that the measurement does not describe any of the objects that are already being tracked. It therefore must correspond to a new/different object.

In this way, each track is given an object ID to distinguish which object it describes. Within the context of track tree diagrams, all of the tracks inside the same track tree have the same object ID. For example, if at some point there are ten separate track trees, this means that ten separate objects are being tracked in the MHT system. When a valid hypothesis is formed, it may turn out that only a few of these objects have compatible tracks.

The hypothesis formation step is formulated as a mixed-integer linear program (MILP) and solved using GLPK. Each track is given an aggregate score that reflects the component score attained from each measurement. The MILP formulation is constructed to select a set of tracks that add up to give the highest score, such that

1. No two tracks have the same object ID.

2. No two tracks have the same measurement index for any scan.

In addition, we extend the formulation with an option to solve for multiple hypotheses, rather than just one. The algorithm will return the "M best" hypotheses, in descending order of score. This enables tracks to be preserved from alternate hypotheses that may be very close in score to the best.

12.2.6 Track Pruning

The N-scan track pruning is carried out every step using the last n scans of data. We employ a pruning method in which the following tracks are preserved:

- Tracks with the "N" highest scores

- Tracks that are included in the "M best" hypotheses

- Tracks that have both (1) the object ID and (2) the first "P" measurements found in the "M best" hypotheses

We use the results of hypothesis formation to guide track pruning. The parameters N, M, and P can be tuned to improve performance. The objective of pruning is to reduce the number of tracks as much as possible while not removing any tracks that should be part of the actual true hypothesis.

The second item listed earlier is to preserve all tracks included in the "M best" hypotheses. Each of these is a full path through a track tree, which is clear. The third item listed is similar but less constrained. Consider one of the tracks in the "M best" hypotheses. We will preserve this full track. In addition, we will preserve all tracks that stem from scan "P" of this track.

Figure 12.3 provides an example of which tracks in a track tree might be preserved. The diagram shows 17 different tracks over 5 scans. The green track represents one of the tracks found in the set of "M best" hypotheses, from the hypothesis formation step. This track would be preserved. The orange tracks all stem from the node in this track at scan 2. These would be preserved if we set $P = 2$ from the preceding description.

12.3   Billiard Ball Kalman Filter

12.3.1   Problem

You want to estimate the trajectory of multiple billiard balls. In the billiard ball example, we assume that we have multiple balls moving at once. Let's say we have a video camera placed above the table, and we have software that can measure the position of each ball for each video frame. That software cannot, however, determine the identity of any ball. This is where MHT comes in. We use MHT to develop a set of tracks for the moving balls.

Track preserved - one of the tracks in the "M best" hypotheses

Tracks preserved - first "P" meas. of a track in "M best" hypotheses (P=2)

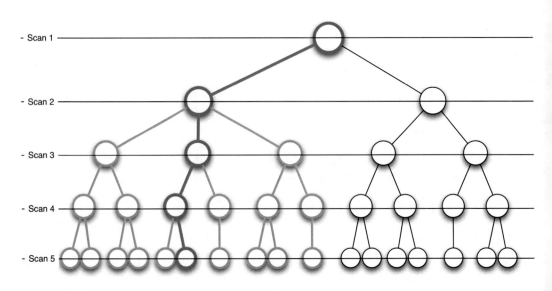

Figure 12.3: *Track pruning example. This shows multiple scans (simultaneous measurements) are how they might be used to remove tracks that do not fit all of the data*

12.3.2 Solution

The solution is to create a linear Kalman Filter.

12.3.3 How It Works

The core estimation algorithm for the MHT system is the Kalman Filter. The Kalman Filter consists of a simulation of the dynamics and an algorithm to incorporate the measurements. For the examples in this chapter, we use a fixed gain, Kalman Filter. The model is

$$x_{k+1} = ax_k + bu_k \tag{12.5}$$

$$y_k = cx_k \tag{12.6}$$

x_k is the state, a column vector that includes position and velocity. y_k is the measurement vector. u_k is the input, the accelerations on the billiard balls. c relates the state to the measurement, y. If the only measurement was the position, then

$$c = \begin{bmatrix} 1 & 0 \end{bmatrix} \tag{12.7}$$

This is a discrete-time equation. Since the second column is zero, it is only measuring the position. Let's assume we have no input accelerations. Also, assume that the time step is τ. Then our equations become

$$\begin{bmatrix} s \\ v \end{bmatrix}_{k+1} = \begin{bmatrix} 1 & \tau \\ 0 & 1 \end{bmatrix} \begin{bmatrix} s \\ v \end{bmatrix}_k \tag{12.8}$$

$$y_k = \begin{bmatrix} 1 & 0 \end{bmatrix} \begin{bmatrix} s \\ v \end{bmatrix}_k \tag{12.9}$$

where s is the position and v is the velocity. $y_k = s$. This says that the new position is the old position plus velocity times time. Our measurement is just the position. If there are no external accelerations, the velocity is constant. If we can't measure acceleration directly, then this is our model. Our filter will estimate the velocity given changes in position.

A track, in this case, is a sequence of s. MHT assigns measurements, y, to the track. If we know that we have only one object and that our sensor is measuring the track accurately, and doesn't have any false measurements or possibility of missing measurements, we can use the Kalman Filter directly.

The KFBilliardsDemo simulates billiard balls. It includes two functions to represent the dynamics. The first is RHSBilliards which is the right-hand side of the billiard ball dynamics, which were just given earlier. This computes the position and velocity given external accelerations. The function BilliardCollision applies conservation of momentum whenever a ball hits a bumper. Balls can't collide with other balls. The first part of the script is the simulation that generates a measurement vector for all of the balls. The second part of the script initializes one Kalman Filter per ball. This script perfectly assigns measurements to each track. The function KFPredict is the prediction step, that is, the simulation of the ball motion. It uses the linear model described earlier. KFUpdate incorporates the measurements. MHTDistance is just for information purposes. The initial positions and velocity vectors of the balls are random. The script fixes the seed for the random number generator to make every run the same, which is handy for debugging. If you comment out this code, each run will be different.

Here, we initialize the ball positions.

KFBilliardsDemo.m

```
31  % The number of balls and the random initial position and velocity
32  d       = struct('nBalls',3,'xLim', [-1 1], 'yLim', [-1 1]);
33  sigP    = 0.4; % 1 sigma noise for the position
34  sigV    = 1; % 1 sigma noise for the velocity
35  sigMeas = 0.00000001; % 1 sigma noise for the measurement
36
37  % Set the initial state for  2 sets of position and velocity
38  x  = zeros(4*d.nBalls,1);
39  rN = rand(4*d.nBalls,1);
40
41  for k = 1:d.nBalls
42      j       = 4*k-3;
```

```
43   x(j  ,1) = sigP*(rN(j  ) - 0.5);
44   x(j+1,1) = sigV*(rN(j+1) - 0.5);
45   x(j+2,1) = sigP*(rN(j+2) - 0.5);
46   x(j+3,1) = sigV*(rN(j+3) - 0.5);
47 end
```

We then simulate them. Their motion is a straight line unless they collide with a bumper.

KFBilliardsDemo.m

```
60  % Sensor measurements
61  nM      = 2*d.nBalls;
62  y   = zeros(nM,n);
63  iY  = zeros(nM,1);
64
65  for k = 1:d.nBalls
66    j = 2*k-1;
67    iY(j  )      = 4*k-3;
68    iY(j+1)      = 4*k-1;
69  end
70
71  for k = 1:n
72
73    % Collisions
74    x = BilliardCollision( x, d );
75
76    % Plotting
77    xP(:,k)      = x;
78
79    % Integrate using a 4th Order Runge-Kutta integrator
80    x = RungeKutta(@RHSBilliards, 0, x, dT, d );
81
82    % Measurements with Gaussian random noise
83    y(:,k) = x(iY) + sigMeas*randn(nM,1);
84
85  end
```

We then process the measurements through the Kalman Filter. `KFPredict` predicts the next position of the balls, and `KFUpdate` incorporates measurements. The prediction step does not know about collisions.

KFBilliardsDemo.m

```
110  %% Implement the Kalman Filter
111
112  % Covariances
113  r0      = sigMeas^2*[1;1];     % Measurement covariance
114  q0      = [1;60;1;60];     % The baseline plant covariance diagonal
115  p0      = [0.1;1;0.1;1];       % Initial state covariance matrix
         diagonal
116
```

```
117   % Plant model
118   a          = [1 dT;0 1];
119   b          = [dT^2/2;dT];
120   zA         = zeros(2,2);
121   zB         = zeros(2,1);
122
123   % Create the Kalman Filter data structures. a is for two balls.
124   for k = 1:d.nBalls
125     kf(k) = KFInitialize( 'kf', 'm', x0(4*k-3:4*k), 'x', x0(4*k-3:4*k)
          ,...
126                          'a', [a zA;zA a], 'b', [b zB;zB b],'u'
                              ,[0;0],...
127                          'h', [1 0 0 0;0 0 1 0], 'p', diag(p0), ...
128                          'q', diag(q0),'r', diag(r0) );
129   end
130
131   % Size arrays for plotting
132   pUKF = zeros(4*d.nBalls,n);
133   xUKF = zeros(4*d.nBalls,n);
134   t    = 0;
135
136   for k = 1:n
137     % Run the filters
138     for j = 1:d.nBalls
139
140       % Store for plotting
141       i          = 4*j-3:4*j;
142       pUKF(i,k)  = diag(kf(j).p);
143       xUKF(i,k)  = kf(j).m;
144
145       % State update
146       kf(j).t    = t;
147       kf(j)      = KFPredict( kf(j) );
148
149       % Incorporate the measurements
150       i          = 2*j-1:2*j;
151       kf(j).y    = y(i,k);
152       kf(j)      = KFUpdate( kf(j) );
153     end
154
155     t = t + dT;
156
157   end
```

The results of the Kalman Filter demo are shown in Figures 12.4, 12.5, and 12.6. The covariances and states for all balls are plotted, but we only show one here. The covariances always follow the same trend with time. As the filter accumulates measurements, it adjusts the covariances based on the ratio between the model covariance, that is, how accurate the model is assumed to be, and the measurement covariances. The covariances are not related to actual measurements at all. The Kalman Filter errors are shown in Figure 12.6. They are large

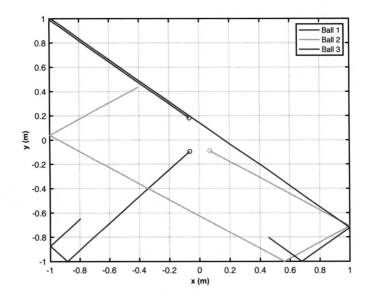

Figure 12.4: *The four balls on the billiard table*

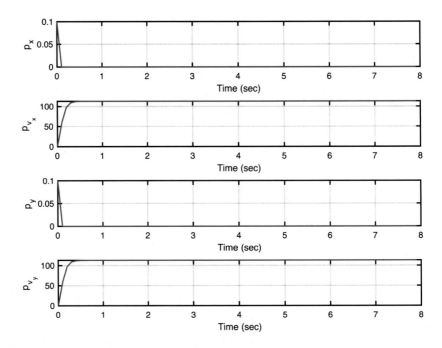

Figure 12.5: *The filter covariances*

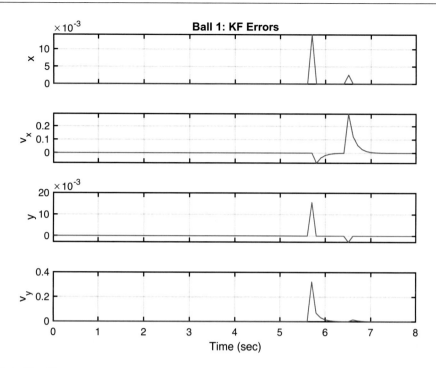

Figure 12.6: *The filter errors*

whenever the ball hits a bumper since the models do not include collisions with the bumpers. They rapidly decrease because our measurements have little noise.

The following code, excerpted from the preceding demo, is a specialized drawing code to show the billiards on the table. It calls `plot` for each ball. Colors are taken from the array c which are blue, green, red, cyan, magenta, yellow, and black. You can run this from the command line once you have computed xP and yP, which are the x and y positions of the balls. The code uses the legend handles to associate the balls with the tracks in the plot in the legend. It manually sets the limits (`gca` is a handle to the current axes).

KFBilliardsDemo.m

```
87   % Plot the simulation results
88   NewFigure( 'Billiard Balls' )
89   c   = 'bgrcmyk';
90   kX = 1;
91   kY = 3;
92   s   = cell(1,d.nBalls);
93   l = [];
94   for k = 1:d.nBalls
95     plot(xP(kX,1),xP(kY,1),['o',c(k)])
96     hold on
97     l(k)   = plot(xP(kX,:),xP(kY,:),c(k));
98     kX     = kX + 4;
```

```
99    kY    = kY + 4;
100   s{k}  = sprintf('Ball %d',k);
101 end
102
103 xlabel('x (m)');
104 ylabel('y (m)');
105 set(gca,'ylim',d.yLim,'xlim',d.xLim);
106 legend(1,s)
107 grid on
```

You can change the covariances, `sigP`, `sigV`, `sigMeas`, in the script and see how it impacts the errors and the covariances.

12.4   Billiard Ball MHT

12.4.1 Problem

You want to estimate the trajectory of multiple billiard balls.

12.4.2 Solution

The solution is to create an MHT system with a linear Kalman Filter. This example involves billiard balls bouncing off of the bumpers of a billiard table. The model does not include the bumper collisions.

12.4.3 How It Works

The following code adds the MHT functionality. It first runs the demo, just like in the preceding example, and then tries to sort the measurements into tracks. It only has two balls. When you run the demo, you will see the GUI (Figure 12.7) and the tree (Figure 12.8) change as the simulation progresses. We only include the MHT code in the following listing.

MHTBilliardsDemo.m

```
135 % Create the track data data structure
136 mhtData = MHTInitialize('probability false alarm', 0.001,...
137                         'probability of signal if target present',
                                0.999,...
138                         'probability of signal if target absent',
                                0.001,...
139                         'probability of detection', 1, ...
140                         'measurement volume', 1.0, ...
141                         'number of scans', 3, ...
142                         'gate', 0.2,...
143                         'm best', 2,...
144                         'number of tracks', 1,...
145                         'scan to track function',@ScanToTrackBilliards
                                ,...
146                         'scan to track data',struct('r',diag(r0),'p',
                                diag(p0)),...
147                         'distance function',@MHTDistance,...
```

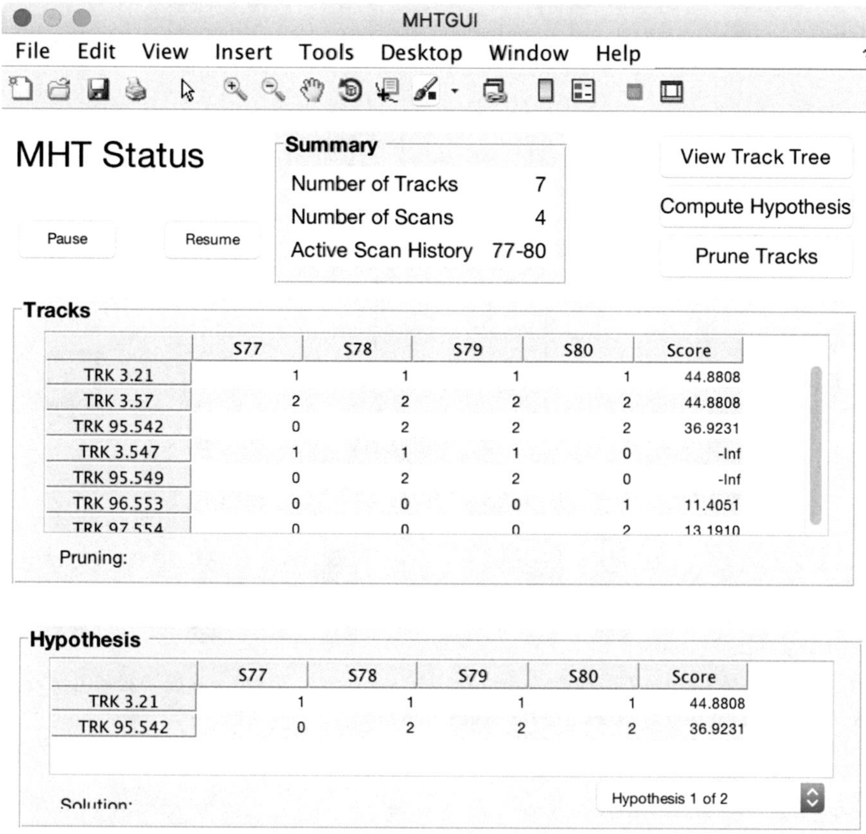

MHT Status

Pause   Resume

Summary

Number of Tracks	7
Number of Scans	4
Active Scan History	77-80

View Track Tree

Compute Hypothesis

Prune Tracks

Tracks

	S77	S78	S79	S80	Score
TRK 3.21	1	1	1	1	44.8808
TRK 3.57	2	2	2	2	44.8808
TRK 95.542	0	2	2	2	36.9231
TRK 3.547	1	1	1	0	-Inf
TRK 95.549	0	2	2	0	-Inf
TRK 96.553	0	0	0	1	11.4051
TRK 97.554	0	0	0	2	13.1910

Pruning:

Hypothesis

	S77	S78	S79	S80	Score
TRK 3.21	1	1	1	1	44.8808
TRK 95.542	0	2	2	2	36.9231

Solution: Hypothesis 1 of 2

Figure 12.7: *The MHT GUI*

```
148                         'hypothesis scan last', 0,...
149                         'filter data',kf(1),...
150                         'prune tracks', 1,...
151                         'remove duplicate tracks across all trees'
                                ,1,...
152                         'average score history weight',0.01,...
153                         'filter type','kf');
154
155   % Create the tracks
156   for k = 1:d.nBalls
157           trk(k) = MHTInitializeTrk( kf(k) );
158   end
159
160   % Size arrays
161   b = MHTTrkToB( trk );
162
163   %% Initialize MHT GUI
164   MHTGUI;
165   MLog('init')
```

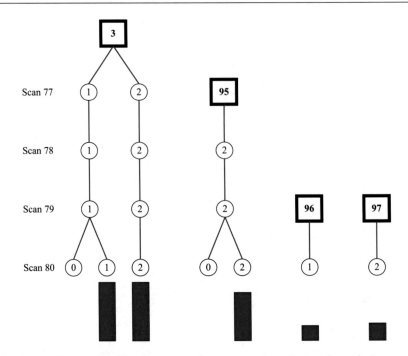

Figure 12.8: *The MHT tree. The blue bars give the score assigned to each track. Longer is better. The numbers in the framed black boxes are the track numbers*

```
166    MLog('name','Billiards Demo')
167    TOMHTTreeAnimation( 'initialize', trk );
168    TOMHTTreeAnimation( 'update', trk );
169
170    t = 0;
171
172    for k = 1:n
173
174      % Get the measurements - zScan.data
175      z = reshape( y(:,k), 2, d.nBalls );
176      zScan = AddScan( z(:,1) );
177      for j = 2:size(z,2)
178        zScan = AddScan( z(:,j),[],zScan);
179      end
180
181      % Manage the tracks and generate hypotheses
182      [b, trk, sol, hyp] = MHTTrackMgmt( b, trk, zScan, mhtData, k, t );
183
184      % Update MHTGUI display
185      if( ~isempty(zScan) && graphicsOn )
186        if (treeAnimationOn)
187          TOMHTTreeAnimation( 'update', trk );
188        end
189        MHTGUI(trk,sol,'hide');
```

```
190      drawnow
191    end
192
193    t = t + dT;
194  end
195
196  % Show the final GUI
197  if (~treeAnimationOn)
198    TOMHTTreeAnimation( 'update', trk );
199  end
200  if (~graphicsOn)
201    MHTGUI(trk,sol,'hide');
202  end
203  MHTGUI;
```

The parameter pairs in MHTInitialize are described in Table 12.2.

Figure 12.7 shows the MHT GUI. This shows the GUI at the end of the simulation. The table shows scans on the x-axis and tracks on the y-axis (vertical). Each track is numbered xxx. yyy where xxx is the track and yyy is the tag. Every track is assigned a new tag number. For example, 95.542 is track 95 and tag 542 means it is the 542nd track generated. The numbers in the table show the measurements associated with the track and the scan. TRK 3.21 and TRK 3.57 are duplicates. In both cases, one measurement per scan is associated with the TRK. Their scores are the same because they are consistent. We can only pick one or the other for our hypothesis. TRK 95.542 doesn't get a measurement from scan 77, but for the rest of the scans, it gets measurement 2. Scans 77 through 80 are active. A scan is a set of four position measurements. The summary shows there are seven active tracks, but we know (but the software does not necessarily) that there are only four balls in play. The number of scans is the ones currently in use to determine valid tracks. There are two active hypotheses.

Figure 12.8 shows the decision tree. You can see that with scan 80, two new tracks are created. This means MHT thinks that there could be as many as four tracks. However, at this point, only two tracks, 3 and 95, have multiple measurements associated with them.

Figure 12.9 shows the information window. This shows the MHT algorithm's thinking. It gives the decisions made with each scan.

The demo shows that the MHT algorithm correctly associates measurements with tracks.

12.5   One-Dimensional Motion

12.5.1 Problem

You want to estimate the position of an object moving in one direction with unknown accelerations.

Table 12.2: *MHT parameters*

Term	Definition
'probability false alarm'	The probability that a measurement is spurious
'probability of signal if target present'	The probability of getting a signal if the target is present
'probability of signal if target absent'	The probability of getting a signal if the target is absent
'probability of detection'	Probability of detection of a target
'measurement volume'	Scales the likelihood ratio
'number of scans'	The number of scans to consider in hypothesis formulation
'gate'	The size of the gate
'm best'	Number of hypotheses to consider
'number of tracks'	Number of tracks to maintain
'scan to track function'	Pointer to the scan to track function. This is custom for each application
'scan to track data'	Data for the scan to track function
'distance function'	Pointer for the MHT distance function. Different definitions are possible
'hypothesis scan last'	The last scan used in a hypothesis
'prune tracks'	Prune tracks if true
'filter type'	Type of Kalman Filter
'filter data'	Data for the Kalman Filter
'remove duplicate tracks across all trees'	If true, removes duplicate tracks from all trees
'average score history weight'	A number to multiple the average score history
'create track'	If entered, it will create a track instead of using an existing track

12.5.2 Solution

The solution is to create a linear Kalman Filter with an acceleration state.

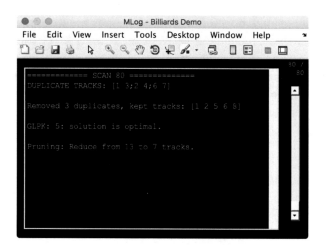

Figure 12.9: *The MHT information window. It tells you what the MHT algorithm is thinking*

12.5.3 How It Works

In this demo, we have a model of objects that includes an unknown acceleration state:

$$\begin{bmatrix} s \\ v \\ a \end{bmatrix}_{k+1} = \begin{bmatrix} 1 & \tau & \frac{1}{2}\tau^2 \\ 0 & 1 & \tau \\ 0 & 0 & 1 \end{bmatrix} \begin{bmatrix} s \\ v \\ a \end{bmatrix}_k \tag{12.10}$$

$$y_k = \begin{bmatrix} 1 & 0 & 0 \end{bmatrix} \begin{bmatrix} s \\ v \\ a \end{bmatrix}_k \tag{12.11}$$

where s is position, v is velocity, and a is acceleration. $y_k = s$. τ is the time step. The input to the acceleration state is the time rate of change of acceleration.

The function `DoubleIntegratorWithAccel` creates the matrices shown earlier:

```
>> [a, b]  = DoubleIntegratorWithAccel( 0.5 )

a =
    1.0000     0.5000     0.1250
         0     1.0000     0.5000
         0          0     1.0000

b =
     0
     0
     1
```

with $\tau = 0.5$ second.

We will set up the simulation so that one object has no acceleration but starts in front of the other. The other will overtake the first. We want to see if MHT can sort out the trajectories. Passing would happen all the time with autonomous driving.

The following code implements the Kalman Filters for two vehicles. The simulation runs first to generate the measurements. The Kalman Filter runs next. Note that the plot array is updated after the filter update. This keeps it in sync with the simulation.

KF1DDemo.m

```
56  %% Run the Kalman Filter
57  % The covariances
58  r        = r(1,1);
59  q        = diag([0.5*aRand*dT^2;aRand*dT;aRand].^2 + q0);
60
61  % Create the Kalman Filter data structures
62  d1   = KFInitialize( 'kf', 'm', [0;0;0], 'x', [0;0;0], 'a', a, 'b', b
        , 'u',0,...
63                      'h', h(1,1:3), 'p', diag(p0), 'q', q, 'r', r );
64  d2   = d1;
65  d1.m = x(1:3,1) + sqrt(p0).*rand(3,1);
66  d2.m = x(4:6,1) + sqrt(p0).*rand(3,1);
67  xE   = zeros(6,n);
68
69  for k = 1:n
70    d1        = KFPredict( d1 );
71    d1.y      = z(1,k);
72    d1        = KFUpdate( d1 );
73
74    d2        = KFPredict( d2 );
75    d2.y      = z(2,k);
76    d2        = KFUpdate( d2 );
77
78    xE(:,k) = [d1.m;d2.m];
79  end
```

We use PlotSet with the argument 'plot set' to group inputs and the argument 'legend' to put legends on each plot. 'plot set' takes a cell array of $1 \times n$ arrays, and 'legend' takes a cell array of cell arrays as inputs. We don't need to numerically integrate the equations of motion because the state equations have already done that. You can always propagate a linear model in this fashion. We set the model noise matrix using aRand but don't input any random accelerations. As written, our model is perfect, which is never true in a real system, hence the need for model uncertainty.

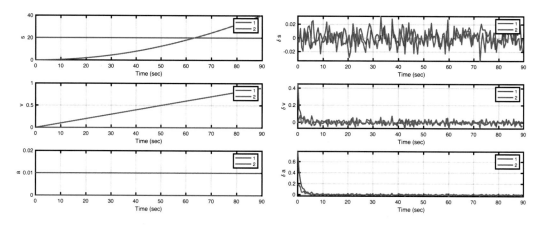

Figure 12.10: *The object states and filter errors*

Figure 12.10 shows the states and the errors. The filters track all three states for both objects pretty well. The acceleration and velocity estimates converge within ten seconds. It does a good job of estimating the fixed disturbance acceleration despite only having a position, s, measurement.

12.6   One-Dimensional MHT

The next problem is one in which we need to associate measurements with a track.

12.6.1   Problem

You want to estimate the position of an object moving in one direction with measurements that need to be associated with a track.

12.6.2   Solution

The solution is to create an MHT system with the Kalman Filter as the state estimator.

12.6.3   How It Works

The MHT code is shown in the following listing. We append the MHT software to the script shown earlier. The Kalman Filters are embedded in the MHT software. We first run the simulation and gather the measurements and then process them in the MHT code.

MHT1DDemo.m

```
69   % Initialize the MHT parameters
70   [mhtData, trk] = MHTInitialize( 'probability false alarm', 0.001,...
71                                    'probability of signal if target
                                        present', 0.999,...
72                                    'probability of signal if target absent
                                        ', 0.001,...
73                                    'probability of detection', 1, ...
74                                    'measurement volume', 1.0, ...
```

```
75                                          'number of scans', 3, ...
76                                          'gate', 0.2,...
77                                          'm best', 2,...
78                                          'number of tracks', 1,...
79                                          'scan to track function',@ScanToTrack1D
                                               ,...
80                                          'scan to track data',struct('v',0),...
81                                          'distance function',@MHTDistance,...
82                                          'hypothesis scan last', 0,...
83                                          'prune tracks', true,...
84                                          'filter type','kf',...
85                                          'filter data', f,...
86                                          'remove duplicate tracks across all
                                               trees',true,...
87                                          'average score history weight',0.01,...
88                                          'create track', '');
89
90   % Size arrays
91   m                    = zeros(3,n);
92   p                    = zeros(3,n);
93   scan                 = cell(1,n);
94   b                    = MHTTrkToB( trk );
95
96   TOMHTTreeAnimation( 'initialize', trk );
97   TOMHTTreeAnimation( 'update', trk );
98
99   % Initialize the MHT GUI
100  MHTGUI;
101  MLog('init')
102  MLog('name','MHT 1D Demo')
103
104  t = 0;
105
106  for k = 1:n
107
108     % Get the measurements
109     zScan = AddScan( z(1,k) );
110     zScan = AddScan( z(2,k), [], zScan );
111
112     % Manage the tracks
113     [b, trk, sol, hyp] = MHTTrackMgmt( b, trk, zScan, mhtData, k, t );
114
115     % Update MHTGUI display
116     MHTGUI(trk,sol,'update');
117
118     % A guess for the initial velocity of any new track
119     for j = 1:length(trk)
120        mhtData.fScanToTrackData.v = mhtData.fScanToTrackData.v + trk(j).
                m(1);
121     end
122     mhtData.fScanToTrackData.v = mhtData.fScanToTrackData.v/length(trk);
```

350

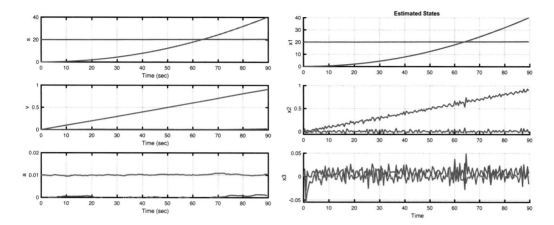

Figure 12.11: *The MHT object states and estimated states. The colors are switched between plots*

```
123
124     % Animate the tree
125         TOMHTTreeAnimation( 'update', trk );
126         drawnow;
127     t = t + dT;
128     end
```

Figure 12.11 shows the states and the errors. The MHT hypothesized tracks are a good fit for the data.

Figure 12.12 shows the MHT GUI and the tree. Track 1 contains only measurements from object 2. Track 2 contains only measurements from object 1. Tracks 354 and 360 are spurious tracks. Track 354 has one measurement of 1 for scan 177, but none for the following scan. Track 360 was created on scan 180 and has just one measurement. Tracks 1 and 2 have the same score. The results show that the MHT software has successfully sorted out the measurements and assigned them correctly. At this point, at the end of the sim, four scans are active.

12.7 Summary

This chapter demonstrated the fundamentals of Multiple Hypothesis Testing. Table 12.3 lists the functions and scripts included in the companion code. Table 12.4 lists the code MHT implementation functions.

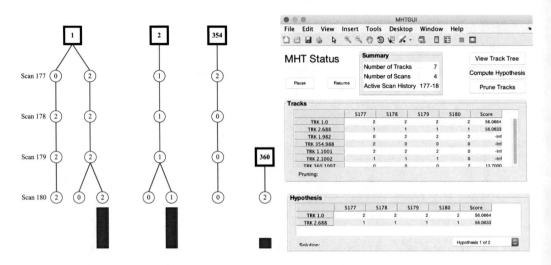

Figure 12.12: *The GUI and MHT tree. The tree shows the MHT decision process*

Table 12.3: *Chapter code listing*

File	Description
BilliardCollision	Billiard ball collision model
DoubleIntegratorWithAccel	Plant model for a double integrator with an acceleration state
KF1DDemo	One-dimensional Kalman Filter demo with two vehicles with random accelerations
KFBilliardsDemo	Billiard demo using a Kalman Filter to estimate the ball states
MHT1DDemo	One-dimensional MHT demo with two vehicles with random accelerations
MHTBilliardsDemo	MHT billiard demo
RHSBilliards	Billiard ball dynamical model
ScanToTrackBilliards	Initializes a new track

Table 12.4: *MHT code listing*

File	Description
AddScan	Adds a scan to the data
CheckForDuplicateTracks	Looks through the recorded tracks for duplicates
MHTDistanceUKF	Computes the MHT distance
MHTGUI.fig	Saved layout data for the MHT Graphical User Interface
MHTGUI	GUI for the MHT software
MHTHypothesisDisplay	Displays hypotheses in a GUI
MHTInitialize	Initializes the MHT algorithm
MHTInitializeTrk	Initializes a track
MHTLLRUpdate	Updates the log-likelihood ratio
MHTMatrixSortRows	Sorts rows in the MHT
MHTMatrixTreeConvert	Converts to and from a tree format for the MHT data
MHTTrackMerging	Merges MHT tracks
MHTTrackMgmt	Manages MHT tracks
MHTTrackScore	Computes the total score for the track
MHTTrackScoreKinematic	Computes the kinematic portion of the track score
MHTTrackScoreSignal	Computes the signal portion of the track score
MHTTreeDiagram	Draws an MHT tree diagram
MHTTrkToB	Converts tracks to a B matrix
PlotTracks	Plots object tracks
Residual	Computes the residual
TOMHTTreeAnimation	Track-Oriented MHT tree diagram animation
TOMHTAssignment	Assigns a scan to a track
TOMHTPruneTracks	Prunes the tracks

CHAPTER 13

■ ■ ■

Autonomous Driving with MHT

In this chapter, we will apply the MHT techniques from the previous chapter to the interesting problem of autonomous driving. As with MHT, this chapter falls in the Learning portion of our taxonomy. Consider a primary car that is driving along a highway at variable speeds. It carries a radar that measures the azimuth, range, and range rate. Cars pass the primary car, some of which change lanes from behind the car and cut in front. The multiple hypothesis system tracks all cars around the primary

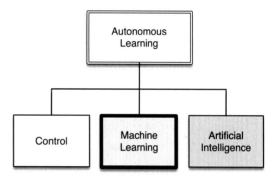

car. At the start of the simulation, there are no cars in the radar field of view. One car passes and cuts in front of the radar car. The other two just pass in their lanes. You want to accurately track all cars that your radar can see.

There are two elements to this problem. One is to model the motion of the tracked automobiles using measurements to improve your estimate of each automobile's location and velocity. The second is to systematically assign measurements to different tracks. A track should represent a single car, but the radar is just returning measurements on echoes; it doesn't know anything about the source of the echoes.

You will solve the problem by first implementing a Kalman Filter to track one automobile. We need to write measurement and dynamics functions that will be passed to the Kalman Filter, and we need a simulation to create the measurements. Then we will apply the Multiple Hypothesis Testing (MHT) techniques developed in the previous chapter to this problem.

We'll do the following things in this chapter:

1. Model the automobile dynamics

2. Model the radar system

3. Write the control algorithms

M. Paluszek, S. Thomas, *MATLAB Machine Learning Recipes*,
https://doi.org/10.1007/978-1-4842-9846-6_13

4. Implement visualization to let us see the maneuvers in 3D

5. Implement the Unscented Kalman Filter

6. Implement MHT

13.1   Automobile Dynamics

13.1.1 Problem

We need to model the car dynamics. We will limit this to a planar model in two dimensions. We are modeling the location of the car in x/y and the angle of the wheels which allows the car to change direction.

13.1.2 Solution

Write a right-hand-side function that can be called by `RungeKutta`.

13.1.3 How It Works

Much like with the radar, we will need two functions for the dynamics of the automobile. `RHSAutomobile` is used by the simulation. `RHSAutomobile` has the full dynamic model including the engine and steering model. Aerodynamic drag, rolling resistance, and side force resistance (the car doesn't slide sideways without resistance) are modeled. `RHSAutomobile` handles multiple automobiles. An alternative would be to have a one-automobile function and call `RungeKutta` once for each automobile. The latter approach works in all cases, except when you want to model collisions. In many types of collisions, two cars collide and then stick, effectively becoming a single car. A real tracking system would need to handle this situation. Each vehicle has six states. They are

1. x position

2. y position

3. x velocity

4. y velocity

5. Angle about vertical

6. Angular rate about vertical

   The velocity derivatives are driven by the forces and the angular rate derivative by the torques. The planar dynamics model is illustrated in Figure 13.1 [33]. Unlike the reference we constrain the rear wheels to be fixed and the angles for the front wheels to be the same.

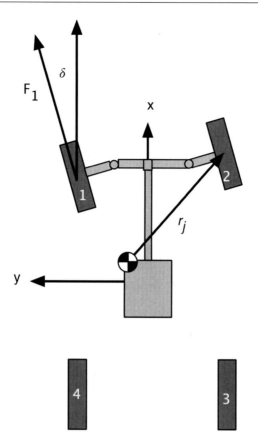

Figure 13.1: *Planar automobile dynamical model*

The dynamical equations are written in the rotating frame:

$$m(\dot{v}_x - 2\omega v_y) = \sum_{k=1}^{4} F_{k_x} - qC_{D_x}A_x u_x \tag{13.1}$$

$$m(\dot{v}_y + 2\omega v_x) = \sum_{k=1}^{4} F_{k_y} - qC_{D_y}A_y u_y \tag{13.2}$$

$$I\dot{\omega} = \sum_{k=1}^{4} r_k^{\times} F_k \tag{13.3}$$

where the dynamic pressure is

$$q = \frac{1}{2}\rho\sqrt{v_x^2 + v_y^2} \tag{13.4}$$

and

$$v = \begin{bmatrix} v_x \\ v_y \end{bmatrix} \qquad (13.5)$$

The unit vector is

$$u = \frac{\begin{bmatrix} v_x \\ v_y \end{bmatrix}}{\sqrt{v_x^2 + v_y^2}} \qquad (13.6)$$

The normal force is mg where g is the acceleration of gravity. The force at the tire contact point, where the tire touches the road, for tire k is

$$F_{t_k} = \begin{bmatrix} T/\rho - F_r \\ -F_c \end{bmatrix} \qquad (13.7)$$

where ρ is the radius of the tire and F_r is the rolling friction and is

$$F_r = f_0 + K_1 v_{t_x}^2 \qquad (13.8)$$

where v_{t_x} is the velocity in the tire frame in the rolling direction. For front-wheel drive cars, the torque, T, is zero for the rear wheels. The contact friction is

$$F_c = \mu_c mg \frac{v_{t_y}}{|v_t|} \qquad (13.9)$$

This is the force perpendicular to the normal rolling direction of the wheel, that is, into or out of the paper in Figure 13.2. The velocity term ensures that the friction force does not cause limit cycling. That is, when the y velocity is zero, the force is zero. μ_c is a constant for the tires.

The transformation from the tire to the body frame is

$$c = \begin{bmatrix} \cos\delta & -\sin\delta \\ \sin\delta & \cos\delta \end{bmatrix} \qquad (13.10)$$

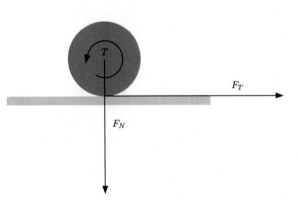

Figure 13.2: *Wheel force and torque*

where δ is the steering angle so that

$$F_k = cF_{t_k} \tag{13.11}$$

$$v_t = c^T \begin{bmatrix} v_x \\ v_y \end{bmatrix} \tag{13.12}$$

The kinematical equation that relates yaw angle and yaw angular rate is

$$\dot{\theta} = \omega \tag{13.13}$$

and the inertial velocity V, the velocity needed to tell you where the car is going, is

$$V = \begin{bmatrix} \cos\theta & -\sin\theta \\ \sin\theta & \cos\theta \end{bmatrix} v \tag{13.14}$$

We'll show you the dynamics simulation when we get to the graphics part of the chapter in Section 13.4.

13.2 Automobile Radar

13.2.1 Problem

The sensor utilized for this example will be the automobile radar. The radar measures the azimuth, range, and range rate. We need two functions: one for the simulation and the second for use by the Unscented Kalman Filter.

13.2.2 Solution

Build a radar model in a MATLAB function. The function will use analytical derivations of range and range rate.

13.2.3 How It Works

The radar model is extremely simple. It assumes the radar measures the line-of-site range, range rate, and azimuth, the angle from the forward axis of the car. The model skips all the details of radar signal processing and outputs those three quantities. A simple model is always the best when you start a project. Later on, you will need to add a very detailed model that has been verified against test data to demonstrate that your system works as expected.

The position and velocity of the radar are entered through the data structure. This does not model the signal-to-noise ratio of a radar. The power received by a radar goes as $\frac{1}{r^4}$. In this model, the signal goes to zero at the maximum range that is specified in the function. The range is found from the difference in position between the radar and the target. If δ is the difference, we write

$$\delta = \begin{bmatrix} x - x_r \\ y - y_r \\ z - z_r \end{bmatrix} \tag{13.15}$$

The range is then

$$\rho = \sqrt{\delta_x^2 + \delta_y^2 + \delta_z^2} \tag{13.16}$$

The delta velocity is

$$\nu = \begin{bmatrix} v_x - v_{x_r} \\ v_y - v_{y_r} \\ v_z - v_{z_r} \end{bmatrix} \tag{13.17}$$

In both equations, the subscript r denotes the radar. The range rate is

$$\dot{\rho} = \frac{\nu^T \delta}{\rho} \tag{13.18}$$

The AutoRadar function handles multiple targets and can generate radar measurements for an entire trajectory. This is convenient because you can give it your trajectory and see what it returns. This gives you a physical feel for the problem without running a simulation. It also allows you to be sure the sensor model is doing what you expect! This is important because all models have assumptions and limitations. It may be that the model isn't suitable for your application. For example, this model is two-dimensional. If you are concerned about your system getting confused about a car driving across a bridge above your automobile, this model will not be useful in testing that scenario.

Notice that the function has a built-in demo and, if there are no outputs, will plot the results. Adding demos to your code is a nice way to make your functions more user-friendly to other people using your code and even to you when you encounter the code again several months after writing the code! We put the demo in a subfunction because it is long. If the demo is one or two lines, a subfunction isn't necessary. Just before the demo function is the function defining the data structure.

The second function, AutoRadarUKF, is the same core code, but designed to be compatible with the Unscented Kalman Filter. We could have used AutoRadar, but this is more convenient. The transformation matrix, cITOC (inertial to car transformation), is two-dimensional since the simulation is in a flat world.

AutoRadarUKF.m

```
 1   %% AUTORADARUKF Radar model for the auto UKF
19   function y = AutoRadarUKF( x, d )
20
21   s        = sin(d.theta);
22   c        = cos(d.theta);
```

```
23  cIToC    = [c s;-s c];
24  dR       = cIToC*x(1:2);
25  dV       = cIToC*x(3:4);
26
27  rng      = sqrt(dR'*dR);
28  y        = [rng; dR'*dV/rng; atan(dR(2)/dR(1))];
```

The radar returns the range, range rate, and the azimuth angle of the target. Even though we are using radar as our sensor, there is no reason why you couldn't use a camera, laser range finder, or sonar instead. The limitation of the algorithms and software provided in this book is that it will only handle one sensor. You can get software from Princeton Satellite Systems that expand this to multiple sensors. For example, cars carry radar, cameras, and lidar. You might want to integrate all of their measurements. Figure 13.3 shows the internal radar demo. The target car is weaving in front of the radar. It is receding at a steady velocity, but the weave introduces a time-varying range rate.

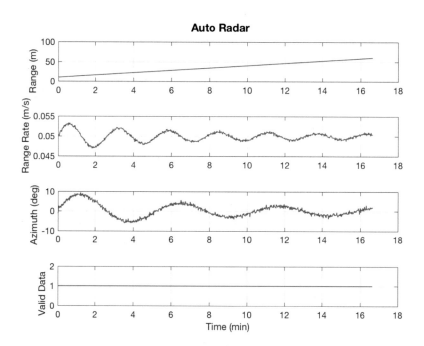

Figure 13.3: *Built-in radar demo. The target is weaving in front of the radar and accelerating away*

13.3   Passing Control

13.3.1 Problem

To have something interesting for our radar to measure, we need our cars to perform some maneuvers. We will develop an algorithm for a car to change lanes.

13.3.2 Solution

The cars are driven by steering controllers that execute basic automobile maneuvers. The throttle (accelerator pedal) and steering angle can be controlled. Multiple maneuvers can be chained together. This provides a challenging test for the MHT system. The first function is for autonomous passing, and the second performs the lane change.

13.3.3 How It Works

The `AutomobilePassing` function implements passing control by pointing the wheels at the target. It generates a steering angle demand and torque demand. Demand is what we want the steering to do. In a real automobile, the hardware will attempt to meet the demand, but there will be a time lag before the wheel angle or motor torque meets the wheel angle or torque demand commanded by the controller. In many cases, you are passing the demand to another control system that will try and meet the demand. The algorithms are quite simple. They don't care if anyone gets in the way. They also don't have any control over avoiding another vehicle. The code assumes that the lane is empty. Don't try this with your car!

The state is defined by the `passState` variable in the `passer` data structure. Before passing, the `passState` is 0. During the passing, it is 1. When it returns to its original lane, the state is set to 0.

AutomobilePassing.m

```
37  function passer = AutomobilePassing( passer, passee, dY, dV, dX, gain )
44  % Lead the target unless the passing car is in front
45  if( passee.x(1) + dX > passer.x(1) )
46      xTarget = passee.x(1) + dX;
47  else
48      xTarget = passer.x(1) + dX;
49  end
50
51  % This causes the passing car to cut in front of the car being passed
52  if( passer(1).passState == 0 )
53      if( passer.x(1) > passee.x(1) + 2*dX )
54          dY = 0;
55          passer(1).passState = 1;
56      end
57  else
58      dY = 0;
59  end
60
61  % Control calculation
```

```
62  target        = [xTarget;passee.x(2) + dY];
63  theta         = passer.x(5);
64  dR            = target - passer.x(1:2);
65  angle         = atan2(dR(2),dR(1));
66  err           = angle - theta;
67  passer.delta  = gain(1)*(err + gain(3)*(err - passer.errOld));
68  passer.errOld = err;
69  passer.torque = gain(2)*(passee.x(3) + dV - passer.x(3));
```

The second function performs a lane change. It implements lane change control by pointing the wheels at the target. The function generates a steering angle demand and a torque demand. The default gains work reasonably well. You should always supply defaults that make sense.

AutomobileLaneChange.m

```
33  function passer = AutomobileLaneChange( passer, dX, y, v, gain )

35  % Default gains
36  if( nargin < 5 )
37    gain = [0.05 80 120];
38  end

39
40  % Lead the target unless the passing car is in front
41  xTarget       = passer.x(1) + dX;

42
43  % Control calculation
44  target        = [xTarget;y];
45  theta         = passer.x(5);
46  dR            = target - passer.x(1:2);
47  angle         = atan2(dR(2),dR(1));
48  err           = angle - theta;
49  passer.delta  = gain(1)*(err + gain(3)*(err - passer.errOld));
50  passer.errOld = err;
51  passer.torque = gain(2)*(v - passer.x(3));
```

13.4   Automobile Animation

13.4.1 Problem

We want to visualize the cars as they maneuver.

13.4.2 Solution

Read in a file in .obj format. Display it using MATLAB's patch function.

13.4.3 How It Works

We create a function to read in .obj files. We then write a function to draw and animate the model.

13.4.4 Solution

The first part is to find an automobile model. A good resource is TurboSquid (www.turbosquid. com). You will find thousands of models. We need .obj format and prefer low polygon count. Ideally, we want models with triangles. In the case of the model found for this chapter, it had rectangles, so we converted them to triangles using a Macintosh application, Cheetah3D (www. cheetah3d.com). An OBJ model comes with an obj file, an mtl file (material file), and images for textures. We will only use the obj file.

LoadOBJFile, covered in Chapter 3, loads the file and puts it into a data structure. The data structure uses the g field of the OBJ file to break the file into components. In this case, the components are the four tires and the rest of the car. The demo is just LoadOBJFile('MyCar.obj'). You do need the extension, .obj. The car is shown in Figure 13.4.

The image is generated with one call to patch per component.

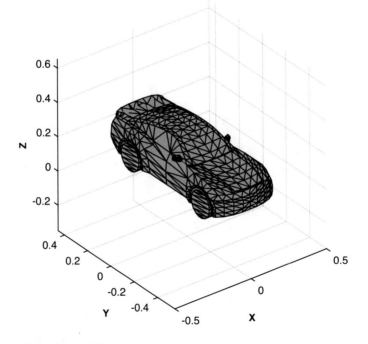

Figure 13.4: *Automobile 3D model*

The first part of `DrawComponents` initializes and updates the model. We save, and return, pointers to the patches so that we only have to update the vectors with each call.

DrawComponents.m

```
33  switch( lower(action) )
34    case 'initialize'
35
36      n = length(g.component);
37      h = zeros(1,n);
38
39      for k = 1:n
40        h(k) = DrawMesh(g.component(k) );
41      end
42
43    case 'update'
44      UpdateMesh(h,g.component,x);
45
46    otherwise
47      warning('%s not available',action);
48  end
```

The mesh is drawn with a call to `patch`. `patch` has many options that are worth exploring. We use the minimal set. We make the edges black to make the model easier to see. The Phong reflection model is an empirical lighting model. It includes diffuse and specular lighting.

DrawComponents.m

```
51  function h = DrawMesh( m )
52
53  h = patch( 'Vertices', m.v, 'Faces',    m.f, 'FaceColor', m.color,...
54             'EdgeColor',[0 0 0],'EdgeLighting', 'phong',...
55             'FaceLighting', 'phong');
```

Updating is done by rotating the vertices around the z-axis and then adding the x and y positional offsets. The input array is [x;y; yaw]. We then set the new vertices. The function can handle an array of positions, velocities, and yaw angles.

DrawComponents.m

```
59  function UpdateMesh( h, c, x )
60
61  for j = 1:size(x,2)
62    for k = 1:length(c)
63      cs       = cos(x(3,j));
64      sn       = sin(x(3,j));
65      b        = [cs -sn 0 ;sn cs 0;0 0 1];
66      v        = (b*c(k).v')';
67      v(:,1)   = v(:,1) + x(1,j);
68      v(:,2)   = v(:,2) + x(2,j);
69      set(h(k),'vertices',v);
```

```
70    end
71  end
```

The graphics demo `AutomobileDemo` implements passing control. `AutomobileInitialize` reads in the OBJ file. The following code sets up the graphics window:

AutomobileDemo.m

```
33  % Set up the figure
34  NewFigure( 'Car Passing' )
35  axes('DataAspectRatio',[1 1 1],'PlotBoxAspectRatio',[1 1 1] );
36
37  h = [];
38  h(1,:) = DrawComponents( 'initialize', d.car(1).g );
39  h(2,:) = DrawComponents( 'initialize', d.car(2).g );
40
41  xlabel('X (m)')
42  ylabel('Y (m)')
43  zlabel('Z (m)')
44
45  set(gca,'ylim',[-4 4],'zlim',[0 2]);
46
47  grid on
48  view(3)
49  rotate3d on
```

During each pass through the simulation loop, we update the graphics. We call `DrawComponents` once per car along with the stored `patch` handles for each car's components. We adjust the limits so that we maintain a tight focus on the two cars. We could have used the camera fields in the axis data structure for this too. We call `drawnow` after setting the new `xlim` for smooth animation. The graphing portion of the loop is shown as follows:

AutomobileDemo.m

```
69  for k = 1:n
70    % Draw the cars
71    pos1 = x([1 2]);
72    pos2 = x([7 8]);
73    DrawComponents( 'update', d.car(1).g, h(1,:), [pos1;pi/2 + x( 5)] );
74    DrawComponents( 'update', d.car(2).g, h(2,:), [pos2;pi/2 + x(11)] );
75
76    xlim = [min(x([1 7]))-10 max(x([1 7]))+10];
77    set(gca,'xlim',xlim);
78    drawnow
```

Figure 13.5 shows four points in the passing sequence.

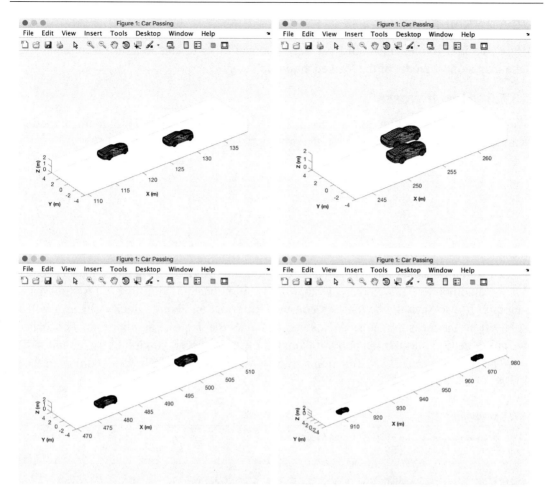

Figure 13.5: *Automobile simulation snapshots showing passing*

13.5 Automobile Simulation and the Kalman Filter

13.5.1 Problem

You want to track a car using radar measurements to track an automobile maneuvering around your car. Cars may appear and disappear at any time. The radar measurement needs to be turned into the position and velocity of the tracked car. In between radar measurements, you want to make your best estimate of where the automobile will be at a given time.

13.5.2 Solution

The solution is to implement an Unscented Kalman Filter to take radar measurements and update a dynamical model of the tracked automobile.

13.5.3 How It Works

We first create the function RHSAutomobileXY with the Kalman Filter dynamical model. The Kalman Filter right-hand side is just the differential equations:

$$\dot{x} = v_x \tag{13.19}$$

$$\dot{y} = v_y \tag{13.20}$$

$$\dot{v}_x = 0 \tag{13.21}$$

$$\dot{v}_y = 0 \tag{13.22}$$

The dot means the time derivative or rate of change with time. These are the state equations for the automobile. This model says that the position change with time is proportional to the velocity. It also says the velocity is constant. Information about velocity changes will come solely from the measurements. We also don't model the angle or angular rate. This is because we aren't getting information about it from the radar. However, you might try including it!

The RHSAutomobileXY function is shown as follows. It models the dynamics of the point mass.

RHSAutomobileXY.m

```
13  function xDot = RHSAutomobileXY ( ~, x, ~ )
19  xDot = [x(3:4);0;0];
```

The demonstration simulation is the same simulation used to demonstrate the multiple hypothesis system tracking. This simulation just demonstrates the Kalman Filter. Since the Kalman Filter is the core of the package, it must work well before adding the measurement assignment part.

MHTDistanceUKF finds the MHT distance for use in gating computations using UKF. The MHT distance is the distance between the observation and predicted locations. The measurement function is of the form h(x,d) where d is the UKF data structure. MHTDistanceUKF uses sigma points. The code is similar to UKFUpdate. As the uncertainty gets smaller, the residual must be smaller to remain within the gate.

MHTDistanceUKF.m

```
27  function [k, del] = MHTDistanceUKF ( d )
28
29  % Get the sigma points
30  pS      = d.c*chol(d.p)';
31  nS      = length(d.m);
32  nSig    = 2*nS + 1;
33  mM      = repmat(d.m,1,nSig);
```

```
34   if ( length(d.m) == 1 )
35       mM = mM';
36   end
37
38   x         = mM + [zeros(nS,1) pS -pS];
39
40   [y, r]    = Measurement( x, d );
41   mu        = y*d.wM;
42   b         = y*d.w*y' + r;
43   del       = d.y - mu;
44   k         = del'*(b\del);
45
46   function [y, r] = Measurement( x, d )
47   %% MHTDistanceUKF>Measurement
48   %        Measurement from the sigma points
49
50   nSigma    = size(x,2);
51   lR        = length(d.r);
52   y         = zeros(lR,nSigma);
53   r         = d.r;
54   iR        = 1:lR;
55
56   for j = 1:nSigma
57       f                 = feval( d.hFun, x(:,j), d.hData );
58       y(iR,j)           = f;
59       r(iR,iR)          = d.r;
60   end
```

The simulation UKFAutomobileDemo uses a car data structure to contain all of the car information. A MATLAB function AutomobileInitialize takes parameter pairs and builds the data structure. This is a lot cleaner than assigning the individual fields in your script. It will return a default data structure if nothing is entered as an argument.

The first part of the demo is the automobile simulation. It generates the measurements of the automobile positions to be used by the Kalman Filter. The second part of the demo processes the measurements in the UKF to generate the estimates of the automobile track. You could move the code that generates the simulated data into a separate file if you were reusing the simulation results repeatably.

The results of the script are shown in Figure 13.6 to Figure 13.8.

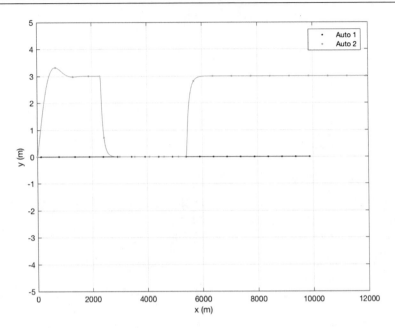

Figure 13.6: *Automobile trajectories*

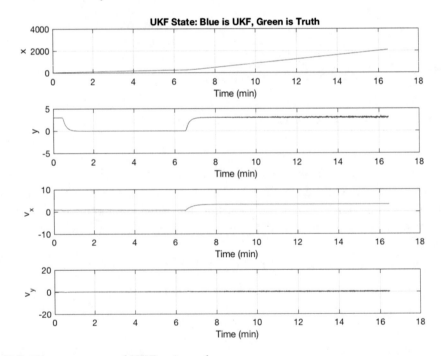

Figure 13.7: *The true states and UKF estimated states*

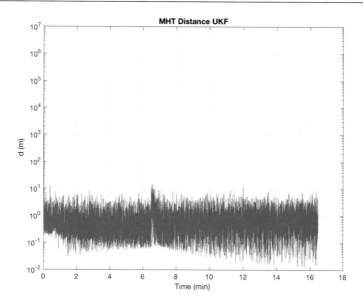

Figure 13.8: *The MHT distance between the automobiles during the simulation. Notice the spike in the distance when the automobile maneuver starts*

13.6 Automobile Target Tracking

13.6.1 Problem

We need to demonstrate target tracking for automobiles.

13.6.2 Solution

Build an automobile simulation with target tracking. This is the script `MHTAutomobileDemo` which will utilize all the pieces created in this chapter, including `AutomobilePassing`, `AutoRadar`, `AutoRadarUKF`, `RHSAutomobile`, `RHSAutomobileXY`, `MHTDistanceUKF`, and `MHTGUI`.

13.6.3 How It Works

The simulation is for a two-dimensional model of automobile dynamics. The primary car is driving along a highway at variable speeds. It carries a radar. Many cars pass the primary car, some of which change lanes from behind the car and cut in front. The MHT system tracks all cars. At the start of the simulation, there are no cars in the radar field of view. One car passes and cuts in front of the radar car. The other two just pass in their lanes. This is a good test of track initiation.

The radar, covered in the first recipe of the chapter, measures the range, range rate, and azimuth in the radar car frame. The model generates those values directly from the target and the tracked cars' relative velocities and positions. The radar signal processing is not modeled, but the radar has field-of-view and range limitations. See `AutoRadar`.

371

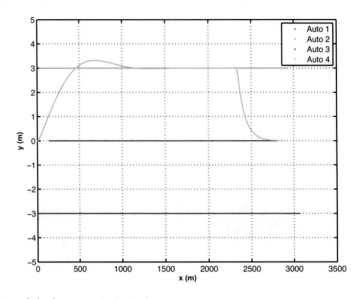

Figure 13.9: *Automobile demo car trajectories*

The cars are driven by steering controllers that execute automobile maneuvers. The throttle (accelerator pedal) and steering angle can be controlled. Multiple maneuvers can be chained together. This provides a challenging test for the MHT system. You can try different maneuvers and add additional maneuver functions of your own.

The Unscented Kalman Filter described in Chapter 4 is used in this demo since the radar is a highly nonlinear measurement. The UKF dynamical model, RHSAutomobileXY, is a pair of double integrators in the inertial frame relative to the radar car. The model accommodates steering and throttle changes by making the plant covariance, both position and velocity, larger than would be expected by analyzing the relative accelerations. An alternative would be to use Interactive Multiple Models (IMM) with a "steering" model and "acceleration" model. This added complication does not appear to be necessary. A considerable amount of uncertainty would be retained even with IMM since a steering model would be limited to one or two steering angles. The script implementing the simulation with MHT is MHTAutomobileDemo. There are four cars in the demo; car 4 will be passing, as shown in Figure 13.9. Figure 13.10 shows the radar measurement for car 3, which is the last car tracked. The MHT system handles vehicle acquisition well. The MHT GUI in Figure 13.11 shows a hypothesis with three tracks at the end of the simulation. This is the expected result.

Figure 13.12 shows the final tree. There are several redundant tracks. These tracks can be removed since they are clones of other tracks. This does not impact the hypothesis generation.

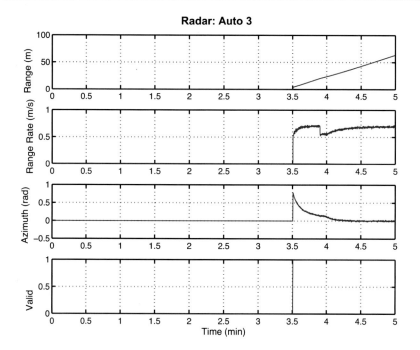

Figure 13.10: *Automobile demo radar measurement for car 3*

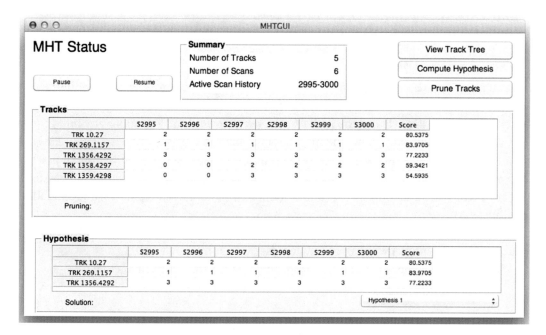

Figure 13.11: *The MHT GUI shows three tracks. Each track has consistent measurements*

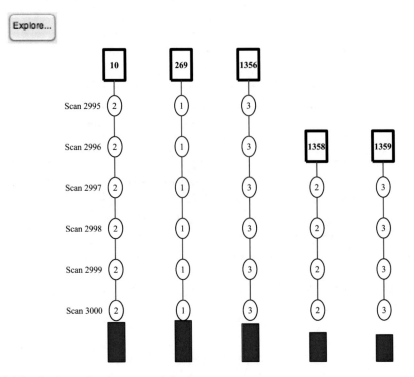

Figure 13.12: *The final tree for the automobile demo*

13.7 Summary

This chapter has demonstrated an automobile tracking problem. The automobile has a radar system that detects cars in its field of view. The system accurately assigns measurements to tracks and successfully learns the path of each neighboring car. You started by building an Unscented Kalman Filter to model the motion of an automobile and to incorporate measurements from a radar system. This was demonstrated in a simulated script. You then built a script that incorporates Track-Oriented Multiple Hypothesis Testing to assign measurements taken by the radar of multiple automobiles. This allows our radar system to autonomously and reliably track multiple cars.

You also learned how to make simple automobile controllers. The two controllers steer the automobiles and allow them to pass other cars.

Table 13.1 lists the functions and scripts included in the companion code.

Table 13.1: *Chapter Code Listing*

File	Description
AutoRadar	Automobile radar model for simulation
AutoRadarUKF	Automobile radar model for the UKF
AutomobileDemo	Demonstrates automobile animation
AutomobileInitialize	Initializes the automobile data structure
AutomobileLaneChange	Automobile control algorithm for lane changes
AutomobilePassing	Automobile control algorithm for passing
DrawComponents	Draws a 3D model
MHTAutomobileDemo	Demonstrates the use of Multiple Hypothesis Testing for automobile radar systems
RHSAutomobile	Automobile dynamical model for simulation
RHSAutomobileXY	Automobile dynamical model for the UKF
UKFAutomobileDemo	Demonstrates the UKF for an automobile

CHAPTER 14

■ ■ ■

Spacecraft Attitude Determination

Many spacecraft use star cameras to determine their orientation. A star camera takes an image of the star field. The first step is to determine what in the image is a star. Bright areas in the image are aggregated into "blobs" that are assumed to be stars. A star camera usually is slightly defocused so that a star image is smeared over multiple pixels. The next step is to find the centroid of each blob. Once this is done, the patterns seen in the image can be compared with the image created by an on-

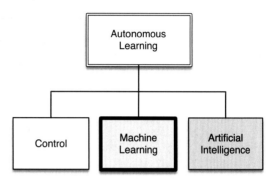

board star catalog. The star catalog produces a celestial sphere. We rotate this catalog until the portion of the catalog that would be seen by the camera matches the star camera image. This then produces an attitude or orientation estimate. Figure 14.1 shows a typical image.

In this chapter, we will build an attitude determination system that uses machine learning. This will work on the entire image, once the star-centroiding process is complete. The attitude determination will not need to explicitly name stars; instead, it will use the whole star pattern to determine the orientation. We will only do single-axis attitude determination. The reader can expand this to three axes using the same approach. All of the tools described can be used for three-axis problems. This chapter will use the MathWorks Deep Learning Toolbox.

14.1 Star Catalog

14.1.1 Problem

We need to generate a star catalog for testing the star identification algorithm.

© The Author(s), under exclusive license to APress Media, LLC, part of Springer Nature 2024   377
M. Paluszek, S. Thomas, *MATLAB Machine Learning Recipes*,
https://doi.org/10.1007/978-1-4842-9846-6_14

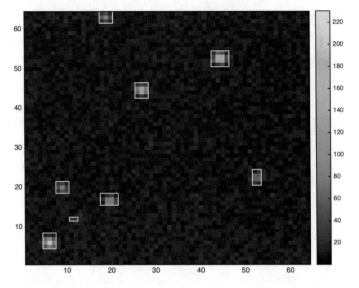

Figure 14.1: *Star image showing star blobs and aggregation into discrete "blobs." The coordinates ar* *pixels. Notice the background noise*

14.1.2 Solution

Build a function, LoadHipparcos, that grabs a subset of the Hipparcos star catalog. We wil use an input visual magnitude to limit the number of stars returned, as the full catalog ha 117,955 stars, most too dim to be seen by a small camera.

14.1.3 How It Works

The Hipparcos star catalog [23] is a well-known star catalog. The catalog was generated b the European Space Agency's Hipparcos space mission which was dedicated to measuring th positions, distances, motions, brightness, and colors of stars. The catalog is contained in a MAT file. The function LoadHipparcos loads the mat-file and saves the stars that are less tha (hence brighter than) the desired visual magnitude. load returns a data structure.

LoadHipparcos.m

```
15  function [rA,dec,vM] = LoadHipparcos( visualMagnitude )
16
17  if( nargin < 1 )
18     visualMagnitude = 6;
19  end
20
21  catalog = load( 'Hipparcos' );
22
23  k      = find( catalog.vM <= visualMagnitude);
24  rA     = catalog.rA(k);
25  dec    = catalog.dec(k);
26  vM     = catalog.vM(k);
```

```
27   nStars = length(catalog.rA);
```

The function has code to plot the star fields. This is only called if no outputs are requested. We used `sprintf` to make a plot title with information about the plots.

LoadHipparcos.m

```
29   % Plot information about the catalog
30   if( nargout == 0 )
31     NewFigure( 'Catalog' );
32     subplot(2,1,1);
33     plot( rA*180/pi, dec*180/pi, '.' )
34     set( gca, 'xlim', [0 360], 'ylim', [-90 90] );
35     xlabel( 'Right Ascension (deg)' );
36     ylabel( 'Declination (deg)' );
37     title( sprintf( 'Hipparcos Star Catalog: %4.0f of %5.0f of the stars
           in the catalog',length(k),nStars ) );
38     grid on;
39
40     x = linspace(min(vM),max(vM),20);
41     j = zeros(1,20);
42     for k = 1:20
43       j(k) = length(find( vM <= x(k) ));
44     end
45
46     subplot(2,1,2)
47     semilogy( x, j )
48     hold on
49     semilogy( visualMagnitude, length(find( vM < visualMagnitude )), 'r*'
           )
50     xlabel( 'Visual Magnitude' );
51     ylabel( 'Number' );
52     title( 'Number of Stars less than a Visual Magnitude' );
53     grid on
54     clear rA
55   end
```

Figure 14.2 shows the stars less than visual magnitude 6 and those less than visual magnitude 1. The plots show the full sky. The catalog has 4559 stars brighter than or equal to visual magnitude 6, so in the left-hand plot, the Milky Way is visible.

Our catalog needs to have enough stars so that no matter where we look in the sky, we get a distinct pattern. At least three stars need to be in the field of view to give three distinct unit vectors, which is the minimum set needed for three-axis attitude determination.

The output right ascension and declination are in the Earth-Centered Inertial (ECI) frame. This is the reference for attitude determination. The ECI frame is shown in Figure 14.3. The Vernal Equinox is the time of year in spring when day and night are the same length, which is also a point in inertial space that the x-axis points toward. The major rotation is about the North Pole. Precession and nutation are small but important for accurately pointing at the Earth from

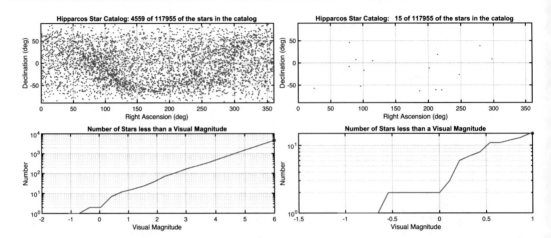

Figure 14.2: *Hipparcos catalog for two different visual magnitudes*

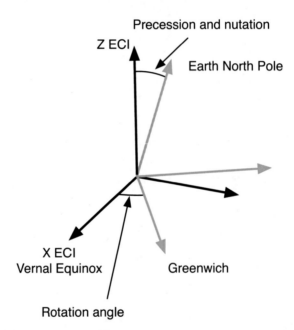

Figure 14.3: *ECI frame. The Earth-fixed frame is green*

space. Greenwich is the Greenwich Meridian which rotates throughout the day. The Earth-fixed coordinates will not be used in this chapter.

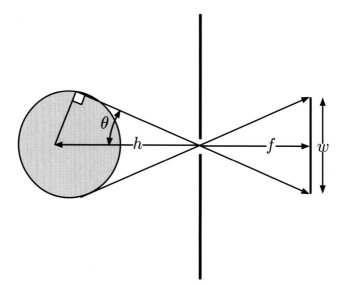

Figure 14.4: *Pinhole camera. You can make a pinhole camera with a piece of cardboard*

14.2 Camera Model

14.2.1 Problem

We need a model of a camera for simulating star identification. The function needs to go from right ascension and declination in the star field to points on the focal plane.

14.2.2 Solution

Build a function, `PinholeCamera`, that models a camera as a pinhole. This model neglects real lens effects, which can be added if needed.

14.2.3 How It Works

Figure 14.4 shows a pinhole camera in two dimensions. Note that the distance doesn't matter, just the angle of the star and the focal length `f`. `w` represents the width of the film or chip capturing the image. The code snippet shows the focal transformation and the screening by width. We need to screen by hemisphere, or we will get stars both in front of and behind the sensor. The index of the stars is returned as well, which can be compared between rotations as stars come into and out of view.

PinholeCamera.m

```
17  function [p,id] = PinholeCamera( rA, dec, f, w )
25  cDec = cos(dec);
26  p    = f*[cos(rA).*cDec;sin(rA).*cDec];
27  j    = sin(dec)>0; % front facing
28  k    = ((abs(p(1,:)) <= w/2) & (abs(p(2,:)) <= w/2)); % within width
```

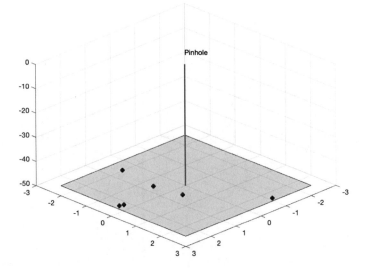

Figure 14.5: `PinholeCamera` output. *This is for all stars brighter than visual magnitude 6 with a focal length of 50 and imager width of 5. The x and y coordinates span the size of the chip, so they are not proportional to the z dimension*

```
29  id    = find(j & k);
30  p     = p(:,id);
```

Figure 14.5 shows the output when no function outputs are requested. It draws the imaging chip and shows where the stars fall on the chip. The following code shows how the plot is created. A patch is used to show the imaging chip area. The focal length is drawn. `plot3` draws the star points. `hold on` prevents the patch from disappearing when `plot3` is called. The celestial sphere will be discussed in the next recipe.

PinholeCamera.m

```
32  % Plot the camera output
33  if( nargout == 0 )
34    NewFigure( 'Focal Plane' );
35    v      = 0.5*w*[1 1 0;1 -1 0;-1 -1 0;-1 1 0];
36    v(:,3) = -f;
37    patch('faces',[1 2 3 4], 'Vertices', v,'facecolor',...
38      [0 0 1],'facealpha',0.2);
39    hold on
40    rotate3d on
41    view([1 1 1]);
42    line([0 0],[0 0],[0 -f],'linewidth',2);
43    text(0,0,0.1*f,'Pinhole')
44    p(3,:) = -f;
45    plot3(p(1,:),p(2,:),p(3,:),'o','LineWidth',2,...
46                          'MarkerEdgeColor','r',...
47                          'MarkerFaceColor','r',...
```

```
48                        'MarkerSize',5)
49    grid on
50    CelestialSphere( rA(id), dec(id) )
51  end
```

14.3   Celestial Sphere

14.3.1 Problem

We want a display showing the orientation of stars. This is useful for general visualization and also debugging of the functions.

14.3.2 Solution

Build a function that displays dots for the stars on a unit sphere using `sphere` and `plot3`.

14.3.3 How It Works

The function `CelestialSphere` uses `sphere` to draw a unit sphere. Display it using interpolated shading – `interp` – to get rid of the grid lines. Make it translucent by setting `alpha` to 0.1. The stars are plotting using plot markers with `plot3`. This function has a demo to display the Hipparcos catalog on the sphere.

CelestialSphere.m

```
13  function CelestialSphere( rA, dec )
14
15  % Demo
16  if( nargin < 1 )
17    [rA,dec] = LoadHipparcos( 6 );
18    CelestialSphere( rA, dec );
19    return
20  end
21
22  cDec = cos(dec);
23  u    = [cos(rA).*cDec;sin(rA).*cDec;sin(dec)];
24
25  NewFigure( 'Celestrial Sphere' );
26
27  sphere(120);
28  colormap('gray');
29  axis equal
30  alpha 0.1
31  shading interp
32  hold on
33  plot3(u(1,:),u(2,:),u(3,:),'.');
34  xlabel('x');
35  ylabel('y');
36  zlabel('z');
```

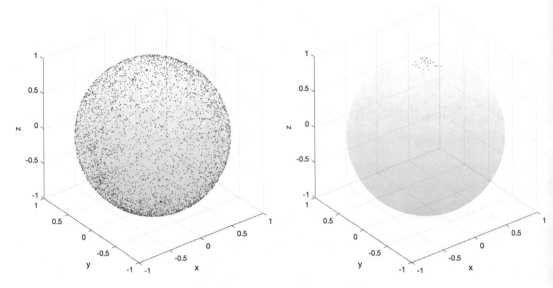

Figure 14.6: *The celestial sphere with the Hipparcos catalog displayed*

Figure 14.6 shows the celestial sphere demo on the left. With a reasonably wide camera aperture, we should always see a distinct pattern. The right-hand plot shows the output of PinholeCamera displayed on the sphere, for a focal length of 50 and an imager width of 15. As expected, we see a small cluster of stars.

14.4 Attitude Simulation of Camera Views

14.4.1 Problem

We need to simulate the attitude of the spacecraft to produce the camera views.

14.4.2 Solution

Build a function, AttitudeSim, whose input is an attitude quaternion and output is the simulated star image.

14.4.3 How It Works

Quaternions are the preferred mathematical representation of satellite attitude for simulation. Propagating a quaternion requires fewer operations than propagating a transformation matrix and avoids singularities that occur with Euler angles. A quaternion has four elements, which correspond to a unit vector a and the angle of rotation ϕ about that vector. The first element is termed the "scalar component" s, and the next three elements are the "vector" components v. Figure 14.7 shows a quaternion.

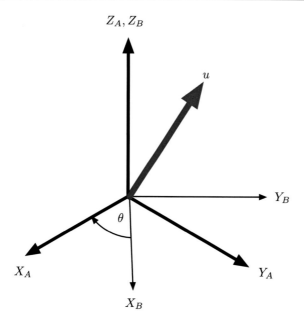

Figure 14.7: *A quaternion is a unit vector and an angle about that vector*

This notation is shown as follows:

$$
\begin{bmatrix} q_0 \\ q_1 \\ q_2 \\ q_3 \end{bmatrix} = \begin{bmatrix} s \\ v_1 \\ v_2 \\ v_3 \end{bmatrix} = \begin{bmatrix} \cos\frac{\phi}{2} \\ a_1 \sin\frac{\phi}{2} \\ a_2 \sin\frac{\phi}{2} \\ a_3 \sin\frac{\phi}{2} \end{bmatrix} = q \tag{14.1}
$$

The "unit" quaternion which represents zero rotation from the initial coordinate frame has a unit scalar component and zero vector components. This is the same convention used on the Space Shuttle, although other conventions are possible.

$$
q_0 = \begin{bmatrix} 1 \\ 0 \\ 0 \\ 0 \end{bmatrix} \tag{14.2}
$$

In order to transform a vector from one coordinate frame a to another b using a quaternion q_{ab}, the operation is

$$
u_b = q_{ab}^T u_a q_{ab} \tag{14.3}
$$

using quaternion multiplication with the vectors defined as quaternions with a scalar part equal to zero, or

$$x_a = \begin{bmatrix} 0 \\ x_a(1) \\ x_a(2) \\ x_a(3) \end{bmatrix} \qquad (14.4)$$

For example, the quaternion

$$\begin{bmatrix} 0.7071 \\ 0.7071 \\ 0.0 \\ 0.0 \end{bmatrix} \qquad (14.5)$$

represents a pure rotation about the x-axis. The first element q_0 is 0.7071 and equals the $\cos(90°/2)$. We cannot tell the direction of rotation from the first element. The second element q_1 is the v_1 component of the unit vector, so $a_1 = 1$ and $a_2 = a_3 = 0$, and the unit vector is

$$v = \begin{bmatrix} 1.0 \\ 0.0 \\ 0.0 \end{bmatrix} \sin(90°/2) \qquad (14.6)$$

Since the sign is positive, the rotation must be a positive 90° rotation.

The catalog outputs the right ascension and declination of the stars. The simplest method to simulate the viewed stars is to convert this to unit vectors, transform with the spacecraft inertial to body quaternion, and then convert back to right ascension and declination. This is done in AttitudeSim. It assumes that the camera is aligned with the spacecraft z-axis.

AttitudeSim.m

```
21  function [rAT, decT, pT, p] = AttitudeSim( q, rA, dec, f, w )
32  cDec = cos(dec);
33  u    = [cos(rA).*cDec;sin(rA).*cDec;sin(dec)];
34
35  uT   = QForm(q,u); % rotated unit vectors
36
37  rAT  = atan2(uT(2,:),uT(1,:));
38  decT = asin(uT(3,:));
```

AttitudeSim has a default output that shows the transformed stars and the catalog stars. Figure 14.8 shows the original inertial frame stars in red and the rotated stars in green.

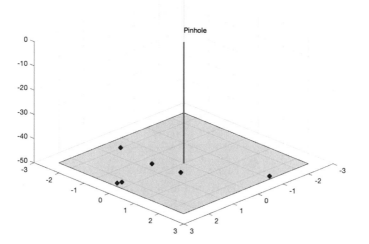

Figure 14.8: *AttitudeSim output for a pure rotation about the z-axis*

14.5 Yaw Angle Rotation

14.5.1 Problem

We want to generate a pixel map from a yaw angle rotation, that is, the angle about the z-axis. This will be a simpler version of the prior recipe for an arbitrary 3D rotation.

14.5.2 Solution

Build a function, YawPixelTransform.m, that computes the pixel map for a single-axis rotation.

14.5.3 How It Works

A yaw rotation, one about the z-axis, produces the transformation matrix:

$$m = \begin{bmatrix} \cos\gamma & \sin\gamma & 0 \\ -\sin\gamma & \cos\gamma & 0 \\ 0 & 0 & 1 \end{bmatrix} \qquad (14.7)$$

The following code implements this transformation.

YawPixelTransform.m

```
20  function pT = YawPixelTransform(yaw,rA,dec,f,w)
21
22  n   = length(yaw);
23  pT  = cell(n,1);
24
25  cDec = cos(dec);
```

```
26   u      = [cos(rA).*cDec;sin(rA).*cDec;sin(dec)];
27
28   for k = 1:n
29     c      = cos(yaw(k));
30     s      = sin(yaw(k));
31     m      = [c s 0;-s c 0;0 0 1]; % about z axis
32     uT     = m*u;
33     rAT    = atan2(uT(2,:),uT(1,:));
34     decT   = asin(uT(3,:));
35     pT{k}  = PinholeCamera(rAT,decT,f,w);
36   end
```

The YawToPixelDemos script runs a demo which plots the original pixels and the transformed pixels. Figure 14.9 shows an example sequence. Note that stars near the edge of the imager may come in and out of the view depending on the angle.

14.6   Yaw Images

14.6.1   Problem

We want to generate images for processing in the neural network. This will be the training data. Here, we are going to generate images for a single-axis rotation using the previous recipe.

14.6.2   Solution

Build a function, YawToImages, that computes the pixel map with YawPixelTransform, displays it in a figure window, and creates images from the figure using getframe and imwrite.

14.6.3   How It Works

We will plot the transformed pixels into a figure as filled plot markers. We can then use frame2im with getframe to extract an image from the figure. The image is then created in grayscale using imwrite with rgb2gray. The demo will create 50 images. You will need to generate a larger database for training the neural net; a separate script performs this task as described in the next recipe.

YawToImages.m

```
20   function YawToImages( yaw, rA, dec, f, w )
21
22   % Demo
23   if( nargin < 1 )
24     [rA,dec] = LoadHipparcos( 6 );
25     n           = 50;
26     yaw         = linspace(0,2*pi,n+1);
27     YawToImages( yaw(1:n), rA, dec, 64, (64/50)*5 );
28     return
29   end
30
31   % Yaw transformed pixels
```

388

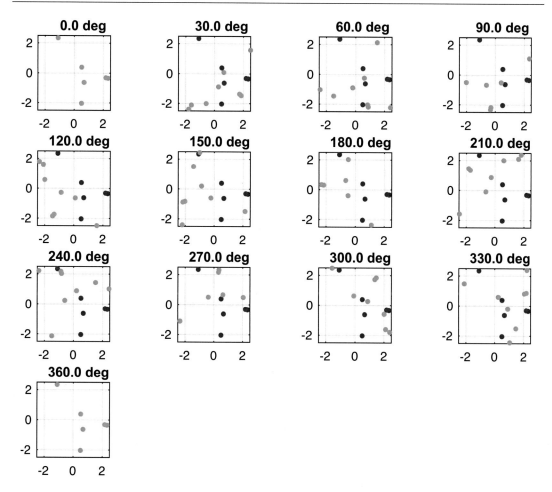

Figure 14.9: `YawPixelTransform.m` *output for a range of yaw angles. The untransformed pixels are shown in red*

```
32  pT   = YawPixelTransform( yaw, rA, dec, f, w );
33
34  % Set up the directory
35  if ~exist('YawImages','dir')
36    warning('Are you in the right folder? No YawImages')
37    mkdir('./','YawImages')
38  end
39  cd YawImages
40  delete *.jpg % Starting from scratch so delete existing images
41
42  n = length(yaw);
43  NewFigure('StarImage');
44
45  for k = 1:n
46    s = sprintf('Yaw %8.2f deg: %d of %d',yaw(k)*180/pi,k,n);
```

```
47    plot(pT{k}(1,:),pT{k}(2,:),'o',...
48    'MarkerEdgeColor','k','MarkerFaceColor','k','MarkerSize',16)
49    title(s);
50    axis off
51    axis square
52    set(gcf,'Color',[1 1 1]);
53    x = frame2im(getframe(gcf));
54    s = sprintf('YawImage%d.jpg',k);
55    x = imresize(x,[f f]);
56    imwrite(rgb2gray(x),s);
57    pause(0.4);
58  end
59  close(gcf);
60
61  save('Label','yaw')
```

An example is built into the function. Figure 14.10 shows the image window. Images are saved as jpegs. The function also saves a Label.mat file with the yaw angles. Note that you must run this function to generate training images before you will be able to run the final neural net training recipe.

Figure 14.10: *YawToImage output for one yaw angle*

14.7   Attitude Determination

14.7.1 Problem

We want to determine the attitude from an image taken by a star camera. We will only do single-axis rotation. This is the penultimate recipe that will combine all the pieces built during this chapter.

14.7.2 Solution

Build a function, `AttitudeDetermination`, that uses regression with a convolutional neural network to determine the attitude of the spacecraft.

14.7.3 How It Works

Convolutional neural networks are well suited for analyzing image data. A regression layer at the end of the network is used to predict numbers associated with images. A convolutional neural net is shown in Figure 14.11. This is also a "deep learning" neural net because it has multiple internal layers, but now the layers are of the three types described earlier. See also Chapter 11.

We can have as many layers as we want in a convolutional neural network. The name comes from the convolution operators used in many of the layers. This example constructs a convolutional neural network architecture, trains a network, and uses the trained network to predict angles of rotated star patterns.

The layers are explained in the following sections.

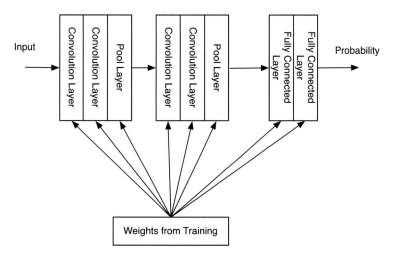

Figure 14.11: *Deep learning convolutional neural net [15]*

batchNormalizationLayer `batchNormalizationLayer` normalizes each input channel across a mini-batch. It automatically divides up the input channel into mini-batches, which are subsets of the entire batch. This reduces the sensitivity to the initialization.

convolution2dLayer `convolution2dLayer` applies sliding convolutional filters to the input. Convolution is the process of highlighting expected features in an image. This layer applies sliding convolutional filters on an image to extract features. You can specify the filters and the stride. Convolution is a matrix multiplication operation. You define the size of the matrices and their contents. For most images, like images of faces, you need multiple filters. Some types of filters are

1. Blurring filter: `ones(3,3)/9`

2. Sharpening filter: `[0 -1 0;-1 5 -1;0 -1 0]`

3. Horizontal Sobel filter for edge detection: `[-1 -2 -1; 0 0 0; 1 2 1]`

4. Vertical Sobel filter for edge detection: `[-1 0 1;-2 0 2;-1 0 1]`

In a convolutional neural network, the weights are computed as part of the training. You don't need to specify a mask. The first argument is the filter size. If it is a scalar, the filter is square. The second argument is the number of filters. We increase the number of filters each time we use the layer.

reluLayer `reluLayer` is a layer that uses the Rectified Linear Unit (ReLU) activation function.

$$f(x) = \begin{cases} x & x >= 0 \\ 0 & x < 0 \end{cases} \tag{14.8}$$

Its derivative is

$$\frac{df}{dx} = \begin{cases} 1 & x >= 0 \\ 0 & x < 0 \end{cases} \tag{14.9}$$

This is very fast to compute. It says that the neuron is only activated for positive values, and the activation is linear for any value greater than zero. You can adjust the activation point with a bias. This code snippet generates a plot of `reluLayer`:

```
x = linspace(-8,8);
y = x;
y(y<0) = 0;
PlotSet(x,y,'x label','Input','y label','reluLayer','plot title','
    reluLayer')
```

Figure 14.12 shows the activation function. An alternative is a leaky reluLayer where the value is not zero for negative input values.

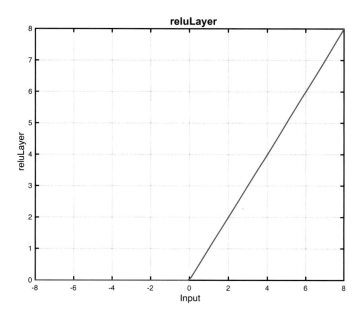

Figure 14.12: *reluLayer*

maxPooling2dLayer `maxPooling2dLayer` creates a layer that breaks the 2D input into rectangular pooling regions and outputs the maximum value of each region. The input `poolSize` specifies the width and height of a pooling region. `poolSize` can have one element (for square regions) or two for rectangular regions. This is a way to reduce the number of inputs that need to be evaluated. Typical images have to be a mega-pixel, and it is not practical to use all pixels as inputs. Furthermore, most images, or two-dimensional entities of any sort, don't have enough information to require finely divided regions. You can experiment with pooling and see how it works for your application. An alternative is `averagePooling2dLayer`.

averagePooling2dLayer `averagePooling2dLayer` creates a layer that breaks the 2D input into rectangular pooling. The average pooling layer performs downsampling by breaking the input into rectangular pooling regions and computing the average of each region. This is an alternative to `maxPooling2dLayer` .

fullyConnectedLayer The fully connected layer connects all of the inputs to the outputs with weights and biases. For example:

```
layer = fullyConnectedLayer(10);
```

creates ten outputs from any number of inputs. You don't have to specify the inputs. Effectively, this is the equation:

$$y = ax + b \qquad (14.10)$$

If there are m inputs and n outputs, b is a column bias matrix of length n and a is n by m.

dropoutLayer　A dropout layer randomly sets input elements to zero based on the input probability.

regressionLayer　A regression layer computes the half-mean-squared-error loss for inputs. It is used to get a scalar output. With a regression layer, we can get output yaw angles that are not specifically associated with a training image.

The images are generated in the following short script. 1000 images at random yaw angles are generated using `YawToImages` from the previous recipe.

RandomYawAngles.m

```
3   [rA,dec] = LoadHipparcos( 6 );
4   n        = 1000;
5   yaw      = rand(1,n)*2*pi;
6   YawToImages( yaw, rA, dec, 64, (64/50)*5 );
```

The training images are stored in a subfolder. They are read here in the following code.

AttitudeDetermination.m

```
5    cd YawImages
6    label = load('Label');
7    cd ..
8
9    n          = length(label.yaw);
10   s          = [64 64 1];
11   nTrain     = 800; % this must be less than the number of images
         available
12
13   % All the data
14   x          = zeros(s(1),s(2),s(3),n);
15   y          = label.yaw';
16
17   %% Put the images in x
18   cd YawImages
19
20   for k = 1:n
21     i = imread(sprintf('YawImage%d.jpg',k));
22     x(:,:,1,k) = i;
23   end
```

The grayscale images are put into a 4D array as follows:

```
x(:,:,1,k) = i;
```

The images are stored in the first two dimensions. The third element tells it that it only has one color channel. The last is the yaw angle index. We break the input data into validation and training data.

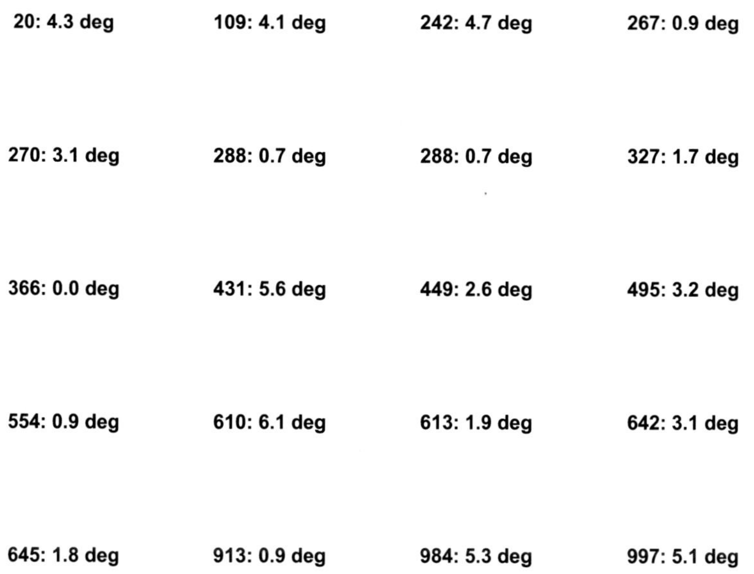

20: 4.3 deg	109: 4.1 deg	242: 4.7 deg	267: 0.9 deg
270: 3.1 deg	288: 0.7 deg	288: 0.7 deg	327: 1.7 deg
366: 0.0 deg	431: 5.6 deg	449: 2.6 deg	495: 3.2 deg
554: 0.9 deg	610: 6.1 deg	613: 1.9 deg	642: 3.1 deg
645: 1.8 deg	913: 0.9 deg	984: 5.3 deg	997: 5.1 deg

Figure 14.13: *Snapshots of star images*

AttitudeDetermination.m

```
25  %% Separate into training and validation sets
26  i           = randperm(n,nTrain);
27
28  xTrain      = x(:,:,:,i);
29  yTrain      = y(i);
30  i           = setxor(1:n,i);
31  xVal        = x(:,:,:,i);
32  yVal        = y(i);
```

Figure 14.13 shows a random sampling of the star images. The stars are very small, but you can see them if you zoom in.

Figure 14.14 shows a histogram of the yaw angles of the images.

Training is done in the following code:

AttitudeDetermination.m

```
52  %% Training setup
53  % This gives the structure of the convolutional neural net
54  layers = [
55      imageInputLayer(s)
56      convolution2dLayer(3,8,'Padding','same')
57      batchNormalizationLayer
58      reluLayer
59      averagePooling2dLayer(2,'Stride',2)
60      convolution2dLayer(3,16,'Padding','same')
61      batchNormalizationLayer
```

395

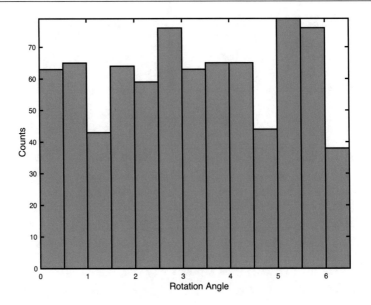

Figure 14.14: *Yaw angle histogram*

```
62        reluLayer
63        averagePooling2dLayer(2,'Stride',2)
64        convolution2dLayer(3,32,'Padding','same')
65        batchNormalizationLayer
66        reluLayer
67        convolution2dLayer(3,32,'Padding','same')
68        batchNormalizationLayer
69        reluLayer
70        dropoutLayer(0.2)
71        fullyConnectedLayer(1)
72        regressionLayer];
73
74  miniBatchSize  = 128;
75  validationFrequency = floor(numel(yTrain)/miniBatchSize);
76  options = trainingOptions('sgdm', ...
77      'MiniBatchSize',miniBatchSize, ...
78      'MaxEpochs',30, ...
79      'InitialLearnRate',1e-3, ...
80      'LearnRateSchedule','piecewise', ...
81      'LearnRateDropFactor',0.1, ...
82      'LearnRateDropPeriod',20, ...
83      'Shuffle','every-epoch', ...
84      'ValidationData',{xVal,yVal}, ...
85      'ValidationFrequency',validationFrequency, ...
86      'Plots','training-progress', ...
87      'Verbose',false);
88
90  %% Train
91  yawAttitudeNet = trainNetwork(xTrain,yTrain,layers,options);
```

396

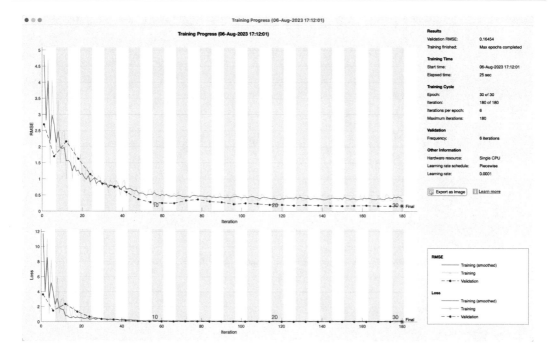

Figure 14.15: *Training GUI*

```
92
93   %% Display
94   disp(layers)
95   disp(options)
```

Figure 14.15 shows the training interface. It converges quickly.
Testing is done in the following code:

AttitudeDetermination.m

```
97   %% Test
98   yPred = predict(yawAttitudeNet,xVal);
99
100  predError = yVal - yPred;
101
102  thr = 10;
103  numCorrect = sum(abs(predError) < thr);
104  numValImages = numel(yVal);
105
106  accuracy = numCorrect/numValImages;
107
108  squares = predError.^2;
109  rmse = sqrt(mean(squares));
110
111  fprintf('Root mean squared error %8.4f\n',rmse);
```

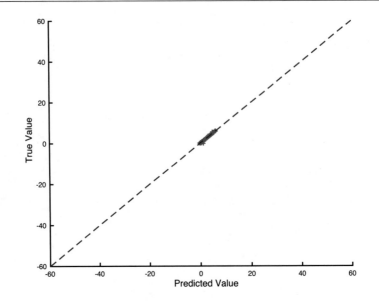

Figure 14.16: *Scatter plot of the results*

```
112  fprintf('Accuracy                    %8.4f\n',accuracy);
113
114  NewFigure('Scatter Plot')
115  scatter(yPred,yVal,'+')
116  xlabel("Predicted Value")
117  ylabel("True Value")
118
119  hold on
120  plot([-60 60], [-60 60],'r--')
```

Figure 14.16 shows the scatter plot.

14.8   Summary

This chapter has demonstrated attitude determination using deep learning. Table 14.1 lists the functions and scripts included in the companion code.

Table 14.1: *Chapter Code Listing*

File	Description
AttitudeDetermination	Trains and tests the neural network
AttitudeSim	Simulates the attitude of the spacecraft
CelestialSphere	Generates a celestial sphere from a star catalog
LoadHipparcos	Generates a star catalog
PinholeCamera	Models a camera as a pinhole to generate pixel maps
RandomYawAngles	Generates random yaw angles for training and testing
YawPixelTransform	Transforms pixels through a yaw rotation
YawToImages	Star tracker frames to image files
YawToPixelsDemo	Computes pixel maps from yaw transformations

CHAPTER 15

■ ■ ■

Case-Based Expert Systems

In this chapter we will introduce case-based expert systems, an example of the artificial intelligence branch of our Autonomous Learning taxonomy. There are two broad classes of expert systems, rule-based and case-based. Rule-based systems have a set of rules that are applied to come to a decision; they are just a more organized way of writing decision statements in computer code.

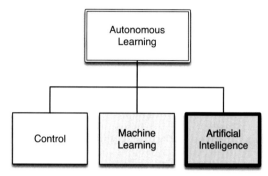

For a spacecraft with six thrusters the jet select logic is shown below. It assumes that thrusters can either be full on or full off. The parameters are chosen so that the thrusters can produce pure rotations about a single axis, either individually or in pairs.

JetSelect.m

```
1  % Force direction
2  u = [0 0 0 0 0 0;1 -1 0 0 0 0;0 0 1 1 1 1];
3
4  % Position vector from the center of mass
5  r = [-10 -10 -10 -10 10 10;1 -1 1 -1 1 -1;0 0 0 0 0 0];
6
7  a = cross(r,u); % Angular acceleration
8
9  aDesired    = [0;10;0];
10 aThreshold  = 0.2;
11
12 % Organize thrusters to get pure couples
13 thrusterPosSet = {[3 5] [3 4] 2};
14 thrusterNegSet = {[4 6] [5 6] 1};
15
16 % Compute pure coupled torques
17 aSetPos = zeros(3,3);
18 aSetNeg = zeros(3,3);
19 for k = 1:3
20   aSetPos(:,k) = sum(a(:,thrusterPosSet{k}),2);
```

```
21      aSetNeg(:,k) = sum(a(:,thrusterNegSet{k}),2);
22   end
23
24   % Jet select
25   useThruster = zeros(1,6);
26   for k = 1:3
27     if( abs(aDesired'*aSetPos(:,k)) > aThreshold )
28        useThruster(thrusterPosSet{k}) = 1;
29     elseif( ( abs(aDesired'*aSetNeg(:,k)) > aThreshold ))
30        useThruster(thrusterNegSet{k}) = 1;
31     end
32   end
33
34   fprintf('Thrusters: %d %d %d %d %d %d\n',useThruster)
35   j = find(useThruster == 1);
36   fprintf('Net angular acceleration: [%5.1f;%5.1f;%5.1f]\n',sum(a(:,j),2))
```

The rules are embodied in the following code snippet from the above script. A set of flags is initialized to zero, and then each is turned on (set to 1) if the desired value is over a threshold.

```
24   % Jet select
25   useThruster = zeros(1,6);
26   for k = 1:3
27     if( abs(aDesired'*aSetPos(:,k)) > aThreshold )
28        useThruster(thrusterPosSet{k}) = 1;
29     elseif( ( abs(aDesired'*aSetNeg(:,k)) > aThreshold ))
30        useThruster(thrusterNegSet{k}) = 1;
31     end
32   end
```

The rules are boolean logic. If there were more than a few thrusters this would become unwieldy. The result of the test case is:

```
>> JetSelect
Thrusters: 0 0 1 1 0 0
Net angular acceleration: [  0.0;  20.0;   0.0]
```

A real system would need a flag to check that the thruster was operating. We were able to organize thrusters into sets which made the logic easier. If the thruster angular acceleration vectors weren't as simple, this would be harder. For example, the Space Shuttle Orbiter thruster locations are shown in Figure 15.1.

Using a search algorithm, such as simplex as was done on the Indostar-1 spacecraft, might be a better choice.

This system had a very simple rule set. Expert systems provide a way of automating a process when decision-making involves hundreds or thousands of rules. Case-based systems decide by example, that is, a set of predefined cases. With its simple rule set it is more of a mapping than a case-based reasoning system.

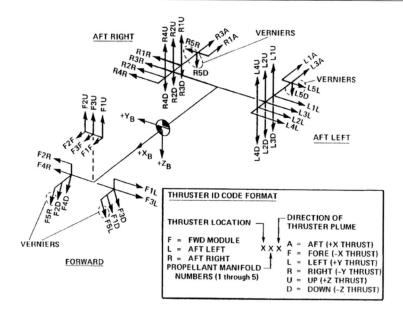

Figure 15.1: *Space Shuttle Orbiter thruster locations and unit vectors. Image courtesy of NASA*

Learning in the context of an expert system depends strongly on the configuration of the expert system. There are three primary methods, which vary in the level of autonomy and the average generalization of the new knowledge of the system.

The least autonomous method of learning is the introduction of **new rule sets** in rule-based expert systems. Learning of this sort can be highly tailored and focused but is done entirely at the behest of external teachers. In general, specific-rule-based systems with extremely general rules tend to have issues with edge cases that require exceptions to their rules. An edge case is a situation that is at the extreme of the operating parameters. For example, operating with a linear actuator in saturation might be an edge case. Driving a car at its maximum speed might also be an edge case. Thus, this type of learning, while easy to manage and implement, is neither autonomous nor generalizable.

The second method is **fact-gathering**. The expert system makes decisions based on the known cause-and-effect relationships along with an evolving model of the world; learning then is broken up into two sub-pieces. Learning new cause-and-effect system rules is very similar to the type of learning described above, requiring external instruction, but can be more generalizable (as it is combined with more general world knowledge than a simple rule-based system might have). Learning new facts, however, can be autonomous and involves the refinement of the expert system's model of reality by increasing the amount of information that can be taken advantage of by automated reasoning systems. The expert system creates new rules based on its experiences.

The third method is **fully autonomous based reasoning**, where actions and their consequences are observed, leading to inferences about what prior information and action combinations lead to what results. For instance, if two similar actions result in positive results, then those priors that are the same in both cases can begin to be inferred as necessary preconditions for a positive result from that action. As additional actions are seen, these inferences can be refined and confidence can increase in the predictions made.

The three methods are listed in increasing difficulty of implementation. Adding rules to a rule-based expert system is straightforward, though rule dependencies and priorities can become complicated. Fact-based knowledge expansion in automated reasoning systems is also fairly straightforward, once suitably generic sensing systems for handling incoming data are set up. The third method, autonomous reasoning, is by far the most difficult; however, rule-based systems can incorporate this type of learning. More general pattern recognition algorithms can be applied to training data (including online, unsupervised training data) to perform this function, learning to recognize, e.g. with a neural network, patterns of conditions that would lead to positive or negative results from a given candidate action. The system can then check possible actions against these learned classification systems to gauge the potential outcome of the candidate's actions.

In this chapter, we will explore case-based reasoning systems. This is a collection of cases with their states and values described by strings. We do not address the problem of having databases with thousands of cases. The code we present would be too slow for a practical system. We will not deal with a system that autonomously learns. However, the code in this chapter can be made to learn by feeding back the results of new cases into the case-based system.

15.1   Building Expert Systems

15.1.1 Problem

We want a tool to build a case-based expert system. Our tool needs to work for small sets of cases.

15.1.2 Solution

Build a function, `BuildExpertSystem`, that accepts parameter pairs to create the case-based expert system. The system is stored as a data structure.

15.1.3 How It Works

The knowledge base consists of states, values, and production rules. There are four parts to a new case: the case name, the states and values, and the outcome. A state can have multiple values.

The state catalog is a list of all of the information that will be available to the reasoning system. It is formatted as states and state values. Only string values are permitted. Cell arrays store all the data.

The default catalog is shown below for the reaction wheel control system. The cell array of acceptable or possible values for each state follows the state definition

```
{
        {'wheel-turning'},          {'yes','no'};
        {'power'},                  {'on','off'};
        {'torque-command'},         {'yes','no'}
}
```

Our database of cases is designed to detect failures. We have three things to check to see if the wheel is working. If the wheel is turning and power is on and there is a torque command then it is working. The wheel can be turned without a torque command or with the power off because it would just be spinning down from prior commands. If the wheel is not turning the possibilities are that there is no torque command or the power is off.

The function takes a data structure as an input d, defining the system, and uses varargin to handle the parameter pair inputs. The header, shown below, lists the parameters accepted. The function can update the data structure if one is passed in, or else create a new one.

BuildExpertSystem.m

```
 1   function d = BuildExpertSystem(d, varargin)
 2
 3   %% BUILDEXPERTSYSTEM Builds or expands a case based expert system.
 4   %% Form
 5   %   d = BuildExpertSystem;                   % default data
 6   %   d = BuildExpertSystem([], varargin)      % create new structure
 7   %   d = BuildExpertSystem( d, varargin)      % update a structure
 8   %% Inputs
 9   %   d               (.) Existing system data structure (can be empty)
10   %   varargin       {:} Parameter pairs
11   %                      id                 (1,1) Numeric index of parameter
12   %                      catalog state name  ''   Name of the system state
13   %                      catalog value      {:}   The possible values for the
                             state
14   %                      case name          ''    Case name
15   %                      case states        {:}   Case states
16   %                      case values        {:}   Case values
17   %                      case outcome       ''    Case outcome
18   %                      match percent      (1,1) Match percent (0-100)
19   %% Outputs
20   %   d               (.) System data structure
```

The next recipe shows a demo using this function. This function also has a default data output that documents the structure.

BuildExpertSystem.m

```
56   % Create a default system to demonstrate the structure
57   function system =  DefaultSystem
58
```

```
59  catalog.state            = {'wheel-turning','power' 'torque-command'};
60  catalog.values           = {{'yes','no'},{'on','off'},{'yes','no'}};
61  system.case              = struct('name','Wheel working',...
62                                    'activeStates',1:3,...
63                                    'values',{{'yes' 'on' 'yes'}},...
64                                    'outcome','working');
65  system.matchPercent      = 100;
66  system.stateCatalogData = catalog;
```

The default data is obtained by calling the function with no inputs.

```
>> d = BuildExpertSystem
d =

  struct with fields:

              case: [1x1 struct]
      matchPercent: 100
    stateCatalogData: [1x1 struct]

>> d.case
ans =

  struct with fields:

            name: 'Wheel working'
    activeStates: [1 2 3]
          values: {'yes'  'on'  'yes'}
         outcome: 'working'

>> d.stateCatalogData
ans =

  struct with fields:

     state: {'wheel-turning'  'power'  'torque-command'}
    values: {{1x2 cell}  {1x2 cell}  {1x2 cell}}
```

15.2   Running an Expert System

15.2.1 Problem

We want to create a case-based expert system and run it.

406

15.2.2 Solution

Build an expert system engine function that implements a case-based reasoning system. It should be designed to handle small numbers of cases and be capable of updating the case database defined in the previous recipe.

15.2.3 How It Works

Once you have defined a few cases from your state catalog, you can test the system. The function `CBREngine` implements the case-based reasoning engine. The idea is to pass it a case, `newCase`, and see if it matches any existing cases stored in the system data structure. For our problem, we think that we have all the cases necessary to detect any failure. We do string matching with the built-in function `strcmpi`, which compares strings ignoring case. We then find the first value that matches.

The algorithm finds the total fraction of the cases that match to determine if the example matches the stored cases. The engine is matching values for states in the new case against values for states in the case database. It weights the results by the number of states. If the new case has more states than an existing case, it biases the result by the number of states in the database case divided by the number of states in the new case. If more than one case matches the new case and the outcomes for the matching cases are different, the outcome is declared "ambiguous." If they are the same, it gives the new case that outcome. The case names make it easier to understand the results. We use `strcmpi` to make string matches case insensitive.

CBREngine.m

```
1   %% CBRENGINE Implements a case-based reasoning engine.
2   % Fits a new case to the existing set of cases.
14  function [outcome, pMatch] = CBREngine( newCase, system )
15
16  % Find the cases that most closely match the given state values
17  pMatch  = zeros(1,length(system.case));
18  pMatchF = length(newCase.state); % Number of states in the new case
19  for k = 1:length(system.case)
20    f = min([1 length(system.case(k).activeStates)/pMatchF]);
21    for j = 1:length(newCase.state)
22      % Does state j match any active states?
23      q = StringMatch( newCase.state(j), system.case(k).activeStates );
24      if( ~isempty(q) )
25        % See if our values match
26        i = strcmpi(newCase.values{j},system.case(k).values{q});
27        if( i )
28          pMatch(k) = pMatch(k) + f/pMatchF;
29        end
30      end
31    end
32  end
33
34  i = find(pMatch == 1);
35  if( isempty(i) )
```

```
36      i = max(pMatch,1);
37    end
38
39    outcome = system.case(i(1)).outcome;
40
41    for k = 2:length(i)
42      if( ~strcmp(system.case(i(k)).outcome,outcome))
43        outcome = 'ambiguous';
44      end
45    end
47
48    function k = StringMatch( testValue, array )
49
50    match = strcmpi(testValue,array);
51    k       = find(match,1);
```

The demo script, ExpertSystemDemo, is relatively short. The first part builds the system. The remaining code runs in some cases. 'id' denotes the index of the following data in its cell array. For example, the first three entries are for the catalog, and they are items 1 through 3. The next three are for cases, and they are items 1 through 4. As BuildExpertSystem goes through the list of parameter pairs, it uses the last id as the index for subsequent parameter pairs.

ExpertSystemDemo.m

```
 1    %% Demo of a Case-Based Expert System
 9    system = BuildExpertSystem( [], 'id',1,...
10                              'catalog state name','wheel-turning',...
11                              'catalog value',{'yes','no'},...
12                              'id',2,...
13                              'catalog state name','power',...
14                              'catalog value',{'on' 'off'},...
15                              'id',3,...
16                              'catalog state name','torque-command',...
17                              'catalog value',{'yes','no'},...
18                              'id',1,...
19                              'case name', 'Wheel operating',...
20                              'case states',{'wheel-turning', 'power', '
                                 torque-command'},...
21                              'case values',{'yes' 'on' 'yes'},...
22                              'case outcome','working',...
23                              'id',2,...
24                              'case name', 'Wheel power ambiguous',...
25                              'case states',{'wheel-turning', 'power', '
                                 torque-command'},...
26                              'case values',{'yes' {'on' 'off'} 'no'},...
27                              'case outcome','working',...
28                              'id',3,...
29                              'case name', 'Wheel broken',...
30                              'case states',{'wheel-turning', 'power', '
                                 torque-command'},...
31                              'case values',{'no' 'on' 'yes'},...
```

```
32                              'case outcome','broken',...
33                              'id',4,...
34                              'case name', 'Wheel turning',...
35                              'case states',{'wheel-turning', 'power'
                                  },...
36                              'case values',{'yes' 'on'},...
37                              'case outcome','working',...
38                              'match percent',80);
39
40   newCase.state  = {'wheel-turning', 'power', 'torque-command'};
41   newCase.values = {'yes','on','no'};
42   newCase.outcome = '';
43
44   [newCase.outcome, pMatch] = CBREngine( newCase, system );
45
46   fprintf(1,'New case outcome: %s\n\n',newCase.outcome);
47
48   fprintf(1,'Case ID Name                              Percentage Match\n');
49   for k = 1:length(pMatch)
50     fprintf(1,'Case %d: %-30s %4.0f\n',k,system.case(k).name,pMatch(k)
           *100);
51   end
```

As you can see, we match two cases, but because their outcome is the same, the wheel is declared working. The wheel power ambiguous is called that because the power could be on or off, hence ambiguous. We could add this new case to the database using `BuildExpertSystem`. We used `fprintf` in the script to print the following results into the command window:

```
>> ExpertSystemDemo
New case outcome: working

Case ID Name                       Percentage Match
Case 1: Wheel working                    67
Case 2: Wheel power ambiguous            67
Case 3: Wheel broken                     33
Case 4: Wheel turning                    44
```

The inputs are

```
newCase.state  = {'wheel-turning', 'power', 'torque-command'};
newCase.values = {'yes','on','no'};
```

This example is for a very small case-based expert system with a binary outcome. Multiple outcomes can be handled without any changes to the code. However, the matching process is slow as it cycles through all the cases. A more robust system, handling thousands of cases, would need some kind of decision tree to cull the cases tested. Suppose we had several different components that we were testing. For example, with a landing gear, we need to know that the tire is not flat, the brakes are working, the gear is deployed, and the gear is locked. If the gear is not deployed, we no longer have to test the brakes or the tires or that the gear is locked.

409

15.3 Summary

This chapter has demonstrated a simple case-based reasoning expert system. The system can be configured to add new cases based on the results of previous cases. An alternative would be a rule-based system. Table 15.1 lists the functions and scripts included in the companion code.

Table 15.1: *Chapter Code Listing*

File	Description
BuildExpertSystem	Function to build a case-based expert system database
CBREngine	Case-based reasoning engine
ExpertSystemDemo	Expert system demonstration

Appendix A

■ ■ ■

A Brief History

A.1 Introduction

In the first chapter of this book, you were introduced to autonomous learning. You saw that autonomous learning could be divided into the areas of machine learning, controls, and artificial intelligence (AI). In this appendix, we will provide some background on how each area evolved. Automatic control predates artificial intelligence. However, we are interested in adaptive or learning control which is a relatively new development and began evolving around the time that artificial intelligence had its foundations. Machine learning is considered an offshoot of artificial intelligence. However, many of the methods used in machine learning came from different fields of study such as statistics and optimization.

A.2 Artificial Intelligence

Artificial intelligence research began shortly after World War II [26]. Early work was based on knowledge of the structure of the brain, propositional logic, and Turing's theory of computation. Warren McCulloch and Walter Pitts created a mathematical formulation for neural networks based on threshold logic. This allowed neural network research to split into two approaches: one centered on biological processes in the brain and the other on the application of neural networks to artificial intelligence. It was demonstrated that any function could be implemented through a set of such neurons and that a neural net could learn. In 1948, Weiner's book *Cybernetics* was published, which described concepts in control, communications, and statistical signal processing. The next major step in neural networks was Hebb's book in 1949, *The Organization of Behavior*, linking connectivity with learning in the brain. His book became a source of learning and adaptive systems. Marvin Minsky and Dean Edmonds built the first neural computer in 1950.

In 1956, Allen Newell and Herbert Simon designed a reasoning program, the Logic Theorist (LT), which worked nonnumerically. The first version was hand-simulated using index cards. It could prove mathematical theorems and even improve on human derivations. It solved 38 of the 52 theorems in *Principia Mathematica*. LT employed a search tree with heuristics to limit the search. LT was implemented on a computer using Information Processing Language (IPL), a programming language that led to Lisp, a programming language that will be discussed later.

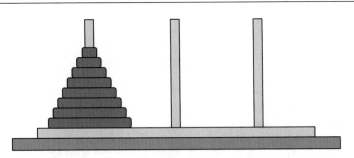

Figure A.1: *Towers of Hanoi. The disks must be moved from the first peg to the last without ever putting a bigger diameter disk on top of a smaller diameter disk*

Blocks World was one of the first attempts to demonstrate general computer reasoning. The blocks world was a micro-world. A set of blocks would sit on a table, some sitting on other blocks. The AI systems could rearrange blocks in certain ways. Blocks under other blocks could not be moved until the block on top was moved. This is not unlike the Towers of Hanoi problem. The blocks world was a spectacular advancement as it showed that a machine could reason at least in a limited environment. Blocks World was an early example of the use of machine vision. The computer had to process an image of the blocks' world and determine what was a block and where they were located.

Blocks World and Newell and Simon's LT were followed up by the General Problem Solver (GPS). It was designed to imitate human problem-solving methods. Within its limited class of puzzles, it could solve them much like a human. While GPS solved simple problems such as the Towers of Hanoi, Figure A.1, it could not solve real-world problems because the search was lost in a combinatorial explosion that represented the enumeration of all choices in a vast decision space.

In 1959, Herbert Gelernter wrote the Geometry Theorem Prover that could prove theorems that were quite tricky. The first game-playing programs were written at this time. In 1958 John McCarthy invented the language Lisp (List Processing) which was to become the main AI language. It is now available as Scheme and Common Lisp. Lisp was introduced only one year after FORTRAN. A typical Lisp expression is

```
1  (defun sqrt-iter (guess x)
2    (if (good-enough-p guess x)
3        guess
4        (sqrt-iter (improve guess x) x)))
```

This computes a square root through recursion. Eventually, dedicated Lisp machines were built, but they went out of favor when general-purpose processors became faster.

Time-sharing was invented at MIT to facilitate AI research. Professor McCarthy created a hypothetical computer program, Advice Taker, a complete AI system that could embody general world information. It would have used a formal language such as predicate calculus. For example, it could come up with a route to the airport from simple rules. Marvin Minsky arrived at MIT in 1958 and began working on micro-worlds. Within these limited domains, AI could solve problems, such as closed-form integrals in calculus.

Minsky wrote the book *Perceptrons* (with Seymour Papert), which was fundamental in the analysis of artificial neural networks. The book contributed to the movement toward symbolic processing in AI. The book noted that single neurons could not implement some logical functions such as exclusive-or and erroneously implied that multilayer networks would have the same issue. It was later found that three-layer networks could implement such functions.

■ **TIP**   Three-layer networks are the minimum to solve most learning problems.

More challenging problems were tried in the 1960s. Limitations in AI techniques became evident. The first language translation programs had mixed results. Trying to solve problems by working through massive numbers of possibilities (such as in chess) ran into computation problems. Human chess play has many forms. Some involve memorization of patterns including openings where the board positions are well defined and end games when the number of pieces is relatively small. Positional play involves seeing patterns on the board through the human brain's ability to process patterns. Someone who is a good positional player will arrange their pieces on the board so that the other player's options are restricted. Localized pattern recognization is seen in mate-in-n problems. Human approaches are not used in computer chess. Computer chess programs have become very capable primarily due to faster processors and the ability to store openings and end games. Multilayer neural networks were discovered in the 1960s but not studied until the 1980s.

In the 1970s, self-organizing maps using competitive learning were introduced [14]. A resurgence in neural networks happened in the 1980s. Knowledge-based systems were also introduced in the 1980s. From Jackson [16]:

> An expert system is a computer program that represents and reasons with knowledge of some specialized subject to solve problems or give advice.

This included expert systems that could store massive amounts of domain knowledge. These could also incorporate uncertainty in their processing. Expert systems are applied to medical diagnoses and other problems. Unlike AI techniques up to this time, expert systems could deal with problems of realistic complexity and attain high performance. They also explain their reasoning. This last feature is critical in their operational use. Sometimes, these are called knowledge-based systems. A well-known open source expert system is CLIPS (C Language Integrated Production System).

Backpropagation for neural networks was reinvented in the 1980s, leading to renewed progress in this field. Studies began both of human neural networks (i.e., the human brain) and the creation of algorithms for effective computational neural networks. This eventually led to deep learning networks in Machine Learning applications.

Advances were made in the 1980s as AI researchers began to apply rigorous mathematical and statistical analysis to develop algorithms. Hidden Markov models were applied to speech. A

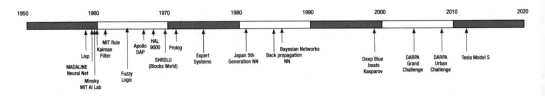

Figure A.2: Artificial intelligence timeline

hidden Markov model is a model with unobserved (i.e., hidden) states. Combined with massive databases, they have resulted in vastly more robust speech recognition. Machine translation has also improved. Data mining, the first form of Machine Learning as it is known today, was developed. Chess programs improved initially through the use of specialized computers, such as IBM's Deep Blue. With the increase in processing power, powerful chess programs that are better than most human players are now available on personal computers.

The Bayesian network formalism was invented to allow for the rigorous application of uncertainty in reasoning problems. In the late 1990s, intelligent agents were introduced. Search engines, bots, and website aggregators are examples of intelligent agents used on the Internet. Figure A.2 gives a timeline of selected events in the history of autonomous systems.

Today, the state of the art in AI includes autonomous cars, speech recognition, planning and scheduling, game playing, robotics, and machine translation. All of these are based on AI technology. They are in constant use today. You can take a PDF document and translate it into any language using Google Translate. The translations are not perfect. One certainly would not use them to translate literature.

Recent advances in AI include IBM's Watson. Watson is a question-answering computing system with advanced natural language processing and information retrieval from massive databases. It defeated champion Jeopardy players in 2011. It is currently being applied to medical problems and many other complex problems. Another advance is ChatGPT that is trained on massive amounts of text available on the Internet. It is a form of generative machine learning. When asked a question or asked to create something, ChatGPT can produce a reasonable output. Here is some haiku it generated:

> Beneath the moon's grace,
> Whispers of the night embrace,,
> Stars in a dark chase.,

> Nature's beauty found,,
> In every sight and each sound,,
> Life's wonders abound.,

A.3 Learning Control

Adaptive or intelligent control was motivated in the 1950s [3] by the problems of aircraft control. Control systems of that time worked very well for linear systems. Aircraft dynamics could be linearized at a particular speed. For example, a simple equation for total velocity in level flight is

$$m\frac{dv}{dt} = T - \frac{1}{2}\rho C_D S v^2 \tag{A.1}$$

This says the mass m times the change in velocity per time, $\frac{dv}{dt}$, equals the thrust T, the force from the aircraft engine, minus the drag. C_D is the aerodynamic drag coefficient, and S is the wetted area (i.e., the area that causes drag such as the wings and fuselage). The thrust is used for control. This is a nonlinear equation in velocity v because of the v^2 term. We can linearize it around a particular velocity v_s so that $v = v_\delta + v_s$ and get

$$m\frac{dv_\delta}{dt} = T - \rho C_D S v_s v_\delta \tag{A.2}$$

This equation is linear in v_δ. We can control velocity with a simple thrust control law:

$$T = T_s - cv_\delta \tag{A.3}$$

where $T_s = \frac{1}{2}\rho C_D S v_s^2$. c is the damping coefficient. ρ is the atmospheric density and is a nonlinear function of altitude. For the linear control to work, the control must be adaptive. If we want to guarantee a certain damping value which is the quantity in parentheses

$$m\frac{dv_\delta}{dt} = -\left(c + \rho C_D S v_s\right)v_\delta \tag{A.4}$$

we need to know ρ, C_D, S, and v_s. This approach leads to a gain scheduling control system where we measure the flight condition – altitude and velocity – and schedule the linear gains based on where the aircraft is in the gain schedule.

In the 1960s, progress was made on adaptive control. State space theory was developed that made it easier to design multi-loop control systems, that is, control systems that controlled more than one state at a time with different control loops. The general space controller is

$$\dot{x} = Ax + Bu \tag{A.5}$$
$$y = Cx + Du \tag{A.6}$$
$$u = -Ky \tag{A.7}$$

where A, B, C, and D are matrices, x is the state, y is the measurement, and u is the control input. A state is a quantity that changes with time that is needed to define what the system is doing. For a point mass that can only move in one direction, the position and velocity make up the two states. If A completely models the system and y contains all of the information about the state vector x, then this system is stable. The full-state feedback would be $x = -Kx$ where K can be computed to have guaranteed phase and gain margins (i.e., tolerance to delays and tolerance to amplification errors). This was a major advance in control theory. Before this multi-loop, systems had to be designed separately and combined very carefully.

Learning control and adaptive control were found to be realizable from a common framework. The Kalman Filter, also known as linear quadratic estimation, was introduced.

Spacecraft required autonomous control since they were often out of contact with the ground or the time delays were too long for effective ground supervision. The first digital autopilots were on the Apollo spacecraft which first flew in 1968 on Apollo 7. Don Eyles's book [10] gives the history of the Lunar Module Digital Autopilot. Geosynchronous communications satellites were automated to the point where one operator could fly a dozen satellites.

Advances in system identification, the process of just determining the parameters of a system (such as the drag coefficient earlier), were made. Adaptive control was applied to real problems. The F-111 aircraft had an adaptive control system. Autopilots have progressed from fairly simple mechanical pilot augmentation systems to sophisticated control systems that can take off, cruise, and land under computer control.

In the 1970s, proofs of adaptive control stability were made. The stability of linear control systems was well established, but adaptive systems are inherently nonlinear. Universally stabilizing controllers were studied. Progress was made in the robustness of adaptive control Robustness is the ability of a system to deal with changes in parameters that were assumed to be known, sometimes due to failures in the systems. It was in the 1970s that digital control became widespread, replacing traditional analog circuits composed of transistors and operational amplifiers.

Adaptive controllers started to appear commercially in the 1980s. Most modern single-loop controllers have some form of adaptation. Adaptive techniques were also found to be useful for tuning controllers.

More recently, there has been a melding of artificial intelligence and control. Expert systems have been proposed that determine what algorithms (not just parameters) to use depending on the environment. For example, during a winged reentry of a glider, the control system would use one system in orbit, a second at high altitudes, a third during high Mach (Mach is the ratio of the velocity to the speed of sound) flight, and a fourth at low Mach numbers and during landing. An F3D Skyknight used the Automatic Carrier Landing System on August 12 1957. This was the first shipboard test of the landing system designed to land aircraft on board autonomously. Naira Hovakimyan University of Illinois Urbana-Champaign (UIUC) and Nhan Nguyen (NASA) were pioneers in this area. Adaptive control was demonstrated on subscale F-18s that controlled and landed the aircraft after most of one wing was lost!

A.4   Machine Learning

Machine learning started as a branch of artificial intelligence. However, many techniques are much older. Thomas Bayes created Bayes' theorem in 1763. Bayes' theorem is

$$P(A_i|B) = \frac{P(B|A_i)P(A_i)}{\sum P(B|A_i)}$$

$$P(A_i|B) = \frac{P(B|A_i)P(A_i)}{P(B)}$$

(A.8)

which is just the probability of A_i given B. This assumes that $P(B) \neq 0$. In the Bayesian interpretation, the theorem introduces the effect of evidence on belief. One technique, regression, was discovered by Legendre in 1805 and Gauss in 1809.

As noted in the section on artificial intelligence, modern Machine Learning began with data mining which is the process of getting new insights from data. In the early days of AI, there was considerable work on machines learning from data. However, this lost favor and in the 1990s was reinvented as the field of machine learning. The goal was to solve practical problems of pattern recognition using statistics. This was greatly aided by the massive amounts of data available online along with the tremendous increase in processing power available to developers. Machine learning is closely related to statistics.

In the early 1990s, Vapnik and coworkers invented a computationally powerful class of supervised learning networks known as support vector machines (SVMs). These networks could solve problems of pattern recognition, regression, and other machine learning problems.

A growing application of Machine Learning is autonomous driving. Autonomous driving makes use of all aspects of autonomous learning including controls, artificial intelligence, and machine learning. Machine vision is used in most systems as cameras are inexpensive and provide more information than lidar, radar, or sonar (which are also useful). It isn't possible to build safe autonomous driving systems without learning through experience. Thus, designers of such systems put their cars on the roads and collect experiences that are used to fine-tune the system.

Other applications include high-speed stock trading and algorithms to guide investments. These are under rapid development and are now available to the consumer. Data mining and machine learning are in use to predict events, both human and natural. Searches on the Internet have been used to track disease outbreaks. If there is a lot of data, and the Internet makes gathering massive data easy, then you can be sure that machine learning techniques are being applied to mine the data.

A.5   Generative Machine Learning

Generative machine learning (ML) models are a class of models that allow you to create new data by modeling the data-generating distribution. For example, a generative model trained on images of human faces would learn what features constitute a realistic human face and how to combine them to generate novel human face images. For a fun demonstration of the power of ML-based human face generation, check out [34].

This is in contrast to a discriminative model that learns an association between a set of labels and the training inputs. Staying with our face example, a discriminative model might predict the age of a person given an image of their face. In this case, the input is the image of the face, and the label is the numerical age. Labels can also be used in generative models.

Generative models are used in a wide variety of applications from drug design to language models for better chatbots and autocomplete features. Generative models are also used in data augmentation to train better discriminative models, especially in situations where training data is difficult or expensive to obtain. Finally, generative models are widely used by artists and composers to inspire or augment their work. ChatGPT, described briefly in this book, is a well-known example of generative machine learning. It is finding applications in many fields from computer coding, as demonstrated in this book, to "creative writing."

A.6   Reinforcement Learning

Reinforcement learning is a machine learning approach in which an intelligent agent learns to take actions to maximize a reward. We will apply this to the design of a Titan landing control system. Reinforcement learning is a tool to approximate solutions that could have been obtained by dynamic programming, but whose exact solutions are computationally intractable [4].

A.7   The Future

Autonomous learning in all its branches is undergoing rapid development today. Many of the technologies are used operationally even in low-cost consumer technology. Virtually, every automobile company in the world and many nonautomotive companies are working to perfect autonomous driving. Military organizations are extremely interested in artificial intelligence and machine learning. Combat aircraft today have systems to take over from the pilot to prevent planes from crashing into the ground.

While completely autonomous systems are the goal in many areas, the meshing of human and machine intelligence is also an area of active research. Much AI research has been done to study how the human mind works. In addition to improving performance by tapping into the extraordinary power of the brain, this work will also enable machine learning systems to mesh more seamlessly with human beings. This is critical for autonomous control involving people but may also allow people to augment their abilities.

This is an exciting time for machine learning! We hope that this book helps you bring your advances to machine learning!

Appendix B

■ ■ ■

Software for Machine Learning

B.1   Autonomous Learning Software

There are many sources for machine learning software. Machine learning encompasses both software to help the user learn from data and software that helps machines learn and adapt to their environment. This book gives you a sampling of software that you can use immediately. However, the software is not designed for industrial applications. This appendix describes software that is available for the MATLAB environment. Both professional and open source MATLAB software are discussed. The book may not cover every available package as new packages are continually becoming available and older packages may become obsolete.

The packages you select for your project depend on your goal and your level of software expertise. Many of the packages in this appendix are research tools that can be used to design, analyze, and improve various types of machine learning systems, but to turn them into deployable, production-quality systems would require compiling, integrating, and testing software that is custom developed for each application. Other packages, such as commercial expert system shells, can be deployed as your application. You'll look for packages that are most compatible with your deployment environment. For example, if your goal is an embedded system, you will need development tools for the embedded processors and packages that are most compatible with that development environment.

This appendix includes software for what is conventionally called "Machine Learning," which are statistics functions that help give us insight into data. Such functions are often used in the context of "big data." It also includes descriptions of packages for other branches of autonomous learning systems such as system identification. System identification is a branch of automatic control that learns about the systems under control, allowing for better and more precise control.

The appendix, for completeness, also covers popular software that is MATLAB compatible but requires extra steps to use it from within MATLAB. Examples include R, Python, and SNOPT. In all cases, it is straightforward to write MATLAB interfaces to these packages. Using MATLAB as a front end can be very helpful and allow you to create integrated packages that include MATLAB, Simulink, and the machine learning package of your choice.

You will note that we include optimization software. Optimization is a tool used as part of machine learning to find the best or "optimal" parameters.

M. Paluszek, S. Thomas, *MATLAB Machine Learning Recipes*,
https://doi.org/10.1007/978-1-4842-9846-6

B.2   Commercial MATLAB Software

B.2.1   MathWorks Products

The MathWorks sells several packages for machine learning. The packages are in the Machine Learning branch of our taxonomy shown in Figure 1.2. The MathWorks products provide high-quality algorithms for data analysis along with graphics tools to visualize the data. Visualization tools are a critical part of any machine learning system. They can be used for data acquisition for example, for image recognition or as part of systems for autonomous control of vehicles or for diagnosis and debugging during development. All of these packages can be integrated with each other and with other MATLAB functions to produce powerful systems for machine learning. The most applicable toolboxes that we will discuss are

- Statistics and Machine Learning Toolbox

- Optimization Toolbox

- Global Optimization Toolbox

- Text Analytics Toolbox

- Deep Learning Toolbox

Statistics and Machine Learning Toolbox

The Statistics and Machine Learning Toolbox provides data analytics methods for gathering trends and patterns from massive amounts of data. Statistical methods do not require a model for analyzing the data. The toolbox functions can be broadly divided into Classification Tools, Regression Tools, and Clustering Tools.

Classification methods are used to place data into different categories. For example, data, in the form of an image, might be used to classify an image of an organ as having a tumor. Classification is used for handwriting recognition, credit scoring, and face identification. Classification methods include support vector machines (SVMs), decision trees, and neural networks.

Regression methods let you build models from current data to predict future data. The models can then be updated as new data becomes available. If the data is only used once to create the model, then it is a batch method. A regression method that incorporates data as it becomes available is recursive.

Clustering finds natural groupings in data. Object recognition is an application of clustering methods. For example, if you want to find a car in an image, you look for data that is associated with the part of an image that is a car. While cars are of different shapes and sizes, they have many features in common.

The toolbox has many functions to support these areas and many that do not fit neatly into these categories. The Statistics and Machine Learning Toolbox is an excellent place to start for professional tools that are seamlessly integrated into the MATLAB environment.

Optimization Toolbox

This toolbox provides tools for optimizing processes. Optimization is closely linked to machine learning. The optimization toolbox functions generally require that the solution start near a global optimum.

Global Optimization Toolbox

This toolbox provides tools for optimizing processes. In this toolbox, methods are included that can find global optimums.

Text Analytics Toolbox

This toolbox provides algorithms for preprocessing, analyzing, and modeling text data. Text data is often used as part of machine learning systems.

Deep Learning Toolbox

The Deep Learning Toolbox allows you to design, build, and visualize convolutional neural networks. You can easily implement models such as GoogLeNet, VGG-16, VGG_19, AlexNet, and ResNet-59. It has extensive capabilities for the visualization and debugging of neural networks. This is important to ensure that your system is behaving properly. It includes several pretrained models. Deep learning is a type of neural net. You can use this when the neural net toolbox functions aren't sufficient for your system. You can use both toolboxes together, along with all other MATLAB toolboxes.

B.2.2 Princeton Satellite Systems Products

Several of our commercial packages provide tools within the purview of autonomous learning.

Core Control Toolbox

The Core Control Toolbox provides the control and estimation functions of our Spacecraft Control Toolbox with general industrial dynamics examples including robotics and chemical processing. The suite of Kalman Filter routines includes conventional filters, Extended Kalman Filters, and Unscented Kalman Filters. The Unscented Kalman Filters have a fast sigma point calculation algorithm. All of the Kalman Filters use a common code format with separate prediction and update functions. This allows the two steps to be used independently. The filters can handle multiple measurement sources that can be changed dynamically, with measurements arriving at different times.

Add-ons for the Core Control Toolbox include Imaging Module and Target Tracking Modules. Imaging includes lens models, image processing, ray tracing, and image analysis tools.

Target Tracking

The Target Tracking Module employs Track-Oriented Multiple Hypothesis Testing. Track Oriented Multiple Hypothesis Testing is a powerful technique for assigning measurements to tracks of objects when the number of objects is unknown or changing. It is essential for accurate tracking of multiple objects.

In many situations, a sensor system must track multiple targets, like in rush hour traffic This leads to the problem of associating measurements with objects or tracks. This is a crucia element of any practical tracking system.

The track-oriented approach recomputes the hypotheses using the newly updated tracks after each scan of data is received. Rather than maintaining, and expanding, hypotheses from scan to scan, the track-oriented approach discards the hypotheses formed on scan $k - 1$. The track that survive pruning are propagated to the next scan k where new tracks are formed, using the new observations, and reformed into hypotheses. The hypothesis formation step is formulated as a mixed-integer linear program (MILP) and solved using GLPK (GNU Linear Programming Kit). Except for the necessity to delete some tracks based on low probability, no information i lost because the track scores that are maintained contain all the relevant statistical data.

The MHT Module uses a powerful track pruning algorithm that does the pruning in one step. Because of its speed, ad hoc pruning methods are not required, leading to more robust and reliable results. The track management software is, as a consequence, quite simple.

All three Kalman Filters in the Core Control Toolbox, including Extended and Unscented can be used independently or as part of the MHT system. The UKF automatically uses sigma points and does not require derivatives to be taken of the measurement functions or linearized versions of the measurement models.

Interactive Multiple Model Systems (IMM) can also be used as part of the MHT system IMM employs multiple dynamic models to facilitate tracking and maneuvering objects. On model might involve maneuvering, while another models constant motion. Measurements ar assigned to all of the models. The Interactive Multiple Model Systems are based on Markovia jump systems.

B.3   Non-MATLAB Products for Machine Learning

There are many products, both open source and commercial, for Machine Learning. We cove some of the more popular open source products. Both Machine Learning and convex optimiza tion packages are discussed.

B.3.1 R

R is open source software for statistical computing. It compiles on MacOS, UNIX, and Windows. It is similar to the Bell Labs S language developed by John Chambers and colleagues. It includes many statistical functions and graphics techniques.

You can use R in batch mode from MATLAB using the `system` command. Write

```
1   system('R CMD BATCH inputfile outputfile');
```

This runs the code in `inputfile` and puts it into `outputfile`. You can then read the `outputfile` into MATLAB.

B.3.2 scikit-learn

scikit-learn is a Machine Learning library for use in Python. It includes a wide variety of tools:

1. Classification

2. Regression

3. Clustering

4. Dimensionality reduction

5. Model selection

6. Preprocessing

scikit-learn is well suited to a wide variety of data mining and data analysis problems.

MATLAB supports the reference implementation of Python and CPython. Mac and Linux users already have Python installed. Windows users need to install a distribution.

B.3.3 LIBSVM

LIBSVM [7] is a library for support vector machines (SVMs). It has an extensive collection of tools for SVMs including extensions by many users of LIBSVM. LIBSVM tools include distributed processing and multicore extensions. The authors are Chih-Chung Chang and Chih-Jen Lin. You can find it at www.csie.ntu.edu.tw/~cjlin/libsvm/.

B.4 Products for Optimization

Optimization tools often are used as part of machine learning systems. Optimizers minimize a cost given a set of constraints on the variables that are optimized. The maximum or minimum values for a variable are one type of constraint. Constraints and costs may be linear or nonlinear.

B.4.1 LOQO

LOQO [31] is a system for solving smooth constrained optimization problems available from Princeton University. The problems can be linear or nonlinear, convex or non-convex, constrained or unconstrained. The only real restriction is that the functions defining the problem must be smooth (at the points evaluated by the algorithm). If the problem is convex, LOQO finds a globally optimal solution. Otherwise, it finds a locally optimal solution near a given starting point.

Once you compile the mex-file interface to LOQO, you must pass it an initial guess and sparse matrices for the problem definition variables. You may also pass in a function handle to provide animation of the algorithm at each iteration of the solution.

B.4.2 SNOPT

SNOPT [11] is a software package for solving large-scale optimization problems (linear and nonlinear programs) hosted at the University of California, San Diego. It is especially effective for nonlinear problems whose functions and gradients are expensive to evaluate. The functions should be smooth but need not be convex. SNOPT is designed to take advantage of the sparsity of the Jacobian matrix, effectively reducing the size of the problem being solved. For optimal control problems, the Jacobian is very sparse because you have a matrix with rows and columns that span a large number of time points, but only adjacent time points can have nonzero entries.

SNOPT makes use of nonlinear functions and gradient values. The solution obtained will be a local optimum (which may or may not be a global optimum). If some of the gradients are unknown, they will be estimated by finite differences. Infeasible problems are treated methodically via elastic bounds. SNOPT allows the nonlinear constraints to be violated and minimizes the sum of such violations. Efficiency is improved in large problems if only some of the variables are nonlinear or if the number of active constraints is nearly equal to the number of variables.

B.4.3 GLPK

GLPK (GNU Linear Programming Kit) solves a variety of linear programming problems. It is part of the GNU project (www.gnu.org/software/glpk/). The most well-known one is solving the linear program:

$$Ax = b \tag{B.1}$$

$$y = cx \tag{B.2}$$

where it is desired to find x that when it is multiplied by A equals b. c is the cost vector that when multiplied by x gives the scalar cost of applying x. If x is the same length as b, the solution is

$$x = A^{-1}b \tag{B.3}$$

Otherwise, we can use GLPK to solve for x that minimizes y. GLPK can solve this problem and others where one or more elements of x have to be an integer or even just 0 or 1.

B.4.4 CVX

CVX [6] is a MATLAB-based modeling system for convex optimization. CVX turns MATLAB into a modeling language, allowing constraints and objectives to be specified using standard MATLAB expression syntax.

In its default mode, CVX supports a particular approach to convex optimization that we call disciplined convex programming. Under this approach, convex functions and sets are built up from a small set of rules from convex analysis, starting from a base library of convex functions and sets. Constraints and objectives that are expressed using these rules are automatically transformed into a canonical form and solved. CVX can be used for free with solvers like SeDuMi or with commercial solvers if a license is obtained from CVX Research.

B.4.5 SeDuMi

SeDuMi [29] is MATLAB software for optimization over second-order cones, currently hosted at Lehigh University. It can handle quadratic constraints. SeDuMi was used in Acikmese [2]. SeDuMi stands for Self-Dual-Minimization. It implements the *self-dual* embedding technique over *self-dual* homogeneous cones. This makes it possible to solve certain optimization problems in one phase. SeDuMi is available as part of YALMIP and as a stand-alone package.

B.4.6 YALMIP

YALMIP is free MATLAB software by Johan Lofberg that provides an easy-to-use interface to other solvers. It interprets constraints and can select the solver based on the constraints. SeDuMi and MATLAB's *fmincon* from the Optimization Toolbox are available solvers. Other available solvers include those for

1. Linear programming

2. Mixed-integer linear programming

3. Quadratic programming

4. Mixed-integer quadratic programming

5. Second-order cone programming

6. Semidefinite programming

7. General nonlinear programming

Mixed-integer problems are problems where only some of the variables are integers. Second-order cone programming is a convex optimization problem. Semidefinite is a subset of cone programming.

B.5 Products for Expert Systems

There are dozens, if not hundreds, of expert system shells. For MATLAB users, the most useful shells are ones for which the C or C++ code is available. It is straightforward to write an interface in C using a .mex file. CLIPS is an expert system shell (www.clipsrules.net/? q=AboutCLIPS). It stands for "C" Language Integrated Production System. It is a rule-based language for creating expert systems. It allows users to implement heuristic solutions easily. It has been used for many applications including

- An Intelligent Training System for Space Shuttle Flight Controllers

- Applications of Artificial Intelligence to Space Shuttle Mission Control

- PI-in-a-Box: A Knowledge-Based System for Space Science Experimentation

- The DRAIR Advisor: A Knowledge-Based System for Materiel Deficiency Analysis

- The Multimission VICAR Planner: Image Processing for Scientific Data

- IMPACT: Development and Deployment Experience of Network Event Correlation Applications

- The NASA Personnel Security Processing Expert System

- Expert System Technology for Nondestructive Waste Assay

- Hybrid knowledge-based system for automatic classification of B-scan images from ultrasonic rail inspection

- An Expert System for Recognition of Facial Actions and Their Intensity

- Development of a Hybrid Knowledge-Based System for Multi-objective Optimization of Power Distribution System Operations

CLIPS is currently maintained by Gary Riley who has written the book *Expert Systems: Principles and Programming*, now in its fourth edition. In the next section, we will learn about mex files to interface CLIPS, and other software, with MATLAB.

B.6 MATLAB mex Files

B.6.1 Problem

CLIPS needs to be connected to MATLAB.

B.6.2 Solution

The solution is to create a mex file to interface MATLAB and CLIPS.

B.6.3 How It Works

Look at the file MEXTest.c. This is a C file with the accompanying header, .h, file.

MEXTest.c

```
1  //
2  //   MEXTest.c
3  //
5
6  #include "MEXTest.h"
7  #include "mex.h"
9
10 void mexFunction( int nlhs, mxArray *plhs[], int nrhs, const mxArray *
       prhs[] )
11 {
12     // Check the arguments
13     if( nrhs != 2 )
14     {
15         mexErrMsgTxt("Two inputs required.");
16     }
17
18     if( nlhs != 1 )
19     {
20         mexErrMsgTxt("One output required.");
21     }
22 }
```

Look at the mex file.

MEXTest.h

```
1  //
2  //   MEXTest.h
3  //
4
5  #ifndef MEXTest_h
6  #define MEXTest_h
7
8  #include <stdio.h>
9
10 #endif /* MEXTest_h */
```

You can edit the files in MATLAB or any text editor.

Now type in the command line:

```
>> mex MEXTest.c
Building with 'Xcode Clang++'.
MEX was completed successfully.
```

mex just calls your development system's compiler and linker. "XCode" is the MacOS development environment. "Clang" is a C/C++ compiler. You end up with the file MEXTest.mexmaci6 if you are using macOS. Typing help MEXTest in the command window does not return anything. If you want help, add a file MEXTest.m such as

```
1  function CLIPS
2  %% Help for the CLIPS.cpp file
```

If you try and run the function, you will get

```
>> MEXTest
Error using MEXTest
Two inputs are required.
```

The CLIPS mex file isn't much more complicated. The CLIPS mex file reads in a rule file and then takes inputs to which the rule is applied. The rule file is Rules.CLP and is

```
1  (defrule troubleshoot-car
2     (wheel-turning no) (power yes) (torque-command yes)
3     =>
4     (cbkFunction))
```

defrule defines a rule. cbkFunction is the callback function. It is called (also known as fired) if the rule troubleshoot-car is true. If all conditions are true, it fires cbkFunction. You can write the rule file using any text editor. To run the mex file, you need to type

```
>> mex CLIPS.c -lCLIPS
Building with 'Xcode Clang++'.
MEX was completed successfully.
```

"-lCLIPS" loads the dynamic library libCLIPS.dylib. This dynamic library was built using XCode on the MacOS from CLIPS source code, which is a set of C files. Now you are ready to run the function. Type the facts into the command window:

```
>> facts = '(wheel-turning no) (power yes) (torque-command yes)'

facts =
     '(wheel-turning no) (power yes) (torque-command yes)'
```

To run the function, call CLIPS with this facts variable:

```
>> CLIPS(facts)

ans =
    'Wheel failed'
```

Now see if it knows when the wheel is working. Change the first fact to "yes":

```
>> facts =  '(wheel-turning yes) (power yes) (torque-command yes)';
>> CLIPS(facts)

ans =
    'Wheel working'
```

You can use this function to create any expert system by making a more elaborate rule base.

Bibliography

[1] A. E. Bryson Jr. *Control of Spacecraft and Aircraft*. Princeton, 1994.

[2] Behcet Acikmese and Scott R. Ploen. Convex Programming Approach to Powered Descent Guidance for Mars Landing. *Journal of Guidance, Control, and Dynamics*, 30(5):1353–1366, 2007.

[3] K. J. Åström and B. Wittenmark. *Adaptive Control Second Edition*. Addison-Wesley, 1995.

[4] D. Bertsekas. *Reinforcement Learning and Optimal Control*. Athena Scientific, 2019.

[5] S.S. Blackman and R.F. Popoli. *Design and Analysis of Modern Tracking Systems*. Artech House, 1999.

[6] S. Boyd. CVX: Matlab Software for Disciplined Convex Programming. http://cvxr.com/cvx/, 2015.

[7] Chih-Chung Chang and Chih-Jen Lin. LIBSVM – A Library for Support Vector Machines. www.csie.ntu.edu.tw/~cjlin/libsvm/, 2015.

[8] Ka Cheok et. al. Fuzzy Logic-Based Smart Automatic. *IEEE Control Systems*, December 1996.

[9] Corinna Cortes and Vladimir Vapnik. Support-Vector Networks. *Machine Learning*, 20:273–297, 1995.

[10] Don Eyles. *Sunburst and Luminary: An Apollo Memoir*. Fort Point Press, 2018.

[11] Philip Gill, Walter Murray, and Michael Saunders. SNOPT 6.0 -Description. www.sbsi-sol-optimize.com/asp/sol_products_snopt_desc.htm, 2013.

[12] J. Grus. *Data Science from Scratch*. O'Reilly, 2015.

[13] Peter Hawke. The Turing Test: Then and Now. Technical report, 2012], month = October, howpublished = www-logic.stanford.edu/seminar/1213/Hawke_TuringTest.pdf.

[14] S. Haykin. *Neural Networks*. Prentice-Hall, 1999.

[15] Matthijs Hollemans. Convolutional neural networks on the iPhone with VGGNet. http://matthijshollemans.com/2016/08/30/vggnet-convolutional-neural-network-iphone/, 2016.

© The Author(s), under exclusive license to APress Media, LLC, part of Springer Nature 2024
M. Paluszek, S. Thomas, *MATLAB Machine Learning Recipes*,
https://doi.org/10.1007/978-1-4842-9846-6

[16] P. Jackson. *Introduction to Expert Systems, Third Edition*. Addison-Wesley, 1999.

[17] Byoung S. Kim and Anthony J. Calise. Nonlinear flight control using neural networks. *Journal of Guidance, Control, and Dynamics*, 20(1):26–33, 1997.

[18] Daniel Liberzon. ECE 517: Nonlinear and Adaptive Control Lecture Notes, November 2021.

[19] J. B. Mueller. *Design and Analysis of Optimal Ascent Trajectories for Stratospheric Airships*. PhD thesis, University of Minnesota, 2013.

[20] J. B. Mueller and G. J. Balas. Implementation and Testing of LPV Controllers for the F/A-18 System Research Aircraft. In *Proceedings*, number 4446 in AIAA-2000. AIAA, August 2000.

[21] Andrew Ng, Jiquan Ngiam, Chuan Yu Foo, Yifan Mai, Caroline Suen, Adam Coates, Andrew Maas, Awni Hannun, Brody Huval, Tao Wang, and Sameep Tandon. UFDL Tutorial. http://ufldl.stanford.edu/tutorial/, 2016.

[22] Nils J. Nilsson. *Artificial Intelligence: A New Synthesis*. Morgran Kaufmann Publishers, 1998.

[23] M. A. C. Perryman, L. Lindegren, J. Kovalevsky, E. Hog, U. Bastian, P. L. Bernacca, M. Creze, F. Donati, M. Grenon, M. Grewing, F. van Leeuwen, H. van der Marel, F. Mignard, C. A. Murray, R. S. Le Poole, H. Schrijver, C. Turon, F. Arenou, M. Froeschle, and C. S. Petersen. The hipparcos catalogue.

[24] Sebastian Raschka. *Python Machine Learning*. [PACKT], 2015.

[25] D. B. Reid. An algorithm for tracking multiple targets. *IEEE Transactions on Automatic Control*, AC=24(6):843–854, December 1979.

[26] S. Russell and P. Norvig. *Artificial Intelligence A Modern Approach Third Edition*. Prentice-Hall, 2010.

[27] S. Sarkka. Lecture 3: Bayesian Optimal Filtering Equations and the Kalman Filter. Technical report, Department of Biomedical Engineering and Computational Science, Aalto University School of Science, February 2011.

[28] L. D. Stone, C. A. Barlow, and T. L. Corwin. *Bayesian Multiple Target Tracking*. Artech House, 1999.

[29] Jos F. Sturm. Using SeDuMi 1.02, a MATLAB toolbox for optimization over symmetric cones. http://sedumi.ie.lehigh.edu/wp-content/sedumi-downloads/usrguide.ps, 1998.

[30] K Terano, Asai T., and M. Sugeno. *Fuzzy Systems Theory and its Applications*. Academic Press, 1992.

[31] R. J. Vanderbvei. LOQO USER'S MANUAL D VERSION 4.05. www.princeton.edu/~rvdb/tex/loqo/loqo405.pdf, September 2013.

[32] M. C. VanDyke, J. L. Schwartz, and C. D. Hall. Unscented Kalman Filtering for Spacecraft Attitude State and Parameter Estimation. *Advances in Astronautical Sciences*, 2005.

[33] Matthew G. Villella. *Nonlinear Modeling and Control of Automobiles with Dynamic Wheel-Road Friction and Wheel Torque Inputs*. PhD thesis, Georgia Institute of Technology, April 2004.

[34] Phillip Wang, 2019.

[35] Peggy S. Williams-Hayes. Flight Test Implementation of a Second Generation Intelligent Flight Control System. Technical Report NASA/TM-2005-213669, NASA Dryden Flight Research Center, November 2005.

Index

A

Activation functions, 202–203, 236, 237
Adaptive control
    designed and implemented, 127
    learning and adaptive, 127
    self-tuning, 128
    taxonomy, 127, 128
Adaptive control systems, xiii, 2, 4, 7, 9–10
Adaptive controllers, 416
Adaptive/intelligent control, 415, 416
Advice Taker, 412
Aerodynamic coefficients, 194
Aerodynamic models, 193, 198
Air traffic control radar systems, 327
Aircraft dynamics, 415
`AircraftSim`, 219–221, 226
`AircraftSimOpenLoop`, 201–202
AlexNet, 322–325, 421
Artificial intelligence (AI)
    advances, 414
    Advice Taker, 412
    Bayesian network, 414
    blocks world, 412
    ChatGPT, 414
    chess, 413, 414
    definition, 17
    expert systems, 18, 413
    GPS, 412
    Hanoi Towers, 412
    intelligent cars, 17–18
    limitations, 413
    Lisp, 412
    LT, 411
    mathematical formulation, 411
    neural network, 411
    Time-sharing, 412
    timeline, 414
    Watson, 414
Artificial neural networks, 413
Aspect ratio, 193, 194
`AttitudeDetermination` function, 391, 399
`AttitudeSim` function, 384, 386, 387, 399
Automatic control systems, 2
Automobile animation
    automobile 3D model, 364
    `AutomobileDemo`, 366
    cars, 363
    `DrawComponents`, 365, 366
    drawnow, 366
    graphics window, 366
    `LoadOBJFile`, 364
    Macintosh application, 364
    mesh, 365
    `.obj` files, 364
    OBJ model, 364
    passing sequence, 366, 367
    patch. patch, 365
    updation, 365
Automobile controllers, 374
Automobile dynamics
    collisions, 356
    contact friction, 358
    dynamic pressure, 357
    dynamical equations, 357
    inertial velocity, 359
    planar model, 356, 357
    `RHSAutomobile` function, 356
    rolling friction, 358
    `RungeKutta` function, 356
    steering angle, 359

Printed in the United States
by Baker & Taylor Publisher Services